STUDENT SOLUTIONS MANUAL

to accompany

FUNCTIONS MODELING CHANGE:
A PREPARATION FOR CALCULUS

Fifth Edition

Eric Connally
Harvard University Extension School

Deborah Hughes-Hallett
University of Arizona

Andrew M. Gleason
Harvard University
et al.

Prepared by
Elliot J. Marks

WILEY

Cover Photo © Patrick Zephyr / Patrick Zephyr Nature Photography

This material is based upon work supported by the National Science Foundation under Grant No. DUE-9352905. Opinions expressed are those of the authors and not necessarily those of the Foundation.

To order books or for customer service, please call 1-800-CALL-WILEY (225-5945).

ISBN-13 978-1-118-94163-8

10 9 8 7 6 5 4 3 2 1

Table of Contents

CHAPTER ONE

Solutions for Section 1.1

Skill Refresher

S1. Finding the common denominator we get $c + \frac{1}{2}c = \frac{2c+c}{2} = \frac{3c}{2} = \frac{3}{2}c$.

S5. $\left(\frac{1}{2}\right) - 5(-5) = \frac{1}{2} + 25 = \frac{51}{2}$.

S9. The figure is a parallelogram, so $A = (-2, 8)$.

Exercises

1. $m = f(v)$.

5. Appropriate axes are shown in Figure 1.1.

p, pressure (lbs/in^2)

v, volume (in^3)

Figure 1.1

9. **(a)** Since the vertical intercept is $(0, 40)$, we have $f(0) = 40$.
(b) Since the horizontal intercept is $(2, 0)$, we have $f(2) = 0$.

13. Since $f(0) = f(4) = f(8) = 0$, the solutions are $x = 0, 4, 8$.

17. Here, y is a function of x, because any particular x value gives one and only one y value. For example, if we input the constant a as the value of x, we have $y = a^4 - 1$, which is one particular y value.

However, some values of y lead to more than one value of x. For example, if $y = 15$, then $15 = x^4 - 1$, so $x^4 = 16$, giving $x = \pm 2$. Thus, x is not a function of y.

21. We apply the vertical-line test. As you can see in Figure 1.2, there is a vertical line meeting the graph in more than one point. Thus, this graph fails the vertical-line test and does not represent a function.

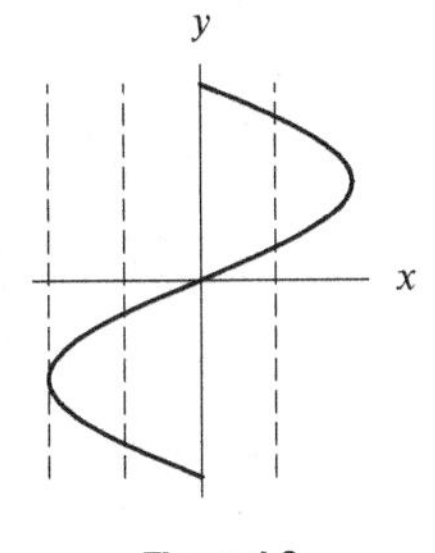

Figure 1.2

25. We apply the vertical-line test. As you can see in Figure 1.3, there is no vertical line that meets the graph at more than one point, so this graph represents y as a function of x.

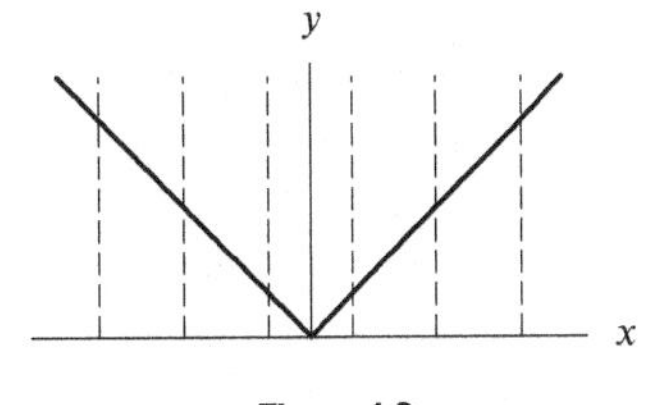

Figure 1.3

Problems

29. A possible graph is shown in Figure 1.4

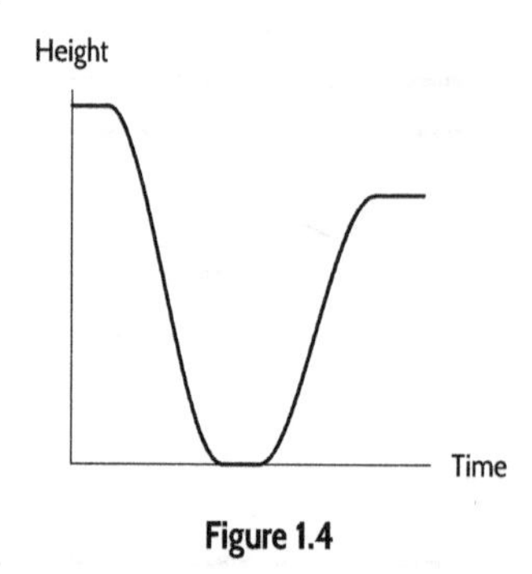

Figure 1.4

33. **(a)** Since the person starts out 5 miles from home, the vertical intercept on the graph must be 5. Thus, (i) and (ii) are possibilities. However, since the person rides 5 mph away from home, after 1 hour the person is 10 miles from home. Thus, (ii) is the correct graph.

(b) Since this person also starts out 5 miles from home, (i) and (ii) are again possibilities. This time, however, the person is moving at 10 mph and so is 15 miles from home after 1 hour. Thus, (i) is correct.

(c) The person starts out 10 miles from home so the vertical intercept must be 10. The fact that the person reaches home after 1 hour means that the horizontal intercept is 1. Thus, (v) is correct.

(d) Starting out 10 miles from home means that the vertical intercept is 10. Being half way home after 1 hour means that the distance from home is 5 miles after 1 hour. Thus, (iv) is correct.

(e) We are looking for a graph with vertical intercept of 5 and where the distance is 10 after 1 hour. This is graph (ii). Notice that graph (iii), which depicts a bicyclist stopped 10 miles from home, does not match any of the stories.

37. Figure 1.5 shows the tank.

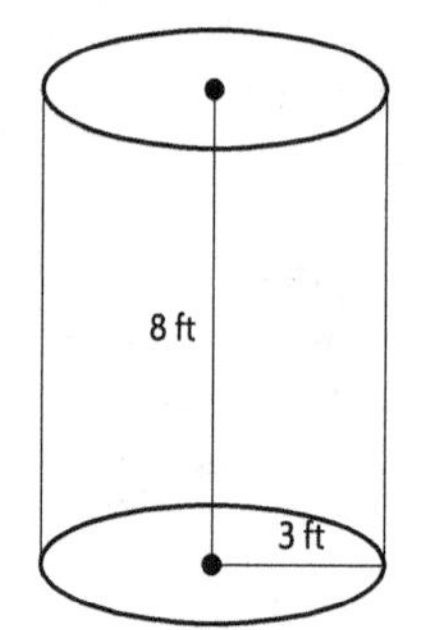

Figure 1.5: Cylindrical water tank

(a) The volume of a cylinder is equal to the area of the base times the height, where the area of the base is πr^2. Here, the radius of the base is $(1/2)(6) = 3$ ft, so the area is $\pi \cdot 3^2 = 9\pi$ ft^2. Therefore, the capacity of this tank is $(9\pi)8 = 72\pi$ ft^3.

(b) If the height of the water is 5 ft, the volume becomes $(9\pi)5 = 45\pi$ ft^3.

(c) In general, if the height of water is h ft, the volume of the water is $(9\pi)h$. If we let $V(h)$ be the volume of water in the tank as a function of its height, then

$$V(h) = 9\pi h.$$

Note that this function only makes sense for a non-negative value of h, which does not exceed 8 feet, the height of the tank.

41. **(a)** No, in the year 1954 there were two world records; in the year 1981 there were three world records.

(b) Yes, each world record occurred in only one year.

(c) The world record of 3 minutes and 47.33 seconds was set in 1981.

(d) The statement $y(3{:}51.1) = 1967$ tells us that the world record of 3 minutes, 51.1 seconds was set in 1967.

45. **(a)** Adding the male total to the female total gives $x + y$, the total number of applicants.

(b) Of the men who apply, 15% are accepted. So $0.15x$ male applicants are accepted. Likewise, 18% of the women are accepted so we have $0.18y$ women accepted. Summing the two tells us that $0.15x + 0.18y$ applicants are accepted.

(c) The number accepted divided by the number who applied times 100 gives the percentage accepted. This expression is

$$\frac{(0.15)x + (0.18)y}{x+y}(100), \quad \text{or} \quad \frac{15x+18y}{x+y}.$$

Solutions for Section 1.2

Skill Refresher

S1. $\frac{4-6}{3-2} = \frac{-2}{1} = -2.$

S5. $\frac{\frac{1}{2}-(-4)^2-\left(\frac{1}{2}-(5^2)\right)}{-4-5} = \frac{\frac{1}{2}-16-\frac{1}{2}+25}{-9} = \frac{9}{-9} = -1.$

S9. $\frac{x^2-\frac{3}{4}-\left(y^2-\frac{3}{4}\right)}{x-y} = \frac{x^2-\frac{3}{4}-y^2+\frac{3}{4}}{x-y} = \frac{x^2-y^2}{x-y} = \frac{(x-y)(x+y)}{x-y} = x+y.$

Exercises

1. The function is increasing for $x > 0$, since the graph rises there as we move to the right. The function is decreasing for $x < 0$, since the graph falls as we move to the right.

5. **(a)** Let $s = V(t)$ be the sales (in millions) of feature phones in year t. Then

$$\begin{aligned}\text{Average rate of change of } s \text{ from } t = 2010 \text{ to } t = 2012 &= \frac{\Delta s}{\Delta t} = \frac{V(2012)-V(2010)}{2012-2010} \\ &= \frac{914-1079}{2} \\ &= -82.5 \text{ million feature phones/year.}\end{aligned}$$

Let $q = D(t)$ be the sales (in millions) of smartphones in year t. Then

$$\begin{aligned}\text{Average rate of change of } q \text{ from } t = 2010 \text{ to } t = 2012 &= \frac{\Delta q}{\Delta t} = \frac{D(2012)-D(2010)}{2012-2010} \\ &= \frac{661-301}{2} \\ &= 180 \text{ million smartphones/year.}\end{aligned}$$

(b) By the same argument

$$\begin{aligned}\text{Average rate of change of } s \text{ from } t = 2012 \text{ to } t = 2013 &= \frac{\Delta s}{\Delta t} = \frac{V(2012)-V(2013)}{2012-2013} \\ &= \frac{838-914}{1} \\ &= -76 \text{ million feature phones/year.}\end{aligned}$$

$$\begin{aligned}\text{Average rate of change of } q \text{ from } t = 2012 \text{ to } t = 2013 &= \frac{\Delta q}{\Delta t} = \frac{D(2013)-D(2012)}{2013-2012} \\ &= \frac{968-661}{1} \\ &= 307 \text{ million smartphones /year.}\end{aligned}$$

(c) The fact that $\Delta s/\Delta t = -82.5$ tells us that feature phone sales decreased at an average rate of 82.5 million feature phones/year between 2010 and 2012. The fact that the average rate of change is negative tells us that annual sales are decreasing.

The fact that $\Delta s/\Delta t = -76$ tells us that feature phone sales decreased at an average rate of -76 million feature phones/year between 2012 and 2013.

The fact that $\Delta q/\Delta t = 180$ means that smartphone sales increased at an average rate of 180 million players/year between 2010 and 2012. The fact that $\Delta q/\Delta t = 307$ means that smartphone sales increased at an average rate of 307 million smartphones/year between 2012 and 2013.

9. (a) (i) After 2 hours 60 miles had been traveled. After 5 hours, 150 miles had been traveled. Thus on the interval from $t = 2$ to $t = 5$ the value of Δt is

$$\Delta t = 5 - 2 = 3$$

and the value of ΔD is

$$\Delta D = 150 - 60 = 90.$$

(ii) After 0.5 hours 15 miles had been traveled. After 2.5 hours, 75 miles had been traveled. Thus on the interval from $t = 0.5$ to $t = 2.5$ the value of Δt is

$$\Delta t = 2.5 - .5 = 2$$

and the value of ΔD is

$$\Delta D = 75 - 15 = 60.$$

(iii) After 1.5 hours 45 miles had been traveled. After 3 hours, 90 miles had been traveled. Thus on the interval from $t = 1.5$ to $t = 3$ the value of Δt is

$$\Delta t = 3 - 1.5 = 1.5$$

and the value of ΔD is

$$\Delta D = 90 - 45 = 45.$$

(b) For the interval from $t = 2$ to $t = 5$, we see

$$\text{Rate of change} = \frac{\Delta D}{\Delta t} = \frac{90}{3} = 30.$$

For the interval from $t = 0.5$ to $t = 2.5$, we see

$$\text{Rate of change} = \frac{\Delta D}{\Delta t} = \frac{60}{2} = 30.$$

For the interval from $t = 1.5$ to $t = 3$, we see

$$\text{Rate of change} = \frac{\Delta D}{\Delta t} = \frac{45}{1.5} = 30.$$

This suggests that the average speed is 30 miles per hour throughout the trip.

13. They are equal; both are given by

$$\frac{4.9 - 2.9}{6.1 - 2.2}.$$

Problems

17. (a) The coordinates of point A are $(10, 30)$.
The coordinates of point B are $(30, 40)$.
The coordinates of point C are $(50, 90)$.
The coordinates of point D are $(60, 40)$.
The coordinates of point E are $(90, 40)$.

(b) From Figure 1.6, we see that F is on the graph but G is not.

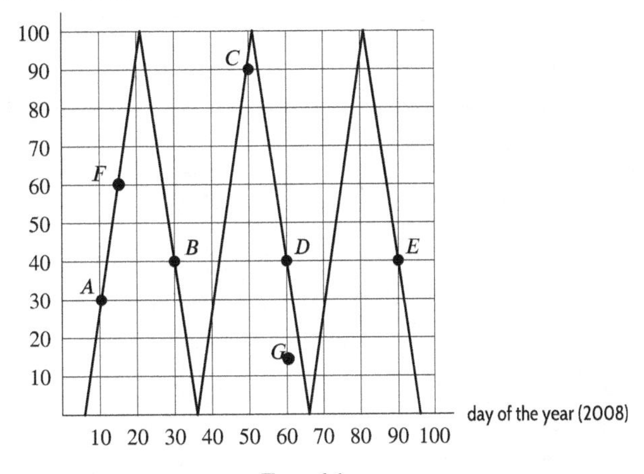

Figure 1.6

(c) The function is increasing from, approximately, days 6 through 21, 36 through 51, and 66 through 81.

(d) The function is decreasing from, approximately, days 22 through 35, 52 through 65, and 82 through 96.

21. A knowledge of when the record was established determines the world record time, so the world record time is a function of the time it was established. Also, when a world record is established it is smaller than the previous world record and occurs later in time. Thus, it is a decreasing function. Because a world record could be established twice in the same year, a knowledge of the year does not determine the world record time, so the world record time is not a function of the year it was established.

25. **(a)** The number of sunspots, s, is a function of the year, t, because knowing the year is enough to uniquely determine the number of sunspots. The graph passes the vertical line test.

(b) When read from left to right, the graph increases from approximately $t = 1964$ to $t = 1969$, from approximately 1971 to 1972, from approximately $t = 1976$ to $t = 1979$, from approximately 1986 to 1989, from approximately 1990 to 1991 and from approximately 1996 to 2000. Thus, s is an increasing function of t on the intervals $1964 < t < 1969$, $1971 < t < 1972$, $1976 < t < 1979$, $1986 < t < 1989$, and $1996 < t < 2000$. For each of these intervals, the average rate of change must be positive.

29. **(a)** Between $(1, 4)$ and $(2, 13)$,

$$\text{Average rate of change } = \frac{\Delta y}{\Delta x} = \frac{13-4}{2-1} = 9.$$

(b) Between (j, k) and (m, n),

$$\text{Average rate of change } = \frac{\Delta y}{\Delta x} = \frac{n-k}{m-j}.$$

(c) Between $(x, f(x))$ and $(x+h, f(x+h))$,

$$\begin{aligned}
\text{Average rate of change } &= \frac{\Delta y}{\Delta x} = \frac{(3(x+h)^2+1)-(3x^2+1)}{(x+h)-x} \\
&= \frac{(3(x^2+2xh+h^2)+1)-(3x^2+1)}{h} \\
&= \frac{3x^2+6xh+3h^2+1-3x^2-1}{h} \\
&= \frac{6xh+3h^2}{h} \\
&= 6x+3h.
\end{aligned}$$

Solutions for Section 1.3

Skill Refresher

S1. We have $f(0) = \frac{2}{3}(0) + 5 = 5$ and $f(3) = \frac{2}{3}(3) + 5 = 2 + 5 = 7$.

S5. To find the y-intercept, we let $x = 0$,

$$\begin{aligned} y &= -4(0) + 3 \\ &= 3. \end{aligned}$$

To find the x-intercept, we let $y = 0$,

$$\begin{aligned} 0 &= -4x + 3 \\ 4x &= 3 \\ x &= \frac{3}{4}. \end{aligned}$$

S9. Combining like terms we get

$$(a - 3)x - ab + a + 3.$$

Hence the constant term is $-ab + a + 3$ and the coefficient is $a - 3$.

Exercises

1. **(a)** Since the slopes are 2 and 3, we see that $y = -2 + 3x$ has the greater slope.
(b) Since the y-intercepts are -1 and -2, we see that $y = -1 + 2x$ has the greater y-intercept.

5. The function h is not linear even though the value of x increases by $\Delta x = 10$ each time. This is because $h(x)$ does not increase by the same amount each time. The value of $h(x)$ increases from 20 to 40 to 50 to 55 taking smaller steps each time.

9. This table could represent a linear function because the rate of change of $p(\gamma)$ is constant. Between consecutive data points, $\Delta\gamma = -1$ and $\Delta p(\gamma) = 10$. Thus, the rate of change is $\Delta p(\gamma)/\Delta\gamma = -10$. Since this is constant, the function could be linear.

As we just observed, the rate of change is

$$\frac{\Delta p(\gamma)}{\Delta \gamma} = \frac{10}{-1} = -10.$$

13. The vertical intercept is 29.99, which tells us that the company charges \$29.99 per month for the phone service, even if the person does not talk on the phone. The slope is 0.05. Since

$$\text{Slope} = \frac{\Delta\text{cost}}{\Delta\text{minutes}} = \frac{0.05}{1},$$

we see that, for each minute the phone is used, it costs an additional \$0.05.

Problems

17. Since the depreciation can be modeled linearly, we can write the formula for the value of the car, V, in terms of its age, t, in years, by the following formula:

$$V = b + mt.$$

Since the initial value of the car is \$23,500, we know that $b = 23{,}500$.

Hence,

$$V = 23{,}500 + mt.$$

To find m, we know that $V = 18{,}823$ when $t = 3$, so

$$\begin{aligned} 18{,}823 &= 23{,}500 + m(3) \\ -4{,}677 &= 3m \\ \frac{-4{,}677}{3} &= m \\ -1559 &= m. \end{aligned}$$

So, $V = 23{,}500 - 1559t$.

21. **(a)** Any line with a slope of 2.1, using appropriate scales on the axes. The horizontal axis should be labeled "days" and the vertical axis should be labeled "inches." See Figure 1.7.

(b) Any line with a slope of -1.3, using appropriate scales on the axes. The horizontal axis should be labeled "miles" and the vertical axis should be labeled "gallons." See Figure 1.8.

Figure 1.7

Figure 1.8

25. **(a)** We see that the population of Country B grows at the constant rate of roughly 2.4 million every ten years. Thus Country B must be Sri Lanka. The population of country A did not change at a constant rate: In the ten years of 1970–1980 the population of Country A grew by 2.7 million while in the ten years of 1980–1990 its population dropped. Thus, Country A is Afghanistan.

(b) The rate of change of Country B is found by taking the population increase and dividing it by the corresponding time in which this increase occurred.Thus

$$\text{Rate of change of population} = \frac{9.9 - 7.5}{1960 - 1950} = \frac{2.4 \text{ million people}}{10 \text{ years}} = 0.24 \text{ million people/year}.$$

This rate of change tells us that on the average, the population of Sri Lanka increases by 0.24 million people every year. The rate of change for the other intervals is the same or nearly the same.

(c) In 1980 the population of Sri Lanka was 14.9 million. If the population grows by 0.24 million every year, then in the eight years from 1980 to 1988

$$\text{Population increase} = 8 \cdot 0.24 \text{ million} = 1.92 \text{ million}.$$

Thus in 1988

$$\text{Population of Sri Lanka} = 14.9 + 1.92 \text{ million} \approx 16.8 \text{ million}.$$

29. **(a)** Since C is 8, we have $T = 300 + 200C = 300 + 200(8) = 1900$. Thus, taking 8 credits costs \$1900.

(b) Here, the value of T is 1700 and we solve for C.

$$\begin{aligned} T &= 300 + 200C \\ 1700 &= 300 + 200C \\ 7 &= C \end{aligned}$$

Thus, \$1, 700 is the cost of taking 7 credits.

(c) Table 1.1 is the table of costs.

Table 1.1

C	1	2	3	4	5	6	7	8	9	10	11	12
T	500	700	900	1100	1300	1500	1700	1900	2100	2300	2500	2700
$\frac{T}{C}$	500	350	300	275	260	250	243	238	233	230	227	225

(d) The largest value for C, that is, 12 credits, gives the smallest value of T/C. In general, the ratio of tuition cost to number of credits is getting smaller as C increases.

(e) This cost is independent of the number of credits taken; it might cover fixed fees such as registration, student activities, and so forth.

(f) The 200 represents the rate of change of cost with the number of credit hours. In other words, one additional credit hour costs an additional \$200.

33. As Figure 1.9 shows, the graph of $y = 2x + 400$ does not appear in the window $-10 \le x \le 10, -10 \le y \le 10$. This is because all the corresponding y-values are between 380 and 420, which are outside this window. The graph can be seen by using a different viewing window: for example, $380 \le y \le 420$.

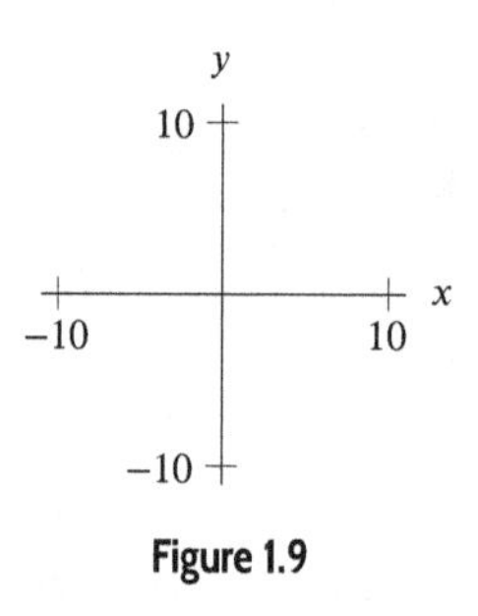

Figure 1.9

Solutions for Section 1.4

Skill Refresher

S1.

$$\begin{aligned} y - 5 &= 21 \\ y &= 26. \end{aligned}$$

S5. We first distribute $\frac{5}{3}(y + 2)$ to obtain:

$$\frac{5}{3}(y + 2) = \frac{1}{2} - y$$

$$\begin{aligned}
\frac{5}{3}y + \frac{10}{3} &= \frac{1}{2} - y \\
\frac{5}{3}y + y &= \frac{1}{2} - \frac{10}{3} \\
\frac{5}{3}y + \frac{3y}{3} &= \frac{3}{6} - \frac{20}{6} \\
\frac{8y}{3} &= -\frac{17}{6} \\
\left(\frac{3}{8}\right)\frac{8y}{3} &= \left(\frac{3}{8}\right)\left(-\frac{17}{6}\right) \\
y &= -\frac{17}{16}.
\end{aligned}$$

S9. We collect all terms involving x and then divide by $2a$:

$$\begin{aligned}
ab + ax &= c - ax \\
2ax &= c - ab \\
x &= \frac{c - ab}{2a}.
\end{aligned}$$

Exercises

1. We have the slope $m = -4$ so

$$y = b - 4x.$$

The line passes through $(7, 0)$ so

$$\begin{aligned}
0 &= b + (-4)(7) \\
28 &= b
\end{aligned}$$

and

$$y = 28 - 4x.$$

5. Since we know the x-intercept and y-intercepts are $(3, 0)$ and $(0, -5)$ respectively, we can find the slope:

$$\text{slope} = m = \frac{-5 - 0}{0 - 3} = \frac{-5}{-3} = \frac{5}{3}.$$

We can then put the slope and y-intercept into the general equation for a line.

$$y = -5 + \frac{5}{3}x.$$

9. We have a V intercept of 2000. Since the value is decreasing by \$500 per year, our slope is -500 dollars per year. So one possible equation is

$$V = 2000 - 500t.$$

13. Since the function is linear, we can choose any two points (from the graph) to find its formula. We use the form

$$p = b + mh$$

to get the price of an apartment as a function of its height. We use the two points (10, 175,000) and (20, 225,000). We begin by finding the slope, $\Delta p/\Delta h = (225{,}000 - 175{,}000)/(20 - 10) = 5000$. Next, we substitute a point into our equation using our slope of 5000 dollars per meter of height and solve to find b, the p-intercept. We use the point (10, 175,000):

$$\begin{aligned}
175{,}000 &= b + 5000 \cdot 10 \\
125{,}000 &= b.
\end{aligned}$$

Therefore,

$$p = 125{,}000 + 5000h.$$

17. Rewriting in slope-intercept form:

$$\begin{aligned} 3x + 5y &= 20 \\ 5y &= 20 - 3x \\ y &= \frac{20}{5} - \frac{3x}{5} \\ y &= 4 - \frac{3}{5}x \end{aligned}$$

21. Writing $y = 5$ as $y = 5 + 0x$ shows that $y = 5$ is the form $y = b + mx$ with $b = 5$ and $m = 0$.

25. Yes. Write the function as

$$g(w) = -\frac{1 - 12w}{3} = -\left(\frac{1}{3} - \frac{12}{3}w\right) = -\frac{1}{3} + 4w,$$

so $g(w)$ is linear with $b = -1/3$ and $m = 4$.

29. The function $h(x)$ is not linear because the 3^x term has the variable in the exponent and is not the same as $3x$ which would be a linear term.

33. These line are parallel because they have the same slope, 2.

Problems

37. We have $f(-3) = -8$ and $f(5) = -20$. This gives $f(x) = b + mx$ where

$$m = \frac{f(5) - f(-3)}{5 - (-3)} = \frac{-20 - (-8)}{8} = -\frac{12}{8} = -1.5.$$

Solving for b, we have

$$\begin{aligned} f(-3) &= b - 1.5(-3) \\ b &= f(-3) + 1.5(-3) \\ &= -8 + 1.5(-3) = -12.5, \end{aligned}$$

so $f(x) = -12.5 - 1.5x$.

41. **(a)** See Figures 1.10 and 1.11.

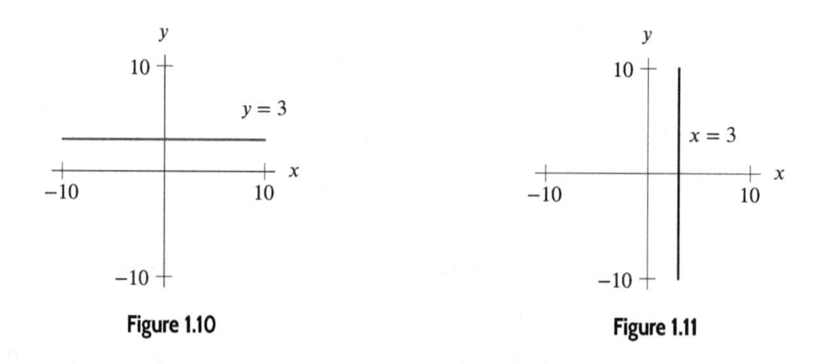

Figure 1.10

Figure 1.11

(b) Yes for $y = 3$: $y = 3 + 0x$. No for $x = 3$, since the slope is undefined, and there is no y-intercept.

45. **(a)** The function $y = f(x)$ is linear because equal spacing between successive input values ($\Delta x = 0.5$) results in equal spacing between successive output values ($\Delta y = 0.58$), so f has a constant rate of change.

(b) A formula for $y = f(x)$ is of the form $y = b + mx$. The slope of this line is

$$m = \frac{\Delta y}{\Delta x} = \frac{-0.64 - (-1.22)}{1.5 - 1} = \frac{0.58}{0.5} = 1.16$$

Thus $y = 1.16x + b$. From the table we see that $(1, -1.22)$ is on the line, so we have

$$\begin{aligned} y &= 1.16x + b \\ -1.22 &= 1.16(1) + b \\ b &= -2.38. \end{aligned}$$

Thus we get $f(x) = 1.16x - 2.38$.

49. We would like to find a table value that corresponds to $n = 0$. The pattern from the table, is that for each decrease of 25 in n, $C(n)$ goes down by 125. It takes four decreases of 25 to get from $n = 100$ to $n = 0$, and $C(100) = 11{,}000$, so we might estimate $C(0) = 11{,}000 - 4 \cdot 125 = 10{,}500$. This means that the fixed cost, before any goods are produced, is \$10,500.

53. **(a)** We are looking at the amount of municipal solid waste, W, as a function of year, t, and the two points are $(1960, 88.1)$ and $(20100, 249.9)$. For the model, we assume that the quantity of solid waste is a linear function of year. The slope of the line is

$$m = \frac{249.9 - 88.1}{2010 - 1960} = \frac{161.8}{50} = 3.236 \frac{\text{millions of tons}}{\text{year}}.$$

This slope tells us that the amount of solid waste generated in the cities of the US has been going up at a rate of 3.236 million tons per year. To find the equation of the line, we must find the vertical intercept. We substitute the point $(1960, 88.1)$ and the slope $m = 3.236$ into the equation $W = b + mt$:

$$\begin{aligned} W &= b + mt \\ 88.1 &= b + (3.236)(1960) \\ 88.1 &= b + 6342.56 \\ -6254.46 &= b. \end{aligned}$$

The equation of the line is $W = -6254.46 + 3.236t$, where W is the amount of municipal solid waste in the US in millions of tons, and t is the year.

(b) How much solid waste does this model predict in the year 2020? We can graph the line and find the vertical coordinate when $t = 2020$, or we can substitute $t = 2020$ into the equation of the line, and solve for W:

$$\begin{aligned} W &= -6254.46 + 3.2368t \\ W &= -6254.46 + (3.236)(2020) = 282.26. \end{aligned}$$

The model predicts that in the year 2020, the solid waste generated by cities in the US will be 282.26 million tons.

57. Point P is on the curve $y = x^2$ and so its coordinates are $(2, 2^2) = (2, 4)$. Since line l contains point P and has slope 4, its equation is

$$y = b + mx.$$

Using $P = (2, 4)$ and $m = 4$, we get

$$\begin{aligned} 4 &= b + 4(2) \\ 4 &= b + 8 \\ -4 &= b \end{aligned}$$

so,

$$y = -4 + 4x.$$

61. **(a)** We know that the equation will be of the form

$$p = b + mt$$

where m is the slope and b is the p-intercept. Since there are 100 minutes in an hour and 40 minutes, two points on this line are $(100, 9)$ and $(50, 4)$. Solving for the slope we get

$$m = \frac{9 - 4}{100 - 50} = \frac{5}{50} = 0.1 \text{ pages/minute}.$$

Thus, we get

$$p = b + 0.1t.$$

Using the point (50, 4) to solve for b, we get

$$\begin{aligned} 4 &= 50(0.1) + b \\ &= 5 + b. \end{aligned}$$

Thus

$$b = -1$$

and

$$p = -1 + 0.1t \quad \text{or} \quad p = 0.1t - 1.$$

Since p must be non-negative, we have $0.1t - 1 \geq 0$, or $t \geq 10$.

(b) In 2 hours there are 120 minutes. If $t = 120$ we get

$$p = 0.1(120) - 1 = 12 - 1 = 11.$$

Thus 11 pages can be typed in two hours.

(c) The slope of the function tells us that you type 0.1 pages per minute.

(d) Solving the equation for time in terms of pages we get

$$\begin{aligned} p &= 0.1t - 1 \\ 0.1t - 1 &= p \\ 0.1t &= p + 1 \\ t &= 10p + 10. \end{aligned}$$

(e) If $p = 15$ and we use the formula from part (d), we get

$$t = 10(15) + 10 = 150 + 10 = 160.$$

Thus it would take 160 minutes, or two hours and forty minutes to type a fifteen page paper.

(f) Answers vary. Sometimes we know the amount of time we have available to type and we could then use $p = f(t)$ to tell us how many pages can be typed in this time. On the other hand, $t = g(p)$ is useful when we know the number of pages we have and want to know how long it will take to type them.

65. (a) We know that $r = 1/t$. Table 1.2 gives values of r. From the table, we see that $\Delta r/\Delta H \approx 0.01/2 = 0.005$, so $r = b + 0.005H$. Solving for b, we have

$$\begin{aligned} 0.070 &= b + 0.005 \cdot 20 \\ b &= 0.070 - 0.1 = -0.03. \end{aligned}$$

Thus, a formula for r is given by $r = 0.005H - 0.03$.

Table 1.2 *Development time t (in days) for an organism as a function of ambient temperature H (in °C)*

H, °C	20	22	24	26	28	30
r, rate	0.070	0.080	0.090	0.100	0.110	0.120

(b) From Problem 64, we know that if $r = b + kH$ then the number of degree-days is given by $S = 1/k$. From part (a) of this problem, we have $k = 0.005$, so $S = 1/0.005 = 200$.

Solutions for Section 1.5

Skill Refresher

S1. Substituting 5 for y in the first equation, we get

$$\begin{aligned} x + 5 &= 3 \\ x &= -2. \end{aligned}$$

S5. Substituting the value of y from the first equation into the second equation, we obtain

$$\begin{aligned} x + 2(2x - 10) &= 15 \\ x + 4x - 20 &= 15 \\ 5x &= 35 \\ x &= 7. \end{aligned}$$

Now we substitute $x = 7$ into the first equation, obtaining $2(7) - y = 10$, hence $y = 4$.

Exercises

1. (a) is (V), because slope is negative, vertical intercept is 0
(b) is (VI), because slope and vertical intercept are both positive
(c) is (I), because slope is negative, vertical intercept is positive
(d) is (IV), because slope is positive, vertical intercept is negative
(e) is (III), because slope and vertical intercept are both negative
(f) is (II), because slope is positive, vertical intercept is 0

5. (a) is (II), since this is the only system where both the bounding lines have a positive slope.
(b) is (IV), since this is the only system where one line has a positive slope and the other has a negative slope.
(c) is (I), since this is the only region bounded below by a horizontal line.
(d) is (III), since this is the only region bounded above by a horizontal line.

Problems

9. Since P is the x-intercept, we know that point P has y-coordinate $= 0$, and if the x-coordinate is x_0, we can calculate the slope of line l using $P(x_0, 0)$ and the other given point $(0, -2)$.

$$m = \frac{-2 - 0}{0 - x_0} = \frac{-2}{-x_0} = \frac{2}{x_0}.$$

We know this equals 2, since l is parallel to $y = 2x + 1$ and therefore must have the same slope. Thus we have

$$\frac{2}{x_0} = 2.$$

So $x_0 = 1$ and the coordinates of P are $(1, 0)$.

13. Let n be the number of drinks he has. Since the BAC goes up linearly by 0.02% per drink, we see that

$$\underbrace{\text{BAC}}_{B} = \underbrace{\text{Starting amount}}_{0.01\%} + \underbrace{\text{Amount per drink}}_{0.02\%} \times \underbrace{\text{Number of drinks}}_{n}$$

$$\text{so} \quad B = 0.01\% + (0.02\%)n.$$

To remain below the legal limit, we require:

$$0.01\% + (0.02\%)n < 0.08\%.$$

Solving for n gives:

$$\begin{aligned} 0.01\% + (0.02\%)n &< 0.08\% \\ (0.02\%)n &< 0.07\% \quad \text{subtract } 0.01\% \text{ from both sides} \\ n &< 3.5. \end{aligned}$$

Thus, he must have fewer than 3.5 drinks, or at most 3 whole drinks.

17. (a) Since $y = f(x)$, to show that $f(x)$ is linear, we can solve for y in terms of A, B, C, and x.

$$\begin{aligned} Ax + By &= C \\ By &= C - Ax, \quad \text{and, since } B \neq 0, \\ y &= \frac{C}{B} - \frac{A}{B}x \end{aligned}$$

Because C/B and $-A/B$ are constants, the formula for $f(x)$ is of the linear form:

$$f(x) = y = b + mx.$$

Thus, f is linear, with slope $m = -(A/B)$ and y-intercept $b = C/B$.
To find the x-intercept, we set $y = 0$ and solve for x:

$$\begin{aligned} Ax + B(0) &= C \\ Ax &= C, \text{ and, since } A \neq 0, \\ x &= \frac{C}{A}. \end{aligned}$$

Thus, the line crosses the x–axis at $x = C/A$.

(b) (i) Since $A > 0, B > 0, C > 0$, we know that C/A (the x-intercept) and C/B (the y-intercept) are both positive and we have Figure 1.12.

(ii) Since only $C < 0$, we know that C/A and C/B are both negative, and we obtain Figure 1.13.

(iii) Since $A > 0, B < 0, C > 0$, we know that C/A is positive and C/B is negative. Thus, we obtain Figure 1.14.

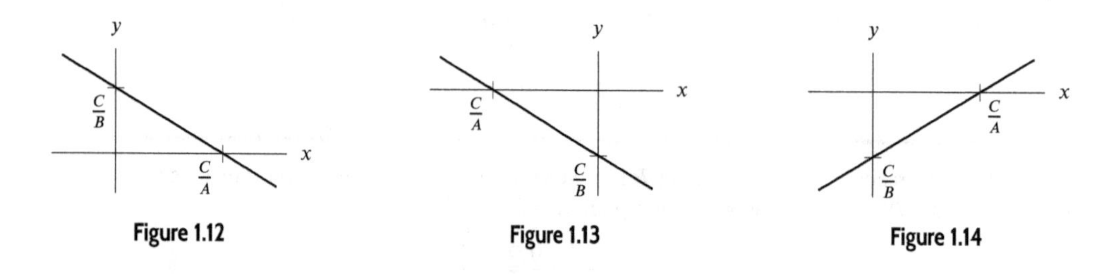

Figure 1.12 Figure 1.13 Figure 1.14

21. The graphs are shown in Figure 1.15.

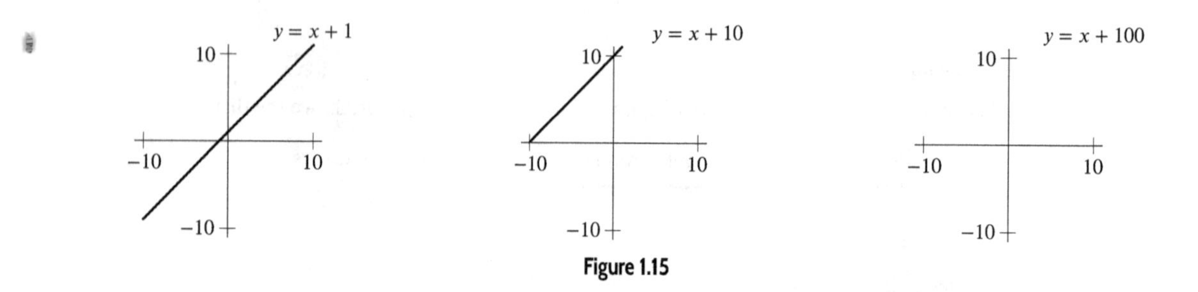

Figure 1.15

(a) As b becomes larger, the graph moves higher and higher up, until it disappears from the viewing rectangle.
(b) There are many correct answers, one of which is $y = x - 100$.

25. **(a)** To have no points in common the lines will have to be parallel and distinct. To be parallel their slopes must be the same, so $m_1 = m_2$. To be distinct we need $b_1 \neq b_2$.
(b) To have all points in common the lines will have to be parallel and the same. To be parallel their slopes must be the same, so $m_1 = m_2$. To be the same we need $b_1 = b_2$.
(c) To have exactly one point in common the lines will have to be nonparallel. To be nonparallel their slopes must be distinct, so $m_1 \neq m_2$.
(d) It is not possible for two lines to meet in just two points.

29. **(a)** When the price of the product went from \$3 to \$4, the demand for the product went down by 200 units. Since we are assuming that this relationship is linear, we know that the demand will drop by another 200 units when the price increases another dollar, to \$5. When $p = 5$, $D = 300 - 200 = 100$. So, when the price for each unit is \$5, consumers will only buy 100 units a week.

(b) The slope, m, of a linear equation is given by

$$m = \frac{\text{change in dependent variable}}{\text{change in independent variable}} = \frac{\Delta D}{\Delta P}.$$

Since quantity demanded depends on price, quantity demanded is the dependent variable and price is the independent variable. We know that when the price changes by \$1, the quantity demand changes by −200 units. That is, the quantity demanded goes down by 200 units. Thus,

$$m = \frac{-200}{1}.$$

Since the relationship is linear, we know that its formula is of the form

$$D = b + mp.$$

We know that $m = -200$, so

$$D = b - 200p.$$

We can find b by using the fact that when $p = 3$ then $D = 500$ or by using the fact that if $p = 4$ then $D = 300$ (it does not matter which). Using $p = 3$ and $D = 500$, we get

$$\begin{aligned} D &= b - 200p \\ 500 &= b - 200(3) \\ 500 &= b - 600 \\ 1100 &= b. \end{aligned}$$

Thus, $D = 1100 - 200p$.

(c) We know that $D = 1100 - 200p$ and $D = 50$, so

$$\begin{aligned} 50 &= 1100 - 200p \\ -1050 &= -200p \\ 5.25 &= p. \end{aligned}$$

At a price of \$5.25, the demand would be only 50 units.

(d) The slope is −200, which means that the demand goes down by 200 units when the price goes up by \$1.

(e) The demand is 1100 when the price is 0. This means that even if you were giving this product away, people would only want 1100 units of it per week. When the price is \$5.50, the demand is zero. This means that at or above a unit price of \$5.50, the company cannot sell this product.

Solutions for Section 1.6

1. These points are very close to a line with negative slope, so r is negative and $r = 1$ is not reasonable. (In fact, $r = -0.998$.)

5. A scatter plot of the data is shown in Figure 1.16. The value $r = 0.9$ is not reasonable. These points are very close to a line with negative slope, so r is negative. (In fact, $r = -0.99$.)

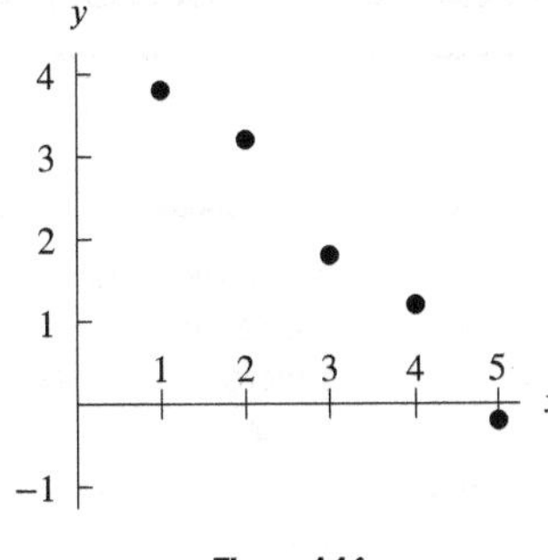

Figure 1.16

9. **(a)** When $n = 700$, we have $p = 4(700) - 2200 = 600$. When 700 people visit the park in a week, the profit is predicted to be $600.

(b) The slope of 4 means that for every weekly visitor, the profit increases by $4. The units are dollars per person.

(c) The vertical intercept of -2200 means that when nobody visits the park, that is, the number of weekly visitors is zero, the park loses $2200.

(d) We solve $4n - 2200 = 0$ to find when the park has a profit of $0. This happens when $n = 550$. With 550 weekly visitors, the park neither makes nor loses money. If more visitors come, it makes money.

13. **(a)** See Figure 1.17.

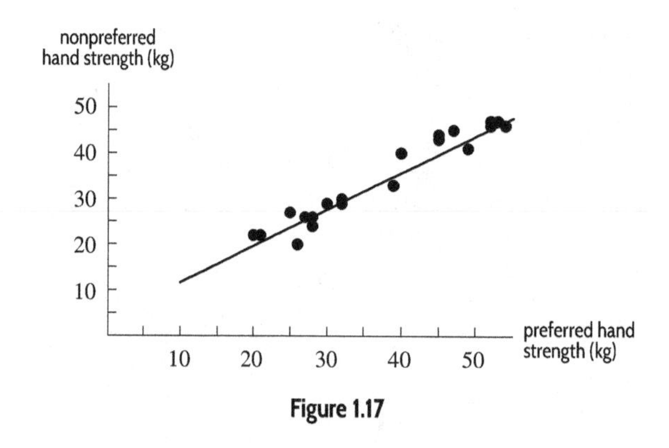

Figure 1.17

(b) Answers vary, but should be close to $y = 3.6 + 0.8x$.

(c) Answers may vary slightly. A possible equation is: $y = 3.623 + 0.825x$.

(d) The preferred hand strength is the independent quantity, so it is represented by x. Substituting $x = 37$ gives

$$y = 3.623 + 0.825(37) \approx 34.$$

So, the nonpreferred hand strength is about 34 kg.

(e) If we predict strength of the nonpreferred hand based on the strength of the preferred hand for values within the observed values of the preferred hand (such as 37), then we are interpolating. However, if we chose a value such as 10, which is below all the actual measurements, and use this to predict the nonpreferred hand strength, then we are extrapolating. Predicting from a value of 100 would be another example of extrapolation. In this case of hand strength, it seems safe to extrapolate; in other situations, extrapolation can be inaccurate.

(f) The correlation coefficient is positive because both hand strengths increase together, so the line has a positive slope. The value of r is close to 1 because the hand strengths lie close to a line of positive slope.

(g) The two clusters suggest that there are two distinct groups of students. These might be men and women, or perhaps students who are involved in college athletics (and therefore in excellent physical shape) and those who are not involved.

Solutions for Chapter 1 Review

Exercises

1. **(a)** No. Because there can be two different points sharing the same x-coordinate. For example, when $x = 0$, $y = 1$ or $y = -1$. So for each value of x, there is not a unique value of y.

(b) Yes. Because on the semi-circle above the x-axis there is only one point for each x-coordinate. Thus, each x-value corresponds to at most one y-value.

5. **(a)** For 1995 to 2005, the rate of change of P_1 is

$$\frac{\Delta P_1}{\Delta t} = \frac{86 - 53}{2005 - 1995} = \frac{33}{10} = 3.3 \text{ hundred people per year,}$$

while for P_2 we have

$$\frac{\Delta P_2}{\Delta t} = \frac{73 - 85}{2005 - 1995} = \frac{-8}{10} = -0.8 \text{ hundred people per year.}$$

(b) For 2000 to 2012,

$$\frac{\Delta P_1}{\Delta t} = \frac{97 - 75}{2012 - 2000} = \frac{22}{12} = 1.83 \text{ hundred people per year,}$$

and

$$\frac{\Delta P_2}{\Delta t} = \frac{69 - 77}{2012 - 2000} = \frac{-8}{12} = -0.75 \text{ hundred people per year.}$$

(c) For 1995 to 2012,

$$\frac{\Delta P_1}{\Delta t} = \frac{97 - 53}{2012 - 1995} = 2.59 \text{ hundred people per year,}$$

and

$$\frac{\Delta P_2}{\Delta t} = \frac{69 - 85}{2012 - 1995} = -0.94 \text{ hundred people per year.}$$

9. This table could not represent a linear function, because the rate of change of $q(\lambda)$ is not constant. Consider the first three points in the table. Between $\lambda = 1$ and $\lambda = 2$, we have $\Delta\lambda = 1$ and $\Delta q(\lambda) = 2$, so the rate of change is $\Delta q(\lambda)/\Delta\lambda = 2$. Between $\lambda = 2$ and $\lambda = 3$, we have $\Delta\lambda = 1$ and $\Delta q(\lambda) = 4$, so the rate of change is $\Delta q(\lambda)/\Delta\lambda = 4$. Thus, the function could not be linear.

13. We know that the function is linear so it is of the form $f(t) = b + mt$. We can choose any two points to find the slope. We use $(5.4, 49.2)$ and $(5.5, 37)$, so

$$m = \frac{37 - 49.2}{5.5 - 5.4} = -122.$$

Thus $f(t)$ is of the form $f(t) = b - 122t$. Substituting the coordinates of the point $(5.5, 37)$ we get

$$37 = b - 122 \cdot 5.5.$$

In other words,

$$b = 37 + 122 \cdot 5.5 = 708.$$

Thus

$$f(t) = 708 - 122t.$$

17. These lines are neither parallel nor perpendicular. They do not have the same slope, nor are their slopes negative reciprocals (if they were, one of the slopes would be negative).

Problems

21. From the table, $r(300) = 120$, which tells us that at a height of 300 m the wind speed is 120 mph.

25. A possible graph is shown in Figure 1.18.

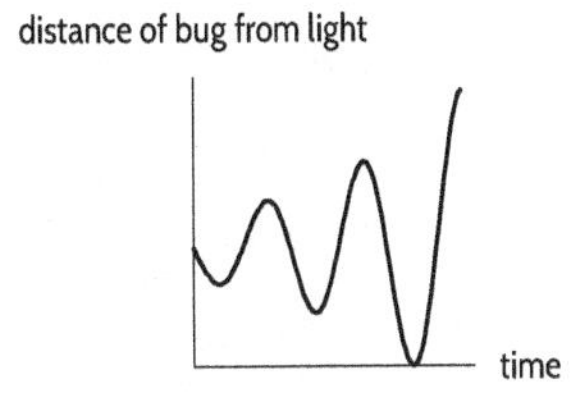

distance of bug from light

time

Figure 1.18

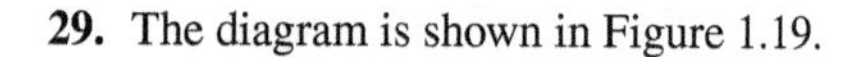

29. The diagram is shown in Figure 1.19.

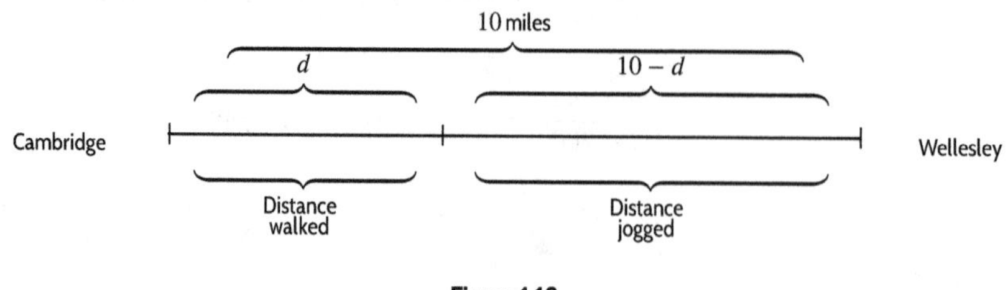

Figure 1.19

The total time the trip takes is given by the equation

$$\text{Total time} = \text{Time walked} + \text{Time jogged}.$$

The distance walked is d, and, since the total distance is 10, the remaining distance jogged is $(10 - d)$. See Figure 1.19. We know that time equals distance over speed, which means that

$$\text{Time walked} = \frac{d}{5} \quad \text{and} \quad \text{Time jogged} = \frac{10-d}{8}.$$

Thus, the total time is given by the equation

$$T(d) = \frac{d}{5} + \frac{10-d}{8}.$$

33. (a) (i) Between $(-1, f(-1))$ and $(3, f(3))$

$$\text{Average rate of change } = \frac{f(3)-f(-1)}{3-(-1)} = \frac{\left(\frac{3}{2}+\frac{5}{2}\right)-\left(\frac{-1}{2}+\frac{5}{2}\right)}{4} = \frac{4-2}{4} = \frac{2}{4} = \frac{1}{2}.$$

(ii) Between $(a, f(a))$ and $(b, f(b))$

$$\text{Average rate of change } = \frac{f(b)-f(a)}{b-a} = \frac{\left(\frac{b}{2}+\frac{5}{2}\right)-\left(\frac{a}{2}+\frac{5}{2}\right)}{b-a} = \frac{\frac{b}{2}+\frac{5}{2}-\frac{a}{2}-\frac{5}{2}}{b-a} = \frac{\frac{b}{2}-\frac{a}{2}}{b-a} = \frac{\frac{1}{2}(b-a)}{b-a} = \frac{1}{2}.$$

(iii) Between $(x, f(x))$ and $(x+h, f(x+h))$

$$\begin{aligned}\text{Average rate of change } &= \frac{f(x+h)-f(x)}{(x+h)-x} = \frac{\left(\frac{x+h}{2}+\frac{5}{2}\right)-\left(\frac{x}{2}+\frac{5}{2}\right)}{(x+h)-x} \\ &= \frac{\frac{x+h}{2}+\frac{5}{2}-\frac{x}{2}-\frac{5}{2}}{x+h-x} = \frac{\frac{x+h-x}{2}}{h} = \frac{\frac{h}{2}}{h} = \frac{1}{2}.\end{aligned}$$

(b) The average rate of change is always $\frac{1}{2}$.

37. (a) $F = 2C + 30$

(b) Since we are finding the difference for a number of values, it would perhaps be easier to find a formula for the difference:

$$\begin{aligned}\text{Difference} &= \text{Approximate value} - \text{Actual value} \\ &= (2C+30) - \left(\frac{9}{5}C + 32\right) = \frac{1}{5}C - 2.\end{aligned}$$

If the Celsius temperature is $-5°$, $(1/5)C - 2 = (1/5)(-5) - 2 = -1 - 2 = -3$. This agrees with our results above.

Similarly, we see that when $C = 0$, the difference is $(1/5)(0) - 2 = -2$ or 2 degrees too low. When $C = 15$, the difference is $(1/5)(15) - 2 = 3 - 2 = 1$ or 1 degree too high. When $C = 30$, the difference is $(1/5)(30) - 2 = 6 - 2 = 4$ or 4 degrees too high.

(c) We are looking for a temperature C, for which the difference between the approximation and the actual formula is zero.

$$\begin{aligned}\frac{1}{5}C - 2 &= 0\\ \frac{1}{5}C &= 2\\ C &= 10\end{aligned}$$

Another way we can solve for a temperature C is to equate our approximation and the actual value.

$$\begin{aligned}\text{Approximation} &= \text{Actual value}\\ 2C + 30 &= 1.8C + 32,\\ 0.2C &= 2\\ C &= 10\end{aligned}$$

So the approximation agrees with the actual formula at 10° Celsius.

41. There are many possible answers. For example, when you buy something, the amount of sales tax depends on the sticker price of the item bought. Let's say Tax = 0.05×Price. This means that the sales tax rate is 5%.

45. (a) If she holds no client meetings, she can hold 30 co-worker meetings. On the other hand, if she holds no co-worker meetings, she can hold 20 client meetings. A graph that describes the relationship is shown in Figure 1.20.

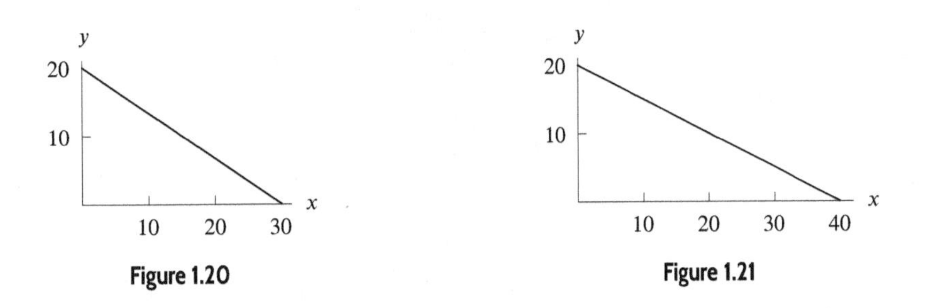

Figure 1.20

Figure 1.21

(b) Since $(0, 20)$ and $(30, 0)$ are on the line, $m = (20 - 0)/(0 - 30) = -(2/3)$. Using the slope intercept form of the line, we have $y = 20 - (2/3)x$.

(c) Since the slope is $-(2/3)$, we know that for every two additional client meetings she must sacrifice three co-worker meetings. Equivalently, for every two fewer client meetings, she gains time for three additional co-worker meetings. The x-intercept is 30. This means that she does not have time for any client meetings at all when she's scheduled 30 co-worker meetings. The y-intercept is 20. This means that she does not have time for any co-worker meetings at all when she's scheduled 20 client meetings.

(d) Instead of 2 hours, co-worker meetings now take $3/2$ hours. If all of her 60 hours are spent in co-worker meetings, she can have $60/(3/2) = 40$ co-worker meetings. The new graph is shown in Figure 1.21. The y-intercept remains at 20. However, the x-intercept is changed to 40. The slope changes, too, from $-(2/3)$ to $-(1/2)$. The new slope is still negative but is less steep because there is less of a decrease in the amount of time available for client meetings due to each extra co-worker meeting.

49. The sloping line has $m = 1$, so its equation is $y = x - 1$. The horizontal line is $y = 3$.

Solving simultaneously gives

$$3 = x - 1 \quad \text{so} \quad x = 4.$$

Thus, the point of intersection is $(4, 3)$.

53. (a) Figure 1.22 shows the data.

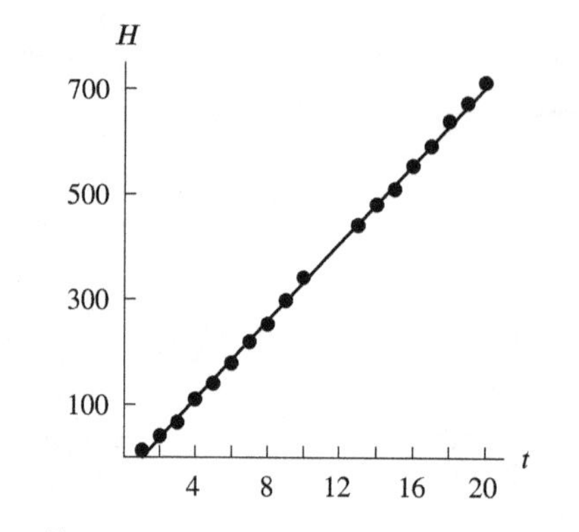

Figure 1.22: Aaron's home-run record from 1954 to 1973

(b) Estimates will vary but the equation $H = 37t - 37$ is typical.

(c) A calculator gives $H = 37.26t - 39.85$, with correlation coefficient $r = 0.9995$, which rounds to $r = 1$. The data set lies very close indeed to the regression line, which has a positive slope. In other words, Aaron's home runs grew at a constant rate over his career.

(d) The slope gives the average number of home runs per year, about 37.

(e) From the answer to part (d) we expect Henry Aaron to hit about 37 home runs in each of the years 1974, 1975, 1976, and 1977. However, the knowledge that Aaron retired in 1976 means that he scored 0 home runs in 1977. Also, people seldom retire at the peak of their abilities, so it is likely that Aaron's performance dropped off in the last few years. In fact he scored 20, 12, and 10 home runs in the years 1974, 1975, and 1976, well below the average of 37.

57. Writing

$$\begin{aligned} y &= \frac{x+k}{z} \\ &= \frac{x}{z} + \frac{k}{z} \\ &= \frac{k}{z} + \frac{1}{z} \cdot x, \end{aligned}$$

and writing

$$\begin{aligned} y &= -3 - \frac{x}{2} \\ &= -3 + \left(-\frac{1}{2}\right) \cdot x, \end{aligned}$$

we see that in order for the coefficients of x to be the same, we have

$$\begin{aligned} \frac{1}{z} &= -\frac{1}{2} \\ z &= -2. \end{aligned}$$

Likewise, for the constant terms to be the same, we have

$$\begin{aligned} \frac{k}{z} &= -3 \\ \frac{k}{-2} &= -3 \qquad \text{because } z = -2 \\ k &= 6, \\ \text{so we have} \quad y &= \frac{x+6}{-2} \qquad \text{because } k = 6, z = -2. \end{aligned}$$

Checking our answer, we see that

$$\begin{aligned} y &= \frac{x+6}{-2} \\ &= \frac{x}{-2} + \frac{6}{-2} \\ &= -3 - \frac{x}{2}. \qquad \text{as required.} \end{aligned}$$

61. **(a)** We know that 75% of David Letterman's 7 million person audience belongs to the nation's work force. Thus

$$\begin{pmatrix} \text{Number of people from the} \\ \text{work force in Dave's audience} \end{pmatrix} = 75\% \text{ of 7 million} = 0.75 \cdot (7 \text{ million}) = 5.25 \text{ million}.$$

Thus the percentage of the work force in Dave's audience is

$$\begin{aligned} \begin{pmatrix} \%\text{ of work force} \\ \text{in audience} \end{pmatrix} &= \left(\frac{\text{People from work force in audience}}{\text{Total work force}}\right) \cdot 100\% \\ &= \left(\frac{5.25}{118}\right) \cdot 100\% = 4.45\%. \end{aligned}$$

(b) Since 4.45% of the work force belongs to Dave's audience, David Letterman's audience must contribute 4.45% of the GDP. Since the GDP is estimated at \$6.325 trillion,

$$\begin{pmatrix} \text{Dave's audience's contribution} \\ \text{to the G.D.P.} \end{pmatrix} = (0.0445) \cdot (6.325 \text{ trillion}) \approx 281 \text{ billion dollars}.$$

(c) Of the contributions by Dave's audience, 10% is estimated to be lost. Since the audience's total contribution is \$281 billion, the "Letterman Loss" is given by

$$\text{Letterman loss } = 0.1 \cdot (281 \text{ billion dollars}) = \$28.1 \text{ billion}.$$

65.
- We know that $\sqrt{3} = 3^{1/2}$, so looking in column $c = 3$, row $r = 2$, we see that $g(8) = 3^{1/2} = \sqrt{3}$. Thus, one possible solution is $n = 8$.
- We know that $\sqrt{3} = 3^{1/2} = 3^{2/4} = \left(3^2\right)^{1/4} = 9^{1/4}$. This means if we look in row $r = 4$, column $c = 9$, we will find another solution. (In fact, if you go to the trouble to write out the table, we find that this solution is $n = 78$.)

STRENGTHEN YOUR UNDERSTANDING

1. False. $f(t)$ is functional notation, meaning that f is a function of the variable t.

5. True. The number of people who enter a store in a day and the total sales for the day are related, but neither quantity is uniquely determined by the other.

9. True. A circle does not pass the vertical line test.

13. True. This is the definition of an increasing function.

17. False. Parentheses must be inserted. The correct ratio is $\dfrac{(10-2^2)-(10-1^2)}{2-1} = -3$.

21. False. Writing the equation as $y = (-3/2)x + 7/2$ shows that the slope is $-3/2$.

25. True. A constant function has slope zero. Its graph is a horizontal line.

29. True. At $y = 0$, we have $4x = 52$, so $x = 13$. The x-intercept is $(13, 0)$.

33. False. Substitute the point's coordinates in the equation: $-3 - 4 \neq -2(4 + 3)$.

37. False. The first line does but the second, in slope-intercept form, is $y = (1/8)x + (1/2)$, so it crosses the y-axis at $y = 1/2$.

41. True. The point $(1, 3)$ is on both lines because $3 = -2 \cdot 1 + 5$ and $3 = 6 \cdot 1 - 3$.

45. True. The slope, $\Delta y/\Delta x$ is undefined because Δx is zero for any two points on a vertical line.

49. False. For example, in children there is a high correlation between height and reading ability, but it is clear that neither causes the other.

53. True. There is a perfect fit of the line to the data.

Solutions to Skills for Chapter 1

1.

$$\begin{aligned} 3x &= 15 \\ \frac{3x}{3} &= \frac{15}{3} \\ x &= 5 \end{aligned}$$

5.

$$\begin{aligned} w - 23 &= -34 \\ w &= -11 \end{aligned}$$

9. The common denominator for this fractional equation is 3. If we multiply both sides of the equation by 3, we obtain:

$$\begin{aligned} 3\left(3t - \frac{2(t-1)}{3}\right) &= 3(4) \\ 9t - 2(t-1) &= 12 \\ 9t - 2t + 2 &= 12 \\ 7t + 2 &= 12 \\ 7t &= 10 \\ t &= \frac{10}{7}. \end{aligned}$$

13. Dividing by w gives $l = A/w$.

17. We collect all terms involving v and then factor out the v.

$$\begin{aligned} u(v+2) + w(v-3) &= z(v-1) \\ uv + 2u + wv - 3w &= zv - z \\ uv + wv - zv &= 3w - 2u - z \\ v(u + w - z) &= 3w - 2u - z \\ v &= \frac{3w - 2u - z}{u + w - z}. \end{aligned}$$

21. Solving for y',

$$\begin{aligned} y'y^2 + 2xyy' &= 4y \\ y'(y^2 + 2xy) &= 4y \\ y' &= \frac{4y}{y^2 + 2xy} \\ y' &= \frac{4}{y + 2x} \text{ if } y \neq 0. \end{aligned}$$

Note that if $y = 0$, then y' could be any real number.

25. We have:

$$\begin{aligned} 3 - 2.1x &< 5 - 0.1x && \\ -2.1x &< 2 - 0.1x && \text{subtract 3 from both sides} \\ -2.1x + 0.1x &< 2 && \text{add } 0.1x \text{ to both sides} \\ -2x &< 2 && \text{add left-hand terms with } x \text{ as a factor} \\ x &> -1. && \text{divide both sides by } -2 \end{aligned}$$

Note that, since we divide by a negative number in the last step, the inequality switches directions. Thus,

$$x > -1.$$

29. We have:

$$
\begin{aligned}
1 + 0.02y &\le 0.001y & \\
1 &\le 0.001y - 0.02y & \text{subtract } 0.02y \text{ from both sides} \\
1 &\le -0.019y & \text{add right-hand terms with } y \text{ as a factor} \\
\frac{1}{-0.019} &\ge y. & \text{divide both sides by } -0.019
\end{aligned}
$$

Note that, since we divide by a negative number in the last step, the inequality switches directions. Thus,

$$\frac{-1}{0.019} \ge y, \quad \text{or, equivalently,} \quad -\frac{1000}{19} \ge y,$$

33. For any two constants a and b we have:

$$
\begin{aligned}
ax - b &> a + 3x & \\
ax &> a + 3x + b & \text{add } b \text{ to both sides} \\
ax - 3x &> a + b & \text{subtract } 3x \text{ from both sides} \\
(a - 3)x &> a + b. & \text{factor out } x \text{ from the left-hand side}
\end{aligned}
$$

The next step is to divide both sides by $a - 3$. However, the inequality must switch directions if $a - 3$ is negative, and must stay the same if $a - 3$ is positive.

(a) If $a > 10$, $a - 3 > 0$. Thus, the inequality stays the same, giving us:

$$x > \frac{a + b}{a - 3}.$$

(b) If $a < 0$, $a - 3 < 0$. Thus, the inequality switches directions, giving us:

$$x < \frac{a + b}{a - 3}.$$

37. Substituting the value of y from the second equation into the first, we obtain

$$
\begin{aligned}
3x - 2(2x - 5) &= 6 \\
3x - 4x + 10 &= 6 \\
-x &= -4 \\
x &= 4.
\end{aligned}
$$

From the second equation, we have

$$y = 2(4) - 5 = 3$$

so

$$y = 3.$$

41. We regard a as a constant. Multiplying the first equation by a and subtracting the second gives

$$
\begin{aligned}
a^2x + ay &= 2a^2 \\
x + ay &= 1 + a^2
\end{aligned}
$$

so, subtracting

$$(a^2 - 1)x = a^2 - 1.$$

Thus $x = 1$ (provided $a \neq \pm 1$). Solving for y in the first equation gives $y = 2a - a(1)$, so $y = a$.

45. Since the y-values of the two lines are equal at the point of intersection, we have:

$$\begin{aligned} 2 + 0.3x &= -5 - 0.5x \\ 2 + 0.8x &= -5 \\ 0.8x &= -7 \\ x &= -\frac{7}{0.8} = -8.75. \end{aligned}$$

We can find the corresponding y-value using either equation:

$$\begin{aligned} \text{First equation:} \quad & y = 2 + 0.3(-8.75) \quad = -0.625 \\ \text{Second equation:} \quad & y = -5 - 0.5(-8.75) = -0.625. \end{aligned}$$

We see that the point of intersection, $(-8.75, -0.625)$, satisfies both equations. The lines and this point are shown in Figure 1.23.

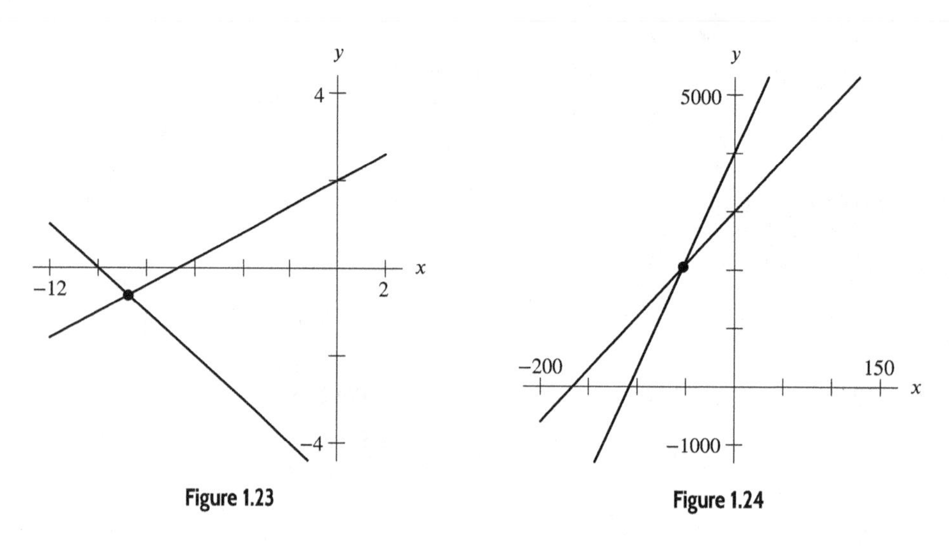

Figure 1.23

Figure 1.24

49. Since the y-values of the two lines are equal at the point of intersection, we have:

$$\begin{aligned} 4000 + 37x &= 3000 + 18x \\ 4000 + 19x &= 3000 \\ 19x &= -1000 \\ x &= \frac{-1000}{19} = -52.6316. \end{aligned}$$

We can find the corresponding y-value using either equation:

$$\begin{aligned} \text{First equation:} \quad & y = 3000 + 18(-1000/19) = 2052.6316 \\ \text{Second equation:} \quad & y = 4000 + 37(-1000/19) = 2052.6316. \end{aligned}$$

We see that the point of intersection, $(-52.6316, 2052.6316)$, satisfies both equations. The lines and this point are shown in Figure 1.24.

53. The radius is 8, so $A = (2, 9)$, $B = (10, 1)$.

CHAPTER TWO

Solutions for Section 2.1

Skill Refresher

S1. $5(x-3) = 5x - 15.$

S5.

$$\begin{aligned} 3\left(1+\frac{1}{x}\right) &= 3\left(\frac{x+1}{x}\right) \\ &= \frac{3x+3}{x}. \end{aligned}$$

S9.

$$\begin{aligned} \frac{21}{z-5} - \frac{13}{z^2-5z} &= 3 \\ \frac{21}{z-5} - \frac{13}{z(z-5)} &= 3 \\ \frac{21z-13}{z(z-5)} &= 3 \\ 21z - 13 &= 3z(z-5) \\ 21z - 13 &= 3z^2 - 15z \\ 3z^2 - 36z + 13 &= 0 \end{aligned}$$

$$\begin{aligned} z &= \frac{-(-36) \pm \sqrt{(-36)^2 - 4(3)(13)}}{2(3)} \\ &= \frac{36 \pm \sqrt{1140}}{6} \\ &= \frac{36 \pm \sqrt{4 \cdot 285}}{6} \\ &= \frac{36 \pm 2\sqrt{285}}{6} \\ &= \frac{18 \pm \sqrt{285}}{3}. \end{aligned}$$

Exercises

1. Substituting -27 for x gives

$$g(-27) = -\frac{1}{2}(-27)^{1/3} = -\frac{1}{2}(-3) = \frac{3}{2}.$$

5. **(a)** Substituting $t = 0$ gives

$$g(0) = \frac{1}{0+2} - 1 = \frac{1}{2} - 1 = -\frac{1}{2}.$$

(b) Setting $g(t) = 0$ and solving gives

$$\begin{aligned}\frac{1}{t+2} - 1 &= 0\\ \frac{1}{t+2} &= 1\\ 1 &= t+2\\ t &= -1.\end{aligned}$$

9.

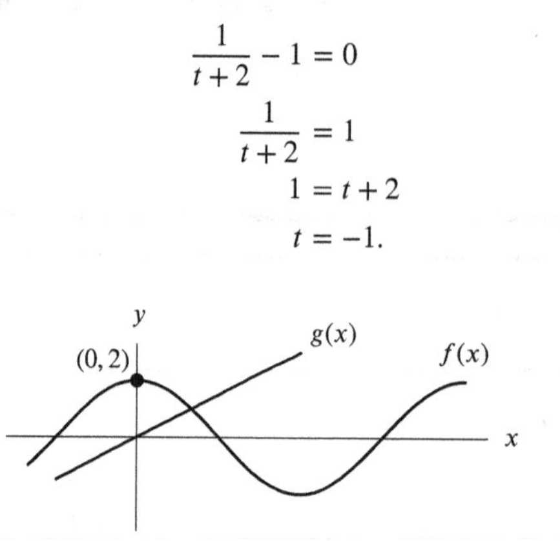

13. The solutions to $g(x) = 2$ are all x values that satisfy $2x + 2 = 2$. Subtracting 2 from both sides, we get $2x = 0$, and dividing by 2, we get $x = 0$. Thus there is only one x value, $x = 0$, with $g(x) = 2$.

Problems

17. The input, t, is the number of months since January 1, and the output, F, is the number of foxes. The expression $g(9)$ represents the number of foxes in the park on October 1. Table 1.3 on page 5 of the text gives $F = 100$ when $t = 9$. Thus, $g(9) = 100$. On October 1, there were 100 foxes in the park.

21. **(a)** $g(100) = 100\sqrt{100} + 100 \cdot 100 = 100 \cdot 10 + 100 \cdot 100 = 11{,}000$
(b) $g(4/25) = 4/25 \cdot \sqrt{4/25} + 100 \cdot 4/25 = 4/25 \cdot 2/5 + 16 = 8/125 + 16 = 16.064$
(c) $g(1.21 \cdot 10^4) = g(12100) = (12100)\sqrt{12100} + 100 \cdot (12100) = 2{,}541{,}000$

25. Since $g(t+h) = 2(t+h) - 1$ we have

$$\frac{g(t+h) - g(t)}{h} = \frac{2(t+h) - 1 - (2t-1)}{h} = \frac{2t + 2h - 1 - 2t + 1}{h} = \frac{2h}{h}.$$

29. **(a)** Substituting $t = 0$ gives $v(0) = 0^2 - 2(0) = 0 - 0 = 0$.
(b) To find when the object has velocity equal to zero, we solve the equation

$$\begin{aligned}t^2 - 2t &= 0\\ t(t-2) &= 0\\ t &= 0 \quad \text{or} \quad t = 2.\end{aligned}$$

Thus the object has velocity zero at $t = 0$ and at $t = 2$.
(c) The quantity $v(3)$ represents the velocity of the object at time $t = 3$. Its units are ft/sec.

33. **(a)** Her tax is \$4650.63 on the first \$77,150 plus 6.65% of income over \$77,150, which is \$88,000 − 77,150 = \$10,850. Thus:

$$\text{Tax owed} = \$4650.63 + 0.0665(\$10{,}850) = \$4650.63 + \$721.53 = \$5372.16.$$

(b) Her taxable income, $T(x)$, is 80% of her total income, or 80% of x. So $T(x) = 0.8x$.
(c) Her tax owed is \$4650.63 plus 6.65% of her taxable income over \$77,150. Since her taxable income is $0.8x$, her taxable income over \$77,850 is $0.8x - 77{,}150$. Therefore,

$$L(x) = 4650.63 + 0.0665(0.8x - 77{,}150),$$

so multiplying out and simplifying, we obtain

$$L(x) = 0.0532x - 479.84$$

(d) Evaluating for $x = \$110{,}000$, we have

$$\begin{aligned} L(110{,}000) &= 0.0532(110{,}000) - 479.84 \\ &= 5372.16. \end{aligned}$$

The values are the same.

37. $r(0.5s_0)$ is the wind speed at a half the height above ground of maximum wind speed.

41. $f(1-a) = \dfrac{a(1-a)}{a+(1-a)} = a(1-a) = a - a^2$

45. This represents the change in average hurricane intensity at average Caribbean Sea surface temperature after CO_2 levels rise to future projected levels.

Solutions for Section 2.2

Skill Refresher

S1. The function is undefined when the denominator is zero. Therefore, $x - 3 = 0$ tells us the function is undefined for $x = 3$.

S5. Adding 8 to both sides of the inequality we get $x > 8$.

S9. $x^2 - 25 > 0$ is true when $x > 5$ or $x < -5$.

Exercises

1. The domain is $1 \le x \le 7$. The range is $2 \le f(x) \le 18$.

5. The graph of $f(x) = 1/x$ for $-2 \le x \le 2$ is shown in Figure 2.1. From the graph, we see that $f(x) = -(1/2)$ at $x = -2$. As we approach zero from the left, $f(x)$ gets more and more negative. On the other side of the y-axis, $f(x) = (1/2)$ at $x = 2$. As x approaches zero from the right, $f(x)$ grows larger and larger. Thus, on the domain $-2 \le x \le 2$, the range is $f(x) \le -(1/2)$ or $f(x) \ge (1/2)$.

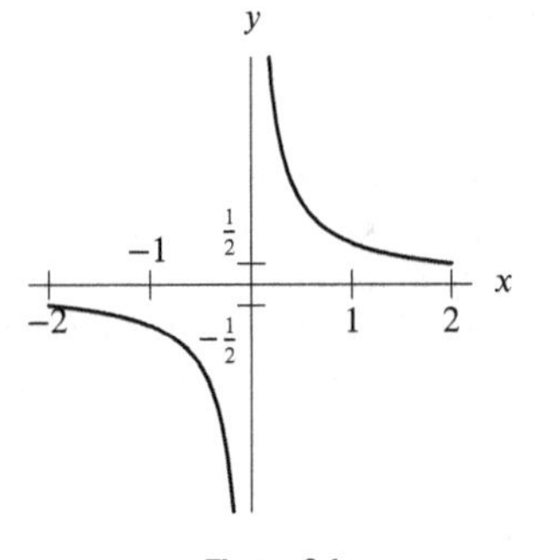

Figure 2.1

9. The domain is all real numbers except those which do not yield an output value. The expression $1/(x+3)$ is defined for any real number x except -3, since for $x = -3$ the denominator of $f(x)$, $x + 3$, is 0 and division by 0 is undefined. Therefore, the domain of $f(x)$ is all real numbers $\neq -3$.

13. To evaluate $f(x)$, we must have $9 + x > 0$. Thus

$$\text{Domain: } x > -9.$$

17. Any number can be squared, so the domain is all real numbers.

21. To evaluate $f(x)$, we must have $x - 4 > 0$. Thus

$$\text{Domain: } x > 4.$$

To find the range, we want to know all possible output values. We solve the equation $y = f(x)$ for x in terms of y. Since

$$y = \frac{1}{\sqrt{x-4}},$$

squaring gives

$$y^2 = \frac{1}{x-4},$$

and multiplying by $x - 4$ gives

$$\begin{aligned} y^2(x-4) &= 1 \\ y^2x - 4y^2 &= 1 \\ y^2x &= 1 + 4y^2 \\ x &= \frac{1+4y^2}{y^2}. \end{aligned}$$

This formula tells us how to find the x-value which corresponds to a given y-value. The formula works for any y except $y = 0$ (which puts a 0 in the denominator). We know that y must be positive, since $\sqrt{x-4}$ is positive, so we have

$$\text{Range: } y > 0.$$

Problems

25. $n(q)$ can be evaluated when $r^2 + a \geq 0$. Note that for all $a \geq 0$, $r^2 + a \geq 0$. Hence the domain of $n(q)$ is all real numbers for any $a \geq 0$.

29. Since the restaurant opens at 2 pm, $t = 0$, and closes at 2 am, $t = 12$, a reasonable domain is $0 \leq t \leq 12$.
Since there cannot be fewer than 0 clients in the restaurant and 200 can fit inside, the range is $0 \leq f(t) \leq 200$.

33. We can put in any number for x except zero, which makes $1/x$ undefined. We note that as x approaches infinity or negative infinity, $1/x$ approaches zero, though it never arrives there, and that as x approaches zero, $1/x$ goes to negative or positive infinity. Thus, the range is all real numbers except a.

37. **(a)** Substituting $t = 0$ into the formula for $p(t)$ shows that $p(0) = 50$, meaning that there were 50 rabbits initially. Using a calculator, we see that $p(10) \approx 131$, which tells us there were about 131 rabbits after 10 months. Similarly, $p(50) \approx 911$ means there were about 911 rabbits after 50 months.

(b) The graph in Figure 2.2 tells us that the rabbit population grew quickly at first but then leveled off at about 1000 rabbits after around 75 months or so. It appears that the rabbit population increased until it reached the island's capacity.

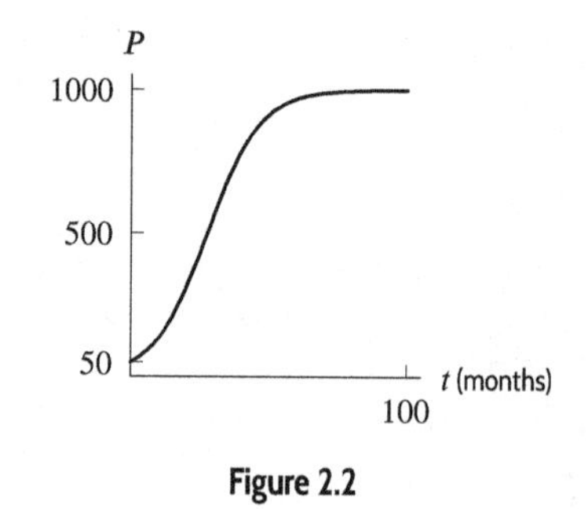

Figure 2.2

(c) From the graph in Figure 2.2, we see that the range is $50 \leq p(t) \leq 1000$. This tells us that (for $t \geq 0$) the number of rabbits is no less than 50 and no more than 1000.

(d) The smallest population occurred when $t = 0$. At that time, there were 50 rabbits. As t gets larger and larger, $(0.9)^t$ gets closer and closer to 0. Thus, as t increases, the denominator of

$$p(t) = \frac{1000}{1 + 19(0.9)^t}$$

decreases. As t increases, the denominator $1 + 19(0.9)^t$ gets close to 1 (try $t = 100$, for example). As the denominator gets closer to 1, the fraction gets closer to 1000. Thus, as t gets larger and larger, the population gets closer and closer to 1000. Thus, the range is $50 \leq p(t) < 1000$.

Solutions for Section 2.3

Skill Refresher

S1. Since the point zero is not included, this graph represents $x > 0$.

S5. Since both end points of the interval are solid dots, this graph represents $x \leq -1$ or $x \geq 2$.

S9. Domain: $-2 \leq x \leq 3$ and range: $-2 \leq x \leq 3$.

Exercises

1. $f(x) = \begin{cases} -1, & -1 \leq x < 0 \\ 0, & 0 \leq x < 1 \\ 1, & 1 \leq x < 2 \end{cases}$ is shown in Figure 2.3.

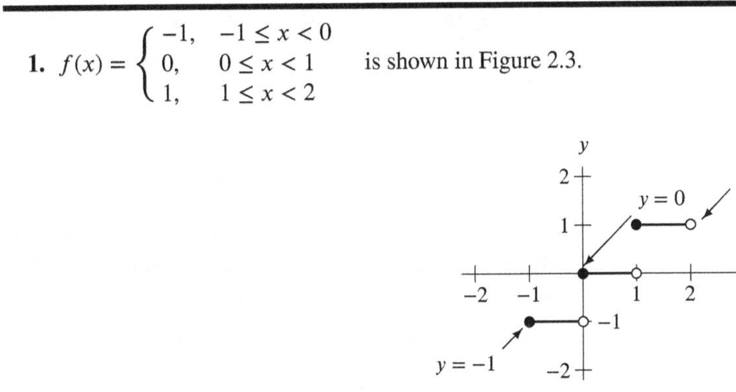

Figure 2.3

5. We find the formulas for each of the lines. For the first, we use the two points we have, $(1, 4)$ and $(3, 2)$. We find the slope: $(2 - 4)/(3 - 1) = -1$. Using the slope of -1, we solve for the y-intercept:

$$\begin{aligned} 4 &= b - 1 \cdot 1 \\ 5 &= b. \end{aligned}$$

Thus, the first line is $y = 5 - x$, and it is for the part of the function where $x < 3$. Notice that we do not use this formula for the value $x = 3$.

We follow the same method for the second line, using the points $(3, \frac{1}{2})$ and $(5, \frac{3}{2})$. We find the slope: $(\frac{3}{2} - \frac{1}{2})/(5 - 3) = \frac{1}{2}$. Using the slope of $\frac{1}{2}$, we solve for the y-intercept:

$$\begin{aligned} \frac{1}{2} &= b + \frac{1}{2} \cdot 3 \\ -1 &= b. \end{aligned}$$

Thus, the second line is $y = -1 + \frac{1}{2}x$, and it is for the part of the function where $x \geq 3$.

Therefore, the function is:

$$y = \begin{cases} 5 - x & \text{for } x < 3 \\ -1 + \frac{1}{2}x & \text{for } x \geq 3. \end{cases}$$

9. Since $G(x)$ is defined for all x, the domain is all real numbers. For $x < -1$ the values of the function are all negative numbers. For $-1 \geq x \geq 0$ the functions values are $4 \geq G(x) \geq 3$, while for $x > 0$ we see that $G(x) \geq 3$ and the values increase to infinity. The range is $G(x) < 0$ and $G(x) \geq 3$.

13. We want to find all numbers x such that $|x - 3| = 7$. That is, we want the distance between x and 3 to be 7. Thus, x must be seven units to the left or seven units to the right of 3; that is, $x = -4$ and $x = 10$.

Problems

17. **(a)** Yes, because every value of x is associated with exactly one value of y.
(b) No, because some values of y are associated with more than one value of x.
(c) Only part (a) leads to a function. Its range is $y = 1, 2, 3, 4$.

21. **(a)** Since zero lies in the interval $-1 \leq x \leq 1$, we find the function value from the formula $f(x) = 3x$. This gives $f(0) = 3 \cdot 0 = 0$. To find $f(3)$, we first note that $x = 3$ lies in the interval $1 < x \leq 5$, so we find the function value from the formula $f(x) = -x + 4$. The result is $f(3) = -3 + 4 = 1$.
(b) By graphing f we can see in Figure 2.4 that the combined domain is $-1 \leq x \leq 5$ and the range is $-3 \leq f(x) \leq 3$.

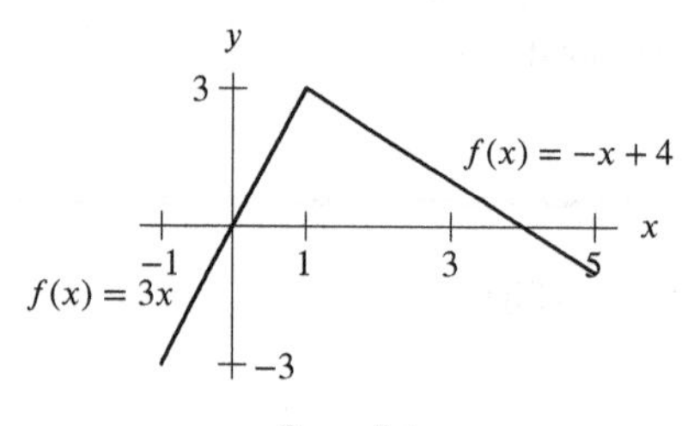

Figure 2.4

25. **(a)** Upon entry, the cost is \$2.50. The tax surcharge of \$0.50 is added to the fare. So, the initial cost will be \$3.00. The cost for the first $1/5$ mile adds \$0.40, giving a fare of \$3.40. For a journey of $2/5$ mile, another \$0.40 is added for a fare of \$3.80. Each additional $1/5$ mile gives an another increment of \$0.40. See Table 2.1.

Table 2.1

Miles	0	0.2	0.4	0.6	0.8	1	1.2	1.4	1.6	1.8	2
Cost	3.00	3.40	3.80	4.20	4.60	5.00	5.40	5.80	6.20	6.60	7.00

(b) The table shows that the cost for a 1.2 mile trip is \$5.40.
(c) From the table, the maximum distance one can travel for 5.80 is 1.4 mile.
(d) See Figure 2.5.

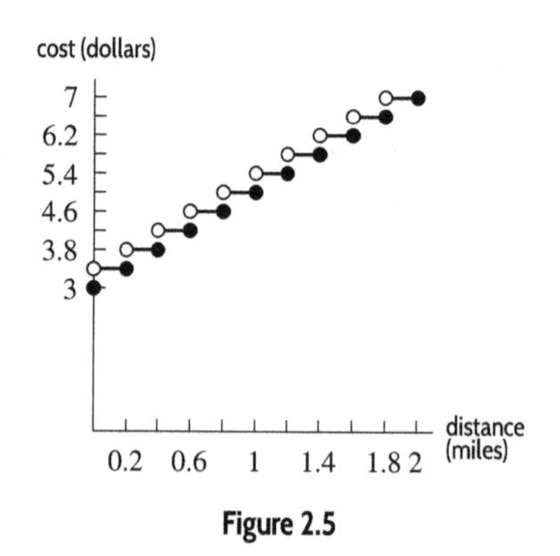

Figure 2.5

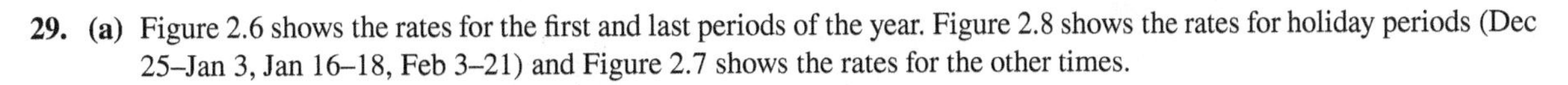

29. **(a)** Figure 2.6 shows the rates for the first and last periods of the year. Figure 2.8 shows the rates for holiday periods (Dec 25–Jan 3, Jan 16–18, Feb 3–21) and Figure 2.7 shows the rates for the other times.

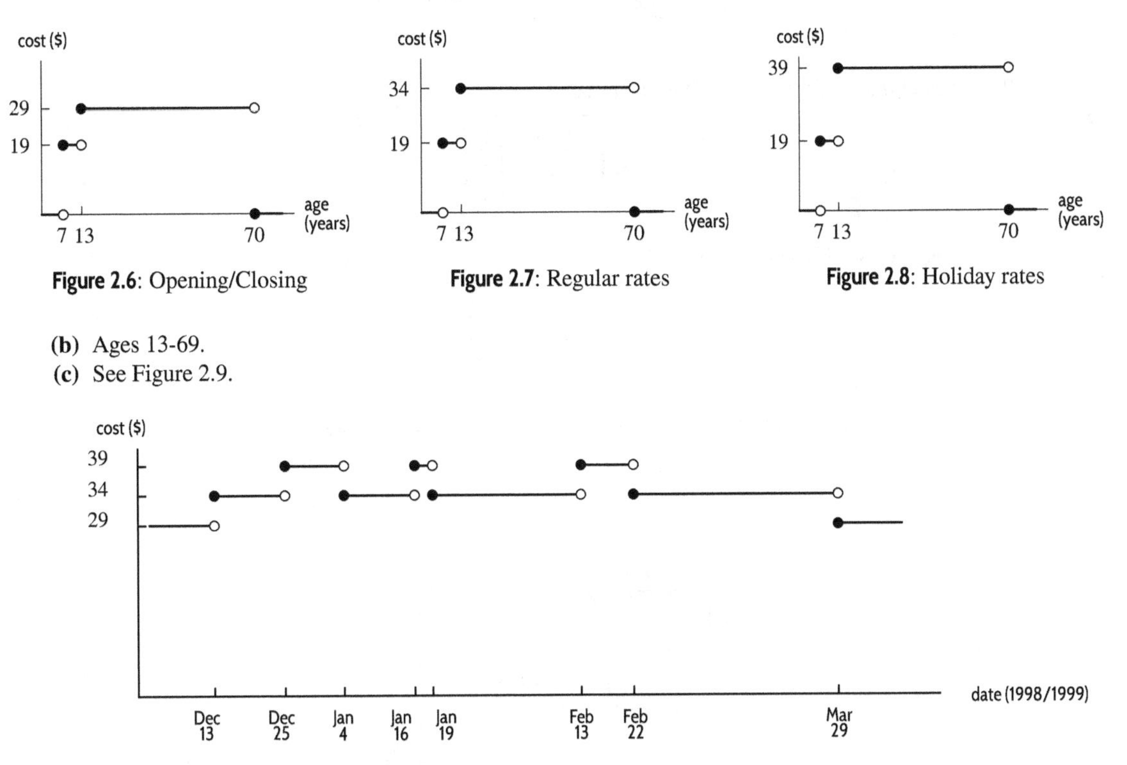

Figure 2.6: Opening/Closing

Figure 2.7: Regular rates

Figure 2.8: Holiday rates

(b) Ages 13-69.

(c) See Figure 2.9.

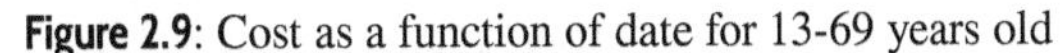

Figure 2.9: Cost as a function of date for 13-69 years old

(d) Rates through 12 December and after 19 March represent early-season and late-season rates, respectively; these are off-peak rates, since it makes economic sense to cut rates when there are fewer skiers. Holiday rates took effect from 25 December through 3 January because of the Christmas/New Year's holiday; they took effect from 16 January through 18 January for Martin Luther King's holiday; they took effect from 13 February through 21 February for the Presidents' Week holiday; it makes economic sense to charge peak rates during the holidays, as more skiers are available to use the facility. Other times represent rates during the heart of the winter skiing season; these are the regular rates.

Solutions for Section 2.4

Skill Refresher

S1. Substituting $x = 4$ into $f(x)$, we have $f(4) = \sqrt{4} = 2$.

S5. The solutions are

$$x = -2 \quad \text{and} \quad x = 2.$$

Exercises

1. **(a)**

x	-1	0	1	2	3
$g(x)$	-3	0	2	1	-1

The graph of $g(x)$ is shifted one unit to the right of $f(x)$.

(b)

x	-3	-2	-1	0	1
$h(x)$	-3	0	2	1	-1

The graph of $h(x)$ is shifted one unit to the left of $f(x)$.

(c)

x	-2	-1	0	1	2
$k(x)$	0	3	5	4	2

The graph $k(x)$ is shifted up three units from $f(x)$.

(d)

x	-1	0	1	2	3
$m(x)$	0	3	5	4	2

The graph $m(x)$ is shifted one unit to the right and three units up from $f(x)$.

5. See Figure 2.10.

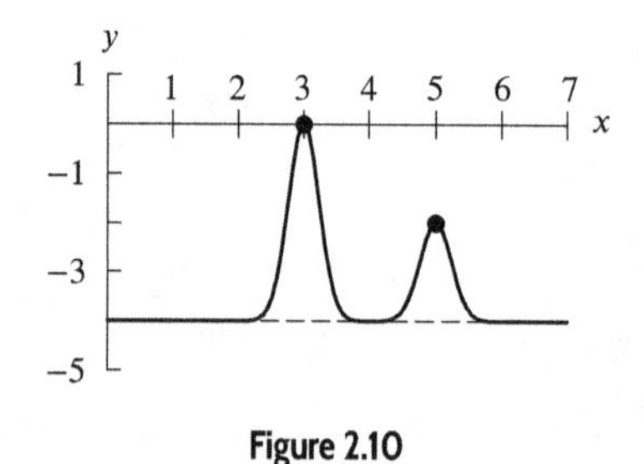

Figure 2.10

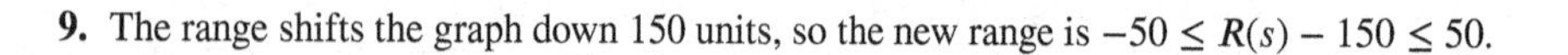

9. The range shifts the graph down 150 units, so the new range is $-50 \leq R(s) - 150 \leq 50$.

Problems

13. **(a)** The translation should leave the x-coordinate unchanged, and shift the y-coordinate up 3; so $y = g(x) + 3$.
(b) The translation should leave the y-coordinate unchanged, and shift the x-coordinate right by 2; so $y = g(x - 2)$.

17. Since $W = s(t + 4)$, at age $t = 3$ months Ben's weight is given by

$$W = s(3 + 4) = s(7).$$

We defined $s(7)$ to be the average weight of a 7-month old baby. At age 3 months, Ben's weight is the same as the average weight of 7-month old babies. Since, on average, a baby's weight increases as the baby grows, this means that Ben is heavier than the average for a 3-month old. Similarly, at age $t = 6$, Ben's weight is given by

$$W = s(6 + 4) = s(10).$$

Thus, at 6 months, Ben's weight is the same as the average weight of 10-month old babies. In both cases, we see that Ben is above average in weight.

21. Since this is an inside change, the graph is four units to the left of $q(z)$. That is, for any given z value, the value of $q(z+4)$ is the same as the value of the function q evaluated four units to the right of z (at $z + 4$).

25. **(a)** There are many possible graphs, but all should show seasonally related cycles of temperature increases and decreases, as in Figure 2.11.

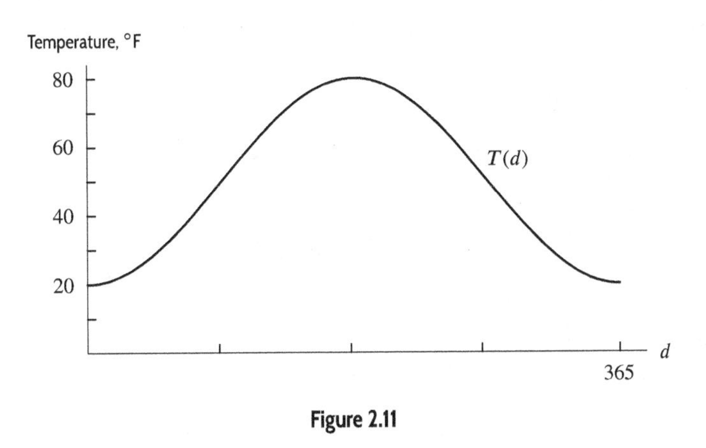

Figure 2.11

(b) While there are a wide variety of correct answers, the value of $T(6)$ is a temperature for a day in early January, $T(100)$ for a day in mid-April, and $T(215)$ for a day in early August. The value for $T(371) = T(365+6)$ should be close to that of $T(6)$.

(c) Since there are usually 365 days in a year, $T(d)$ and $T(d+365)$ represent average temperatures on days which are a year apart.

(d) $T(d+365)$ is the average temperature on the same day of the year a year earlier. They should be about the same value. Therefore, the graph of $T(d+365)$ should be about the same as that of $T(d)$.

(e) The graph of $T(d)+365$ is a shift upward of $T(d)$, by 365 units. It has no significance in practical terms, other than to represent a temperature that is 365° hotter than the average temperature on day d.

Solutions for Section 2.5

Skill Refresher

S1. Adding 4 to both sides and dividing by 3, we get $y = \frac{x+4}{3}$.

S5. Adding 2 to both sides, we get $\sqrt{y} = x + 2$. Squaring both sides, we get

$$y = (x+2)^2 = x^2 + 4x + 4.$$

S9.

$$4x^2 + 4x + 1 - 4 = 4x^2 + 4x - 3.$$

Exercises

1. The domain of $f^{-1}(y)$ is the range of $f(x)$, that is, all numbers between -3 and 2. Thus:

$$\text{Domain of } f^{-1}(y): \ -3 \leq \text{ all real numbers } \leq 2.$$

The range of $f^{-1}(y)$ is the domain of $f(x)$, that is all numbers between -2 and 3. Thus:

$$\text{Range of } f^{-1}(y): \ -2 \leq \text{ all real numbers } \leq 3.$$

5. $f(g(0)) = f(1-0^2) = f(1) = 3 \cdot 1 - 1 = 2.$

9. $f(g(x)) = f(1-x^2) = 3(1-x^2) - 1 = 2 - 3x^2.$

13. $A(f(t))$ is the area, in square centimeters, of the circle at time t minutes.

17. The inverse function, $f^{-1}(T)$, gives the temperature in °F needed if the cake is to bake in T minutes. Units of $f^{-1}(T)$ are °F.

21. Since $y = 2t + 3$, solving for t gives

$$\begin{aligned} 2t + 3 &= y \\ t &= \frac{y-3}{2} \\ f^{-1}(y) &= \frac{y-3}{2}. \end{aligned}$$

25. Since $A = \pi r^2$, solving for r gives

$$\begin{aligned} \frac{A}{\pi} &= r^2 \\ \sqrt{\frac{A}{\pi}} &= r \\ r = f^{-1}(A) &= \sqrt{\frac{A}{\pi}}. \end{aligned}$$

29. We have

$$k(3) = 4(3) - 10 = 12 - 10 = 2.$$

To find $k^{-1}(14)$, we find the value of x so that $k(x) = 14$. If $14 = 4x - 10$, then $24 = 4x$ so $x = 6$. Thus, $k^{-1}(14) = 6$.

33. **(a)** Since the vertical intercept of the graph of f is $(0, b)$, we have $f(0) = b$.
(b) Since the horizontal intercept of the graph of f is $(a, 0)$, we have $f(a) = 0$.
(c) The function f^{-1} goes from y-values to x-values, so to evaluate $f^{-1}(0)$, we want the x-value corresponding to $y = 0$. This is $x = a$, so $f^{-1}(0) = a$.
(d) Solving $f^{-1}(?) = 0$ means finding the y-value corresponding to $x = 0$. This is $y = b$, so $f^{-1}(b) = 0$.

Problems

37. Since $f(A) = A/250$ and $f^{-1}(n) = 250n$, we have

$$\begin{aligned} f^{-1}(f(A)) &= f^{-1}\left(\frac{A}{250}\right) = 250\frac{A}{250} = A. \\ f(f^{-1}(n)) &= f(250n) = \frac{250n}{250} = n. \end{aligned}$$

To interpret these results, we use the fact that $f(A)$ gives the number of gallons of paint needed to cover an area A, and $f^{-1}(n)$ gives the area covered by n gallons. Thus $f^{-1}(f(A))$ gives the area which can be covered by $f(A)$ gallons; that is, A square feet. Similarly, $f(f^{-1}(n))$ gives the number of gallons needed for an area of $f^{-1}(n)$; that is, n gallons.

41. Since we can take the cube root of any number, the domain of $t(a)$ is all real numbers. Since the range of $t(a)$ is the domain of its inverse function, we first compute the inverse of $t(a)$. Let $t(a) = y$. Solving for a gives

$$\begin{aligned} y &= \sqrt[3]{a+1} \\ y^3 &= a + 1 \\ a &= y^3 - 1 \\ t^{-1}(y) &= y^3 - 1. \end{aligned}$$

Since $y^3 - 1$ is defined for any y, the domain of $t^{-1}(y)$ is all real numbers, hence the range of $t(a)$ is all real numbers.

45. **(a)** $G(13)$ is the output corresponding to the input of $t = 13$. So $G(13)$ represents the GDP thirteen years after 2000. This tells us that, in 2013, the gross domestic product was \$16,011.2 billion.
(b) The input to the G^{-1} function is billions of dollars, so its output is a time in years after 2000. Thus, $G^{-1}(13{,}776) = 7$ tells us that, seven years after 2000, the GDP was 13,776 billion dollars. Thus, the GDP was \$13,776 billion in 2007.

49. We can find the inverse function by solving for t in our equation:

$$H = \frac{5}{9}(t - 32)$$
$$\frac{9}{5}H = t - 32$$
$$\frac{9}{5}H + 32 = t.$$

This function gives us the temperature in degrees Fahrenheit if we know the temperature in degrees Celsius.

53. **(a)** $A = f(r) = \pi r^2$
(b) $f(0) = 0$
(c) $f(r+1) = \pi(r+1)^2$. This is the area of a circle whose radius is 1 cm more than r.
(d) $f(r) + 1 = \pi r^2 + 1$. This is the area of a circle of radius r, plus 1 square centimeter more.
(e) Centimeters.

57. **(a)** $f(60) = 30$. A car traveling at 60 km/hr needs 30 meters to stop.
(b) $f(70)$ should be between $f(60) = 30$ and $f(80) = 50$, so we estimate 40 meters.
(c) $f^{-1}(70) = 100$ because $f(100) = 70$. A car that took 70 meters to stop was traveling at 100 km/hr.

Solutions for Section 2.6

Exercises

1. To determine concavity, we calculate the rate of change:

$$\frac{\Delta f(x)}{\Delta x} = \frac{1.3 - 1.0}{1 - 0} = 0.3$$
$$\frac{\Delta f(x)}{\Delta x} = \frac{1.7 - 1.3}{3 - 1} = 0.2$$
$$\frac{\Delta f(x)}{\Delta x} = \frac{2.2 - 1.7}{6 - 3} \approx 0.167.$$

The rates of change are decreasing, so we expect the graph of $f(x)$ to be concave down.

5. The slope of $y = x^2$ is always increasing, so its graph is concave up. See Figure 2.12.

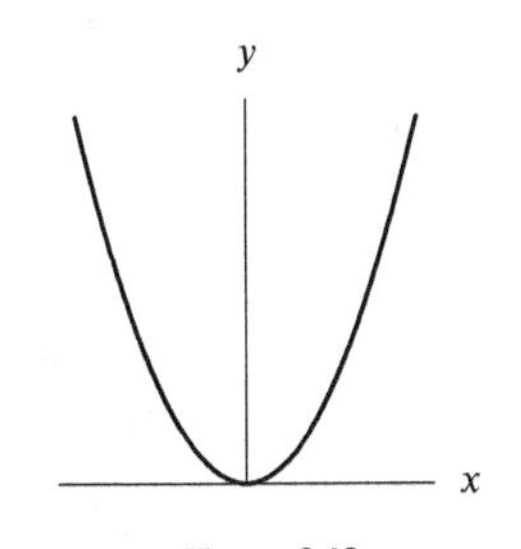

Figure 2.12

9. The rate of change between $x = 12$ and $x = 15$ is

$$\frac{\Delta H(x)}{\Delta x} = \frac{21.53 - 21.40}{15 - 12} \approx 0.043.$$

Similarly, we have

$$\frac{\Delta H(x)}{\Delta x} = \frac{21.75 - 21.53}{18 - 15} \approx 0.073$$
$$\frac{\Delta H(x)}{\Delta x} = \frac{22.02 - 21.75}{21 - 18} \approx 0.090.$$

The rate of change is increasing, so we expect the graph of $H(x)$ to be concave up.

Problems

13. The graph is increasing for all x. It is concave down for $x < o$ and concave up for $x > 0$. Thus we have

(a) Increasing and concave down $x < 0$
(b) Decreasing and concave down nowhere
(c) Increasing and concave up $x > 0$
(d) Decreasing and concave down nowhere

17. Since more and more of the drug is being injected into the body, this is an increasing function. However, since the rate of increase of the drug is slowing down, the graph is concave down.

21. Since new people are always trying the product, it is an increasing function. At first, the graph is concave up. After many people start to use the product, the rate of increase slows down and the graph becomes concave down.

25. **(a)** This is a case in which the rate of decrease is constant, i.e., the change in y divided by the change in x is always the same. We see this in Table (F), where y decreases by 80 units for every decrease of 1 unit in x, and graphically in Graph (IV).

(b) Here, the change in y gets smaller and smaller relative to corresponding changes in x. In Table (G), y decreases by 216 units for a change of 1 unit in x initially, but only decreases by 6 units when x changes by 1 unit from 4 to 5. This is seen in Graph (I), where y is falling rapidly at first, but much more slowly for longer values of x.

(c) If y is the distance from the ground, we see in Table (E) that initially it is changing very slowly; by the end, however, the distance from the ground is changing rapidly. This is shown in Graph (II), where the decrease in y is larger and larger as x gets bigger.

(d) Here, y is decreasing quickly at first, then decreases only slightly for a while, then decreases rapidly again. This occurs in Table (H), where y decreases from 147 units, then 39, and finally by another 147 units. This corresponds to Graph (III).

Solutions for Chapter 2 Review

Exercises

1. To evaluate when $x = -7$, we substitute -7 for x in the function, giving $f(-7) = -\frac{7}{2} - 1 = -\frac{9}{2}$.

5. We have

$$y = f(4) = 4 \cdot 4^{3/2} = 4 \cdot 2^3 = 4 \cdot 8 = 32.$$

Solve for x:

$$\begin{aligned} 4x^{3/2} &= 6 \\ x^{3/2} &= 6/4 \\ x^3 &= 36/16 = 9/4 \\ x &= \sqrt[3]{9/4}. \end{aligned}$$

9. Substituting 4 for t gives

$$P(4) = 170 - 4 \cdot 4 = 154.$$

Similarly, with $t = 2$,

$$P(2) = 170 - 4 \cdot 2 = 162,$$

so

$$P(4) - P(2) = 154 - 162 = -8.$$

13. The expression $x^2 - 9$, found inside the square root sign, must always be non-negative. This happens when $x \geq 3$ or $x \leq -3$, so our domain is $x \geq 3$ or $x \leq -3$.
For the range, the smallest value $\sqrt{x^2 - 9}$ can have is zero. There is no largest value, so the range is $q(x) \geq 0$.

17. Since $m(t)$ is a linear function, the domain of $m(t)$ is all real numbers. For any value of $m(t)$ there is a corresponding value of t. So the range is also all real numbers.

21. **(a)** $2f(x) = 2(1 - x)$.
(b) $f(x) + 1 = (1 - x) + 1 = 2 - x$.
(c) $f(1 - x) = 1 - (1 - x) = x$.
(d) $(f(x))^2 = (1 - x)^2$.
(e) $f(1)/x = (1 - 1)/x = 0$.
(f) $\sqrt{f(x)} = \sqrt{1 - x}$.

25. $P(f(t))$ is the period, in seconds, of the pendulum at time t minutes.

29. $g(f(1)) = g(1^2 + 1) = g(2) = 2 \cdot 2 + 3 = 7$.

33. $g(g(x)) = g(2x + 3) = 2(2x + 3) + 3 = 4x + 9$.

37. The function is defined for all values of x. Since $|x - b| \geq 0$, the range is all numbers greater than or equal to 6.

41. If $P = f(t)$, then $t = f^{-1}(P)$, so $f^{-1}(P)$ gives the time in years at which the population is P million.

45. We find $f(16) = 12 - \sqrt{16} = 8$. If $3 = 12 - \sqrt{x}$, then $\sqrt{x} = 9$, so $x = 81$. Thus, $f^{-1}(3) = 81$

Problems

49. **(a)** Substituting, $q(5) = 3 - (5)^2 = -22$.
(b) Substituting, $q(a) = 3 - a^2$.
(c) Substituting, $q(a - 5) = 3 - (a - 5)^2 = 3 - (a^2 - 10a + 25) = -a^2 + 10a - 22$.
(d) Using the answer to part (b), $q(a) - 5 = 3 - a^2 - 5 = -a^2 - 2$.
(e) Using the answer to part (b) and (a), $q(a) - q(5) = (3 - a^2) - (-22) = -a^2 + 25$.

53. **(a)** Substituting $x = 0$ gives $f(0) = \sqrt{0^2 + 16} - 5 = \sqrt{16} - 5 = 4 - 5 = -1$.
(b) We want to find x such that $f(x) = \sqrt{x^2 + 16} - 5 = 0$. Thus, we have

$$\begin{aligned}
\sqrt{x^2 + 16} - 5 &= 0 \\
\sqrt{x^2 + 16} &= 5 \\
x^2 + 16 &= 25 \\
x^2 &= 9 \\
x &= \pm 3.
\end{aligned}$$

Thus, $f(x) = 0$ for $x = 3$ or $x = -3$.
(c) In part (b), we saw that $f(3) = 0$. You can verify this by substituting $x = 3$ into the formula for $f(x)$:

$$f(3) = \sqrt{3^2 + 16} - 5 = \sqrt{25} - 5 = 5 - 5 = 0.$$

(d) The vertical intercept is the value of the function when $x = 0$. We found this to be -1 in part (a). Thus the vertical intercept is -1.
(e) The graph touches the x-axis when $f(x) = 0$. We saw in part (b) that this occurs at $x = 3$ and $x = -3$.

57. See Figure 2.13.

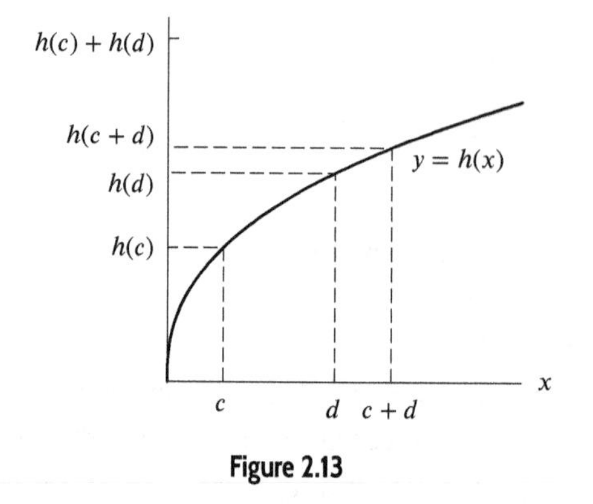

Figure 2.13

61. **(a)** Since the deck is square, $f(s) = s^2$.

(b) Since a can costs \$29.50 and covers 200 ft^2, we know

$$\text{Cost of stain for 1 ft}^2 = \frac{29.50}{200} = 0.1475 \text{ dollars.}$$

Thus

$$\text{Cost of stain for } A \text{ ft}^2 = 0.1475A \text{ dollars,}$$

so

$$C = g(A) = 0.1475A.$$

(c) Substituting for $A = f(s) = s^2$ into g gives

$$C = g(f(s)) = 0.1475s^2.$$

The function $g(f(s))$ gives the cost in dollars of staining a square deck of side s feet.

(d) (i) $f(8) = 8^2 = 64$ square feet; the area of a deck of side 8 feet.

(ii) $g(80) = 0.1475 \cdot 80 = 11.80$ dollars; the cost of staining a deck of area 80 ft^2.

(iii) $g(f(10)) = 0.1475 \cdot 10^2 = 14.75$ dollars; the cost of staining a deck of side 10 feet.

65. **(a)** Since $f(2) = 3$, $f^{-1}(3) = 2$.

(b) Unknown

(c) Since $f^{-1}(5) = 4$, $f(4) = 5$.

69. **(a)**

n	1	2	3	4	5	6	7	8	9	10	11	12
$f(n)$	1	1	2	3	5	8	13	21	34	55	89	144

(b) We note that for every value of n, we can find a unique value for $f(n)$ (by adding the two previous values of the function). This satisfies the definition of function, so $f(n)$ is a function.

(c) Using the pattern, we can figure out $f(0)$ from the fact that we must have

$$f(2) = f(1) + f(0).$$

Since $f(2) = f(1) = 1$, we have

$$1 = 1 + f(0),$$

so

$$f(0) = 0.$$

Likewise, using the fact that $f(1) = 1$ and $f(0) = 0$, we have

$$\begin{aligned} f(1) &= f(0) + f(-1) \\ 1 &= 0 + f(-1) \\ f(-1) &= 1. \end{aligned}$$

Similarly, using $f(0) = 0$ and $f(-1) = 1$ gives

$$\begin{aligned} f(0) &= f(-1) + f(-2) \\ 0 &= 1 + f(-2) \\ f(-2) &= -1. \end{aligned}$$

However, there is no obvious way to extend the definition of $f(n)$ to non-integers, such as $n = 0.5$. Thus we cannot easily evaluate $f(0.5)$, and we say that $f(0.5)$ is undefined.

73.
- A function such as $y = \sqrt{x-4}$ is undefined for $x < 4$, because the input of the square root operation is negative for these x-values.
- A function such as $y = 1/(x-8)$ is undefined for $x = 8$.
- Combining two functions such as these, for example by adding or multiplying them, yields a function with the required domain. Thus, possible formulas include

$$y = \frac{1}{x-8} + \sqrt{x-4} \quad \text{or} \quad y = \frac{\sqrt{x-4}}{x-8}.$$

STRENGTHEN YOUR UNDERSTANDING

1. False. $f(2) = 3 \cdot 2^2 - 4 = 8$.

5. False. $W = (8+4)/(8-4) = 3$.

9. True. A fraction can only be zero if the numerator is zero.

13. False. The domain consists of all real numbers x, $x \neq 3$

17. True. Since f is an increasing function, the domain endpoints determine the range endpoints. We have $f(15) = 12$ and $f(20) = 14$.

21. True. $|x| = |-x|$ for all x.

25. True. If $x < 0$, then $f(x) = x < 0$, so $f(x) \neq 4$. If $x > 4$, then $f(x) = -x < 0$, so $f(x) \neq 4$. If $0 \leq x \leq 4$, then $f(x) = x^2 = 4$ only for $x = 2$. The only solution for the equation $f(x) = 4$ is $x = 2$.

29. True. This looks like the absolute value function shifted right 1 unit and down 2 units.

33. True. To find $f^{-1}(R)$, we solve $R = \frac{2}{3}S + 8$ for S by subtracting 8 from both sides and then multiplying both sides by $(3/2)$.

37. False. Since

$$f(g(x)) = 2\left(\frac{1}{2}x - 1\right) + 1 = x - 1 \neq x,$$

the functions do not undo each other.

41. True. Since the function is concave up, the average rate of change increases as we move right.

45. True. For $x > 0$, the function $f(x) = -x^2$ is both decreasing and concave down.

CHAPTER THREE

Solutions for Section 3.1

Skill Refresher

S1. In this example, we distribute the factors $50t$ and $2t$ across the two binomials $t^2 + 1$ and $25t^2 + 125$, respectively. Thus,

$$\begin{aligned}(t^2+1)(50t) - (25t^2+125)(2t) &= 50t^3 + 50t - (50t^3 + 250t)\\ &= 50t^3 + 50t - 50t^3 - 250t = -200t.\end{aligned}$$

S5. $3x^2 - x - 4 = (3x-4)(x+1)$

S9.

$$\begin{aligned}x^2 + 7x + 6 &= 0\\ (x+6)(x+1) &= 0\\ x+6 = 0 \quad &\text{or} \quad x+1 = 0\\ x = -6 \quad &\text{or} \quad x = -1\end{aligned}$$

Exercises

1. We have:

$$g(t) = 3(t-2)^2 + 7 = 3t^2 - 12t + 19,$$

so $a = 3, b = -12, c = 19$.

5. No. We rewrite the function, giving

$$\begin{aligned}R(q) &= \frac{1}{q^2}(q^2+1)^2\\ &= \frac{1}{q^2}(q^4 + 2q^2 + 1)\\ &= q^2 + 2 + \frac{1}{q^2}\\ &= q^2 + 2 + q^{-2}.\end{aligned}$$

So $R(q)$ is not quadratic since it contains a term with q to a negative power.

9. We solve for r in the equation by factoring

$$\begin{aligned}2r^2 - 6r - 36 &= 0\\ 2(r^2 - 3r - 18) &= 0\\ 2(r-6)(r+3) &= 0.\end{aligned}$$

The solutions are $r = 6$ and $r = -3$.

13. To find the zeros, we solve the equation

$$0 = 9x^2 + 6x + 1.$$

We see that this is factorable, as follows:

$$\begin{aligned}y &= (3x+1)(3x+1)\\ y &= (3x+1)^2.\end{aligned}$$

Therefore, there is only one zero at $x = -\frac{1}{3}$.

17. We solve for r in the equation $Q(r) = 2r^2 - 6r - 36 = 0$ using the quadratic formula with $a = 2$, $b = -6$ and $c = -36$.

$$r = \frac{-(-6) \pm \sqrt{(-6)^2 - 4(2)(-36)}}{2(2)}$$
$$r = \frac{6 \pm \sqrt{36 + 288}}{4}$$
$$r = \frac{6 \pm \sqrt{324}}{4}$$
$$r = \frac{6 \pm 18}{4}.$$

Therefore $r = (6 + 18)/4 = 6$ and $r = (6 - 18)/4 = -3$. The zeros of $Q(r)$ are $r = 6$ and $r = -3$.

21. Without fully multiplying out, we can see that the coefficient of x^2 is 5, so this function has a graph which is concave up.

Problems

25. There is one zero at $x = -2$, so by symmetry the vertex is $(-2, 0)$. We have $y = a(x + 2)^2$. Solving for a, we have

$$a(0 + 2)^2 = 7$$
$$4a = 7$$
$$a = \frac{7}{4},$$

so $y = (7/4)(x + 2)^2$.

29. This is impossible. If the graph is concave down, it opens downward. Then the graph is above the x-axis between the two zeros so could not have a y-intercept of -6.

33. Factoring gives $y = -4cx + x^2 + 4c^2 = x^2 - 4ck + 4c^2 = (x - 2c)^2$. Since $c > 0$, this is the graph of $y = x^2$ shifted to the right $2c$ units. See Figure 3.1.

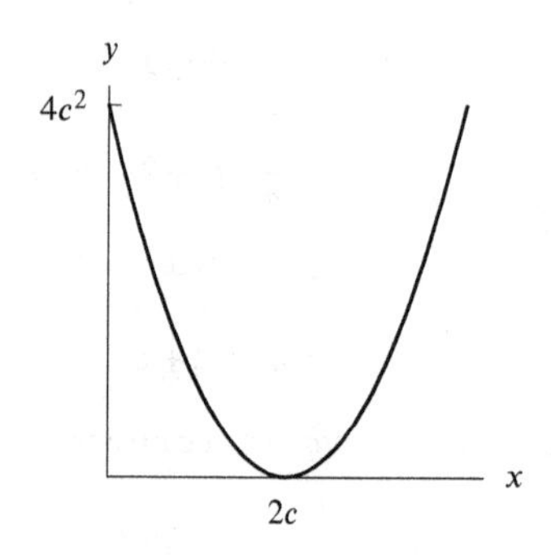

Figure 3.1: $y = -4cx + x^2 + 4c^2$ for $c > 0$

37. Since the parabola has x-intercepts at $x = -1$ and $x = 5$, its formula is:

$$y = a(x + 1)(x - 5).$$

The coordinates $(-2, 6)$ must satisfy the equation, so

$$6 = a(-2 + 1)(-2 - 5).$$

Solving for a gives $a = \frac{6}{7}$. The formula is:

$$y = \frac{6}{7}(x + 1)(x - 5).$$

41. **(a)** At $t = 0$ the snowboarder is 5 meters below the edge of the half-pipe.
(b) We find the zeros of $y = -4.9t^2 + 14t - 5$ using the quadratic formula:

$$t = \frac{-14 \pm \sqrt{14^2 - 4(-4.9)(-5)}}{2(-4.9)}$$
$$t = 0.4184 \text{ or } 2.4387.$$

Thus snowboarder leaves the pipe and flies into the air before she returns to the pipe, so we choose the lower zero. She reaches the air after 0.4184 seconds.

She comes back to the pipe at the second zero, after 2.4387 seconds.

(c) She is in the air from the time she leaves the pipe until the time she returns, from 0.4184 seconds to 2.4387 seconds. Thus, she spends $2.4387 - 0.4184 = 2.0203$ seconds in the air.

45. **(a)** According to the figure in the text, the package was dropped from a height of 5 km.
(b) When the package hits the ground, $h = 0$ and $d = 4430$. So, the package has moved 4430 meters forward when it lands.
(c) Since the maximum is at $d = 0$, the formula is of the form $h = ad^2 + b$ where a is negative and b is positive. Since $h = 5$ at $d = 0$, $5 = a(0)^2 + b = b$, so $b = 5$. We now know that $h = ad^2 + 5$. Since $h = 0$ when $d = 4430$, we have $0 = a(4430)^2 + 5$, giving $a = \frac{-5}{(4430)^2} \approx -0.000000255$. So $h \approx -0.000000255d^2 + 5$.

Solutions for Section 3.2

Skill Refresher

S1. $y^2 - 12y = y^2 - 12y + 36 - 36 = (y - 6)^2 - 36$

S5. Get the variables on the left side, the constants on the right side and complete the square using $(\frac{-6}{2})^2 = 9$.

$$r^2 - 6r = -8$$
$$r^2 - 6r + 9 = 9 - 8$$
$$(r - 3)^2 = 1.$$

Take the square root of both sides and solve for r.

$$r - 3 = \pm 1$$
$$r = 3 \pm 1.$$

So, $r = 4$ or $r = 2$.

S9. Rewrite the equation to equal zero, and factor.

$$n^2 + 4n - 5 = 0$$
$$(n + 5)(n - 1) = 0.$$

So, $n + 5 = 0$ or $n - 1 = 0$, thus $n = -5$ or $n = 1$.

Exercises

1. By comparing $f(x)$ to the vertex form, $y = a(x - h)^2 + k$, we see the vertex is $(h, k) = (1, 2)$. The axis of symmetry is the vertical line through the vertex, so the equation is $x = 1$. The parabola opens upward because the value of a is positive 3.

5. **(a)** See Figure 3.2. For g, we have $a = 1$, $b = 0$, and $c = 3$. Its vertex is at $(0, 3)$, and its axis of symmetry is the y-axis, or the line $x = 0$. This function has no zeros.
(b) See Figure 3.3. For f, we have $a = -2$, $b = 4$, and $c = 16$. The axis of symmetry is the line $x = 1$ and the vertex is at $(1, 18)$. The zeros, or x-intercepts, are at $x = -2$ and $x = 4$. The y-intercept is at $y = 16$.

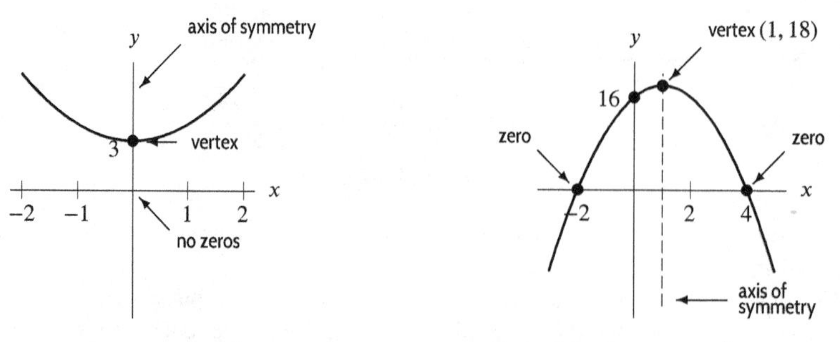

Figure 3.2: $g(x) = x^2 + 3$

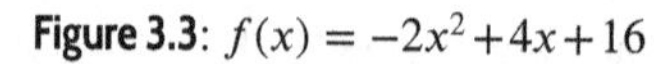
Figure 3.3: $f(x) = -2x^2 + 4x + 16$

9. Since the vertex is $(6, 5)$, we use the form $y = a(x-h)^2 + k$, with $h = 6$ and $k = 5$. We solve for a, substituting in the second point, $(10, 8)$.

$$\begin{aligned} y &= a(x-6)^2 + 5 \\ 8 &= a(10-6)^2 + 5 \\ 3 &= 16a \\ \frac{3}{16} &= a. \end{aligned}$$

Thus, an equation for the parabola is

$$y = \frac{3}{16}(x-6)^2 + 5.$$

13. Letting $y = -3z^2 + 9z - 2$, we have:

$$\begin{aligned} y &= -3z^2 + 9z - 2 && \\ y + 2 &= -3z^2 + 9z && \text{add 2} \\ -\frac{1}{3}(y+2) &= z^2 - 3z && \text{multiply by } -1/3 \\ -\frac{1}{3}(y+2) + \left(\frac{3}{2}\right)^2 &= z^2 - 3x + \left(\frac{3}{2}\right)^2 && \text{complete the square} \\ -\frac{1}{3}(y+2) + \frac{9}{4} &= \left(z - \frac{3}{2}\right)^2 && \text{factor right-hand side} \\ -\frac{1}{3}(y+2) &= \left(z - \frac{3}{2}\right)^2 - \frac{9}{4} && \text{subtract } 9/4 \\ y + 2 &= -3\left(z - \frac{3}{2}\right)^2 + \frac{27}{4} && \text{multiply by } -3 \\ y &= -3\left(z - \frac{3}{2}\right)^2 + \frac{27}{4} - 2 && \text{subtract 2} \\ &= -3\left(z - \frac{3}{2}\right)^2 + \frac{19}{4} && \text{simplify,} \end{aligned}$$

so the vertex is $(3/2, 19/4)$ and the axis of symmetry is $z = 3/2$.

17. $m(n) + 1 = \frac{1}{2}n^2 + 1$

To graph this function, shift the graph of $m(n) = \frac{1}{2}n^2$ one unit up. See Figure 3.4.

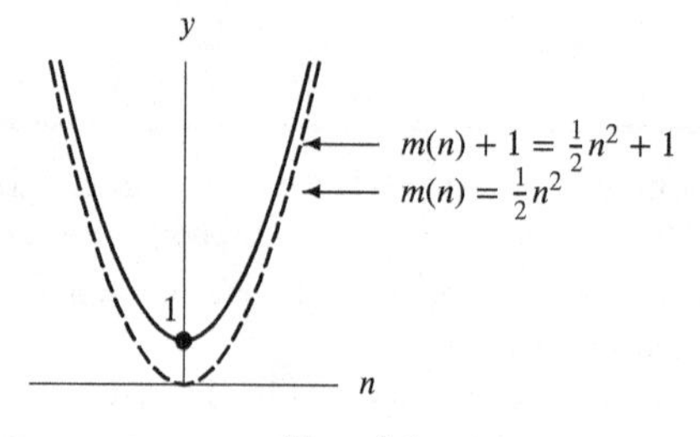

Figure 3.4

21. $m(n) + \sqrt{13} = \dfrac{1}{2}n^2 + \sqrt{13}$

To sketch, shift the graph of $m(n) = \frac{1}{2}n^2$ up by $\sqrt{13}$ units, as in Figure 3.5.

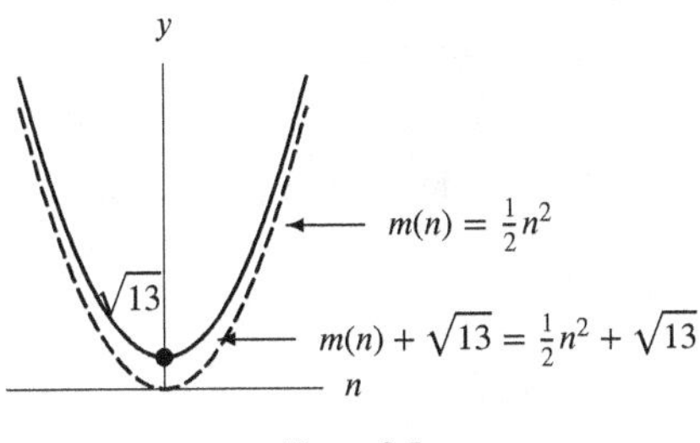

Figure 3.5

Problems

25. We complete the square to write the function in vertex form:

$$\begin{aligned}
y = f(t) &= -16\left(t^2 - \frac{47}{16}t - \frac{3}{16}\right) \\
&= -16\left(t^2 - \frac{47}{16}t + \left(-\frac{47}{32}\right)^2 - \left(-\frac{47}{32}\right)^2 - \frac{3}{16}\right) \\
&= -16\left(\left(t - \frac{47}{32}\right)^2 - \frac{2209}{1024} - \frac{192}{1024}\right) \\
&= -16\left(\left(t - \frac{47}{32}\right)^2 - \frac{2401}{1024}\right) \\
&= -16\left(t - \frac{47}{32}\right)^2 + \frac{2401}{64}.
\end{aligned}$$

Thus, the vertex is at the point $(\frac{47}{32}, \frac{2401}{64})$.

29. Using the vertex form $y = a(x-h)^2 + k$, where $(h,k) = (2,5)$, we have

$$y = a(x-2)^2 + 5.$$

Since the parabola passes through $(1,2)$, these coordinates must satisfy the equation, so

$$2 = a(1-2)^2 + 5.$$

Solving for a gives $a = -3$. The formula is:

$$y = -3(x-2)^2 + 5.$$

33. We have $y = a(x-7)^2 + 3$. Since the point $(3,7)$ is on the curve, we obtain $7 = a(-4)^2 + 3$, so $a = 1/4$. Therefore $(1/4)(x-7)^2 + 3$.

37. We have

$$\begin{aligned}
y &= 0.03x^2 + 1.8x + 2 \\
y - 2 &= 0.03x^2 + 1.8x \\
&= 0.03\left(x^2 + 60x\right) && \text{factor} \\
\frac{y-2}{0.03} &= x^2 + 60x \\
\frac{y-2}{0.03} + (30)^2 &= x^2 + 60x + (30)^2 && \text{complete the square}
\end{aligned}$$

$$\begin{aligned}
\frac{y-2}{0.03}+900 &= (x+30)^2 \qquad \text{factor}\\
\frac{y-2}{0.03} &= (x+30)^2-900\\
y-2 &= 0.03(x+30)^2-0.03(900)\\
y &= 0.03\,(x-(-30))^2-25.
\end{aligned}$$

This is a quadratic function in vertex form with vertex $(h,k)=(-30,-25)$ and $a=0.03$. From the original equation, we see that the y-intercept is $y=2$. Since $a>0$, the graph opens up, and since the vertex lies below the x-axis, there are two x-intercepts. Solving for $y=0$ gives

$$\begin{aligned}
0.03(x+30)^2-25 &= 0\\
0.03(x+30)^2 &= 25\\
(x+30)^2 &= \frac{25}{0.03}\\
&= \frac{2500}{3}\\
x+30 &= \pm\sqrt{\frac{2500}{3}}\\
x &= -30\pm\sqrt{\frac{2500}{3}},
\end{aligned}$$

so the x-intercepts are $x=-58.868$ and $x=-1.133$. See Figure 3.6.

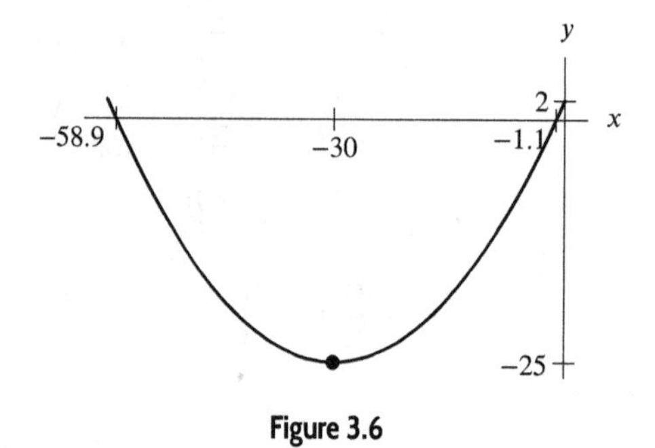

Figure 3.6

41. The graph of $y=x^2-10x+25$ appears to be the graph of $y=x^2$ moved to the right by 5 units. See Figure 3.7. If this were so, then its formula would be $y=(x-5)^2$. Since $(x-5)^2=x^2-10x+25$, $y=x^2-10x+25$ is, indeed, a horizontal shift of $y=x^2$.

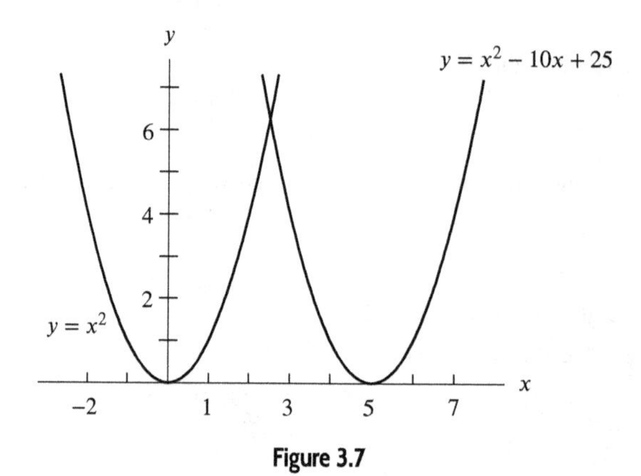

Figure 3.7

45. **(a)** See Figure 3.8.

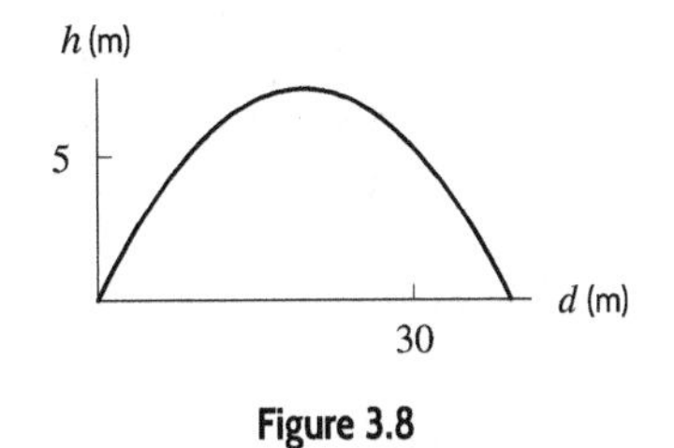

Figure 3.8

(b) When the ball hits the ground $h = 0$, so $h = 0.75d - 0.0192d^2 = d(0.75 - 0.0192d) = 0$ and we get $d = 0$ or $d \approx 39.063$ m. Since $d = 0$ is the position where the kicker is standing, the ball must hit the ground about 39.063 meters from the point where it is kicked.

(c) The path is parabolic and the maximum height occurs at the vertex, which lies on the axis of symmetry, midway between the zeros at $d \approx 19.531$ m. Since $h = 0.75(19.531) - 0.0192(19.531)^2 \approx 7.324$, we know that the ball reaches 7.324 meters above the ground before it begins to fall.

(d) From part (c), the horizontal distance traveled when the ball reaches its maximum height is ≈ 19.531 m.

Solutions for Chapter 3 Review

Exercises

1. Yes. We rewrite the function, giving

$$\begin{aligned} f(x) &= 2(7-x)^2 + 1 \\ &= 2(49 - 14x + x^2) + 1 \\ &= 98 - 28x + 2x^2 + 1 \\ &= 2x^2 - 28x + 99. \end{aligned}$$

So $f(x)$ is quadratic with $a = 2$, $b = -28$ and $c = 99$.

5. To find the zeros, we solve the equation

$$0 = 4x^2 - 4x - 8.$$

We see that this is factorable, as follows:

$$\begin{aligned} 0 &= 4(x^2 - x - 2) \\ 0 &= 4(x-2)(x+1). \end{aligned}$$

Therefore, the zeros occur where $x = 2$ and $x = -1$.

9. Using the quadratic formula, we have

$$\begin{aligned} x &= \frac{-(-2) \pm \sqrt{(-2)^2 - 4(3)(6)}}{2(3)} \\ &= \frac{2 \pm \sqrt{-68}}{6}, \end{aligned}$$

so there are no real-valued zeros.

13. We have $y = a(x-4)^2 - 2$, and if $x = 0$, $y = -3$, so $-3 = a(16) - 2$. Therefore $a = -1/16$, and we have $y = (-1/16)(x-4)^2 - 2$.

17. Since the vertex is $(4, 7)$, we use the form $y = a(x - h)^2 + k$, with $h = 4$ and $k = 7$. We solve for a, substituting in the second point, $(0, 4)$.

$$\begin{aligned} y &= a(x-4)^2 + 7 \\ 4 &= a(0-4)^2 + 7 \\ -3 &= 16a \\ -\frac{3}{16} &= a. \end{aligned}$$

Thus, an equation for the parabola is

$$y = -\frac{3}{16}(x-4)^2 + 7.$$

Problems

21. By inspection the vertex is $(3/4, -2/3)$. The axis of symmetry is $x = 3/4$. The y-intercept occurs when $x = 0$, so $y = 2(3/4)^2 - 2/3 = 11/24$. Since the coefficient a is positive the curve is concave up.

25. In factored form, we have

$$\begin{aligned} y &= 0.3x^2 - 0.6x - 7.2 \\ &= 0.3(x^2 - 2x - 24) \\ &= 0.3(x-6)(x+4). \end{aligned}$$

Therefore, the zeros are at $x = 6$ and $x = -4$. The axis of symmetry is midway between the zeros, so its equation is $x = 1$. The vertex occurs on the axis of symmetry, so substituting $x = 1$ into $y = 0.3(x-6)(x+4)$ gives $y = -7.5$. Hence the vertex is $(1, -7.5)$.

29. Between $x = -2$ and $x = 0$

$$\frac{\Delta f(x)}{\Delta x} = \frac{f(0) - f(-2)}{0 - (-2)} = \frac{(0-1)^2 + 2 - ((-2-1)^2 + 2)}{2} = -4.$$

Between $x = 0$ and $x = 2$

$$\frac{\Delta f(x)}{\Delta x} = \frac{f(2) - f(0)}{2 - 0} = \frac{(2-1)^2 + 2 - ((0-1)^2 + 2)}{2} = 0.$$

Between $x = 2$ and $x = 4$

$$\frac{\Delta f(x)}{\Delta x} = \frac{f(4) - f(2)}{4 - 2} = \frac{(4-1)^2 + 2 - ((2-1)^2 + 2)}{2} = 4.$$

Since rates of change are increasing, the graph of $f(x)$ is concave up.

STRENGTHEN YOUR UNDERSTANDING

1. True. It is of the form $f(x) = a(x - r)(x - s)$ where $a = 1$, $r = 0$ and $s = -2$.

5. False. The time when the object hits the ground is when the height is zero, $s(t) = 0$. The value $s(0)$ gives the height when the object is launched at $t = 0$.

9. False. A quadratic function may have two, one, or no zeros.

13. False. The vertex is located at the point (h, k).

Solutions to Skills for Factoring

1. $2(3x - 7) = 6x - 14$

5. $x(2x+5) = 2x^2 + 5x$

9. $5z(x-2) - 3(x-2) = 5xz - 10z - 3x + 6$

13. $(y+1)(z+3) = yz + 3y + z + 3$

17. $(x-5)6 - 5(1-(2-x)) = 6x - 30 - 5(1-2+x) = 6x - 30 + 5 - 5x = x - 25.$

21. First we square $\sqrt{2x}+1$ and then take the negative of this result. Therefore,

$$\begin{aligned} -\left(\sqrt{2x}+1\right)^2 &= -\left(\sqrt{2x}+1\right)\left(\sqrt{2x}+1\right) = -\left(2x+\sqrt{2x}+\sqrt{2x}+1\right) \\ &= -(2x+2\sqrt{2x}+1) = -2x - 2\sqrt{2x} - 1. \end{aligned}$$

25. $5z - 30 = 5(z-6)$

29. $3u^7 + 12u^2 = 3u^2(u^5+4)$

33. $x^2 - 3x + 2 = (x-2)(x-1)$

37. Can be factored no further.

41. $x^2 + 3x - 28 = (x+7)(x-4)$

45. $x^2 + 2xy + 3xz + 6yz = x(x+2y) + 3z(x+2y) = (x+2y)(x+3z).$

49. We notice that the only factors of 24 whose sum is -10 are -6 and -4. Therefore,

$$B^2 - 10B + 24 = (B-6)(B-4).$$

53. This example is factored as the difference of perfect squares. Thus,

$$\begin{aligned} (t+3)^2 - 16 &= ((t+3)-4)\,((t+3)+4) \\ &= (t-1)(t+7). \end{aligned}$$

Alternatively, we could arrive at the same answer by multiplying the expression out and then factoring it.

57.

$$\begin{aligned} c^2d^2 - 25c^2 - 9d^2 + 225 &= c^2(d^2-25) - 9(d^2-25) \\ &= (d^2-25)(c^2-9) \\ &= (d+5)(d-5)(c+3)(c-3). \end{aligned}$$

61. The common factor is xe^{-3x}. Therefore,

$$x^2e^{-3x} + 2xe^{-3x} = xe^{-3x}(x+2).$$

65. $dk + 2dm - 3ek - 6em = d(k+2m) - 3e(k+2m) = (k+2m)(d-3e).$

69.

$$x = \frac{-3 \pm \sqrt{3^2 - 4(4)(-15)}}{2(4)}$$

$$x = \frac{-3 \pm \sqrt{249}}{8}$$

73.

$$\begin{aligned} -16t^2 + 96t + 12 &= 60 \\ -16t^2 + 96t - 48 &= 0 \\ t^2 - 6t + 3 &= 0 \end{aligned}$$

$$t = \frac{-(-6) \pm \sqrt{(-6)^2 - 4(1)(3)}}{2(1)}$$

$$t = \frac{6 \pm \sqrt{24}}{2} = \frac{6 \pm 2\sqrt{6}}{2}$$

$$t = 3 \pm \sqrt{6}.$$

77.

$$\begin{aligned} N^2 - 2N - 3 &= 2N(N-3) \\ N^2 - 2N - 3 &= 2N^2 - 6N \\ N^2 - 4N + 3 &= 0 \\ (N-3)(N-1) &= 0 \\ N = 3 &\text{ or } N = 1 \end{aligned}$$

81. We rewrite the quadratic equation in standard form and use the quadratic formula. So

$$\begin{aligned} 60 &= -16t^2 + 96t + 12 \\ 16t^2 - 96t + 48 &= 0 \\ t^2 - 6t + 3 &= 0 \\ t &= \frac{-(-6) \pm \sqrt{(-6)^2 - 4(1)(3)}}{2} = \frac{6 \pm \sqrt{36-12}}{2} \\ &= \frac{6 \pm \sqrt{24}}{2} = \frac{6 \pm 2\sqrt{6}}{2} = 3 \pm \sqrt{6}. \end{aligned}$$

85. To find the common denominator, we factor the second denominator

$$\begin{aligned} \frac{2}{z-3} + \frac{7}{z^2 - 3z} &= 0 \\ \frac{2}{z-3} + \frac{7}{z(z-3)} &= 0 \end{aligned}$$

which produces a common denominator of $z(z-3)$. Therefore:

$$\begin{aligned} \frac{2z}{z(z-3)} + \frac{7}{z(z-3)} &= 0 \\ \frac{2z+7}{z(z-3)} &= 0 \\ 2z + 7 &= 0 \\ z &= -\frac{7}{2}. \end{aligned}$$

89. We can solve this equation by squaring both sides.

$$\begin{aligned} \sqrt{r^2 + 24} &= 7 \\ r^2 + 24 &= 49 \\ r^2 &= 25 \\ r &= \pm 5 \end{aligned}$$

93. Multiply by $(x-5)(x-1)$ on both sides of the equation, giving

$$(3x+4)(x-2) = 0.$$

So, $3x + 4 = 0$, or $x - 2 = 0$, that is,

$$x = -\frac{4}{3}, \quad x = 2.$$

97. For a fraction to equal zero, the numerator must equal zero. So, we solve

$$x^2 - 5mx + 4m^2 = 0.$$

Since $x^2 - 5mx + 4m^2 = (x-m)(x-4m)$, we know that the numerator equals zero when $x = 4m$ and when $x = m$. But for $x = m$, the denominator will equal zero as well. So, the fraction is undefined at $x = m$, and the only solution is $x = 4m$.

101. We substitute the expression $4 - x^2$ for y in the second equation.

$$\begin{aligned} y - 2x &= 1 \\ 4 - x^2 - 2x &= 1 \\ -x^2 - 2x + 3 &= 0 \\ x^2 + 2x - 3 &= 0 \\ (x+3)(x-1) &= 0 \end{aligned}$$

$$x = -3 \quad \text{and} \quad y = 4 - (-3)^2 = -5 \quad \text{or}$$
$$x = 1 \quad \text{and} \quad y = 4 - 1^2 = 3$$

105. Solving $y = x^2$ and $y = 15 - 2x$ simultaneously, we have

$$\begin{aligned} x^2 &= 15 - 2x \\ x^2 + 2x - 15 &= 0 \\ (x+5)(x-3) &= 0 \\ x &= -5, 3. \end{aligned}$$

Thus, the points of intersection are $(-5, 25)$, $(3, 9)$.

Solutions to Skills for Completing the Square

1. $x^2 + 8x = x^2 + 8x + 16 - 16 = (x+4)^2 - 16.$

5. We add and subtract the square of half the coefficient of the a-term, $(\frac{-2}{2})^2 = 1$, to get

$$\begin{aligned} a^2 - 2a - 4 &= a^2 - 2a + 1 - 1 - 4 \\ &= (a^2 - 2a + 1) - 1 - 4 \\ &= (a-1)^2 - 5. \end{aligned}$$

9. Completing the square yields

$$x^2 - 2x - 3 = (x^2 - 2x + 1) - 1 - 3 = (x-1)^2 - 4.$$

13. Complete the square and write in vertex form.

$$\begin{aligned} y &= x^2 + 6x + 3 \\ &= x^2 + 6x + 9 - 9 + 3 \\ &= (x+3)^2 - 6. \end{aligned}$$

The vertex is $(-3, -6)$.

17. Complete the square and write in vertex form.

$$\begin{aligned} y &= -x^2 + x - 6 \\ &= -(x^2 - x + 6) \\ &= -\left(x^2 - x + \frac{1}{4} - \frac{1}{4} + 6\right) \\ &= -\left(\left(x - \frac{1}{2}\right)^2 - \frac{1}{4} + 6\right) \\ &= -\left(x - \frac{1}{2}\right)^2 - \frac{23}{4}. \end{aligned}$$

The vertex is $(1/2, -23/4)$.

21. Complete the square and write in vertex form.

$$\begin{aligned} y &= 2x^2 - 7x + 3 \\ &= 2\left(x^2 - \frac{7}{2}x + \frac{3}{2}\right) \\ &= 2\left(x^2 - \frac{7}{2}x + \frac{49}{16} - \frac{49}{16} + \frac{3}{2}\right) \\ &= 2\left(\left(x - \frac{7}{4}\right)^2 - \frac{49}{16} + \frac{3}{2}\right) \\ &= 2\left(x - \frac{7}{4}\right)^2 - \frac{25}{8}. \end{aligned}$$

The vertex is $(7/4, -25/8)$.

25. Complete the square with $(\frac{1}{2})^2 = \frac{1}{4}$ and take the square root of both sides to solve for d.

$$\begin{aligned} d^2 - d + \frac{1}{4} &= \frac{1}{4} + 2 \\ \left(d - \frac{1}{2}\right)^2 &= \frac{9}{4} \\ d - \frac{1}{2} &= \pm\frac{3}{2} \\ d &= \frac{1}{2} \pm \frac{3}{2}. \end{aligned}$$

So $d = 2$ or $d = -1$.

29. Complete the square on the left side.

$$\begin{aligned} 5\left(p^2 + \frac{9}{5}p\right) &= 1 \\ 5\left(p^2 + \frac{9}{5}p + \frac{81}{100}\right) &= 5\left(\frac{81}{100}\right) + 1 \\ 5\left(p + \frac{9}{10}\right)^2 &= \frac{81}{20} + 1 \\ 5\left(p + \frac{9}{10}\right)^2 &= \frac{101}{20}. \end{aligned}$$

Divide by 5 and take the square root of both sides to solve for p.

$$\begin{aligned} \left(p + \frac{9}{10}\right)^2 &= \frac{101}{100} \\ p + \frac{9}{10} &= \pm\sqrt{\frac{101}{100}} \\ p + \frac{9}{10} &= \pm\frac{\sqrt{101}}{10} \\ p &= -\frac{9}{10} \pm \frac{\sqrt{101}}{10}. \end{aligned}$$

33. Set the equation equal to zero, $w^2 + w - 4 = 0$. With $a = 1$, $b = 1$, and $c = -4$, we use the quadratic formula,

$$\begin{aligned} w &= \frac{-b \pm \sqrt{b^2 - 4ac}}{2a} \\ &= \frac{-1 \pm \sqrt{1^2 - 4 \cdot 1 \cdot (-4)}}{2 \cdot 1} \\ &= \frac{-1 \pm \sqrt{1 + 16}}{2} \end{aligned}$$

$$= \frac{-1 \pm \sqrt{17}}{2}.$$

37. Solve by completing the square using $\left(\frac{3}{2}\right)^2 = \frac{9}{4}$.

$$s^2 + 3s + \frac{9}{4} = 1 + \frac{9}{4}$$
$$\left(s + \frac{3}{2}\right)^2 = 1 + \frac{9}{4}$$
$$\left(s + \frac{3}{2}\right)^2 = \frac{13}{4}.$$

Taking the square root of both sides and solving for s,

$$s + \frac{3}{2} = \pm\sqrt{\frac{13}{4}}$$
$$s + \frac{3}{2} = \pm\frac{\sqrt{13}}{2}$$
$$s = -\frac{3}{2} \pm \frac{\sqrt{13}}{2}$$
$$s = \frac{-3 \pm \sqrt{13}}{2}.$$

41. Simplify by dividing by 3 and solve by completing the square.

$$y^2 = 2y + 6$$
$$y^2 - 2y = 6$$
$$y^2 - 2y + 1 = 1 + 6$$
$$(y - 1)^2 = 7.$$

Take the square root of both sides and solve for y to get $y = 1 \pm \sqrt{7}$.

45. Use the quadratic formula with $a = 49$, $b = 70$, $c = 22$, to solve this equation.

$$m = \frac{-70 \pm \sqrt{70^2 - 4 \cdot 49 \cdot 22}}{2 \cdot 49}$$
$$= \frac{-70 \pm \sqrt{4900 - 4312}}{98}$$
$$= \frac{-70 \pm \sqrt{588}}{98}$$
$$= \frac{-70 \pm 14\sqrt{3}}{98}$$
$$= \frac{-5 \pm \sqrt{3}}{7}.$$

CHAPTER FOUR

Solutions for Section 4.1

Skill Refresher

S1. We have $6\% = 0.06$.

Exercises

1. Yes. Writing the function as

$$g(w) = 2\left(2^{-w}\right) = 2\left(2^{-1}\right)^{w} = 2\left(\frac{1}{2}\right)^{w},$$

we have $a = 2$ and $b = 1/2$.

5. No. The base must be a constant.

9. Yes. Writing the function as

$$K(x) = \frac{2^x}{3 \cdot 3^x} = \frac{1}{3}\left(\frac{2^x}{3^x}\right) = \frac{1}{3}\left(\frac{2}{3}\right)^{x},$$

we have $a = 1/3$ and $b = 2/3$.

13. We can rewrite this as

$$\begin{aligned} Q &= 0.0022(2.31^{-3})^t \\ &= 0.0022(0.0811)^t, \end{aligned}$$

so $a = 0.0022$, $b = 0.0811$, and $r = b - 1 = -0.9189 = -91.89\%$.

17. The growth factor per century is 1+ the growth per century. Since the forest is shrinking, the growth is negative, so we subtract 0.80, giving 0.20.

Problems

21. If an investment decreases by 5% each year, we know that only 95% remains at the end of the first year. After 2 years there will be 95% of 95%, or 0.95^2 left. After 4 years, there will be $0.95^4 \approx 0.81451$ or 81.451% of the investment left; it therefore decreases by about 18.549% altogether.

25. **(a)** Since the initial amount is 112.8 and the quantity is decreasing at a rate of 23.4% per year, the formula is $Q = 112.8(1 - 0.234)^t = 112.8(0.766)^t$.

(b) At $t = 10$, we have $Q = 112.8(0.766)^{10} = 7.845$.

29. The population is growing at a rate of 1.9% per year. So, at the end of each year, the population is $100\% + 1.9\% = 101.9\%$ of what it had been the previous year. The growth factor is 1.019. If P is the population of this country, in millions, and t is the number of years since 2014, then, after one year,

$$\begin{aligned} & P = 70(1.019). \\ \text{After two years,}\quad & P = 70(1.019)(1.019) = 70(1.019)^2 \\ \text{After three years,}\quad & P = 70(1.019)(1.019)(1.019) = 70(1.019)^3 \\ \text{After } t \text{ years,}\quad & P = 70\underbrace{(1.019)(1.019)\ldots(1.019)}_{t\text{ times}} = 70(1.019)^t \end{aligned}$$

33. **(a)** We have $C = C_0(1 - r)^t = 100(1 - 0.16)^t = 100(0.84)^t$, so

$$C = 100(0.84)^t.$$

(b) At $t = 5$, we have $C = 100(0.84)^5 = 41.821$ mg

37. Let $D(t)$ be the difference between the oven's temperature and the yam's temperature, which is given by an exponential function $D(t) = ab^t$. The initial temperature difference is 300°F – 0°F = 300°F, so $a = 300$. The temperature difference decreases by 3% per minute, so $b = 1 - 0.03 = 0.97$. Thus,

$$D(t) = 300(0.97)^t.$$

If the yam's temperature is represented by $Y(t)$, then the temperature difference is given by

$$D(t) = 300 - Y(t),$$

so, solving for $Y(t)$, we have

$$Y(t) = 300 - D(t),$$

giving

$$Y(t) = 300 - 300(0.97)^t.$$

41. For $f(x) = 3^x$ on the interval $1 \le x \le 2$ we have

$$\text{Rate of change} = \frac{f(2) - f(1)}{2 - 1} = \frac{3^2 - 3^1}{1} = \frac{9 - 3}{1} = 6.$$

45. **(a)** (i) On the interval $0 \le x \le 1$ we have

$$\text{Rate of change} = \frac{f(1) - f(0)}{1 - 0} = \frac{2^1 - 2^0}{1} = \frac{2 - 1}{1} = 1.$$

(ii) On the interval $1 \le x \le 2$ we have

$$\text{Rate of change} = \frac{f(2) - f(1)}{2 - 1} = \frac{2^2 - 2^1}{1} = \frac{4 - 2}{1} = 2.$$

(iii) On the interval $2 \le x \le 3$ we have

$$\text{Rate of change} = \frac{f(3) - f(2)}{3 - 2} = \frac{2^3 - 2^2}{1} = \frac{8 - 4}{1} = 4.$$

(iv) On the interval $3 \le x \le 4$ we have

$$\text{Rate of change} = \frac{f(4) - f(3)}{4 - 3} = \frac{2^4 - 2^3}{1} = \frac{16 - 8}{1} = 8.$$

(b) The rate of change is increasing by a factor of 2 over consecutive intervals. This means for the interval $19 \le x \le 20$, we expect a rate of change of $2^{19} = 524{,}288$ since the interval $19 \le x \le 20$ is 19 intervals past $0 \le x \le 1$. Checking, we get, as predicted,

$$\text{Rate of change} = \frac{f(20) - f(19)}{20 - 19} = \frac{2^{20} - 2^{19}}{1} = \frac{1{,}048{,}576 - 524{,}288}{1} = 524{,}288.$$

49. We have

$$\begin{aligned} y &= 5(0.5)^{t/3} \\ &= 5\left(2^{-1}\right)^{\frac{1}{3}\cdot t} \\ &= 5 \cdot 2^{-\frac{1}{3}\cdot t} \\ &= 5\left(4^{1/2}\right)^{-\frac{1}{3}\cdot t} \\ &= 5 \cdot 4^{-\frac{1}{6}\cdot t}, \end{aligned}$$

so $a = 5, k = -1/6$.

53. (a) Since $N(0)$ gives the number of teams remaining in the tournament after no rounds have been played, we have $N(0) = 64$. After 1 round, half of the original 64 teams remain in the competition, so

$$N(1) = 64\left(\frac{1}{2}\right).$$

After 2 rounds, half of these teams remain, so

$$N(2) = 64\left(\frac{1}{2}\right)\left(\frac{1}{2}\right).$$

And, after r rounds, the original pool of 64 teams has been halved r times, so that

$$N(r) = 64\underbrace{\left(\frac{1}{2}\right)\left(\frac{1}{2}\right)\cdots\left(\frac{1}{2}\right)}_{\text{Pool halved } r \text{ times}},$$

giving

$$N(r) = 64\left(\frac{1}{2}\right)^r.$$

The graph of $y = N(r)$ is given in Figure 4.1. The domain of N is $0 \le r \le 6$, for r an integer. A curve has been dashed in to help you see the overall shape of the function.

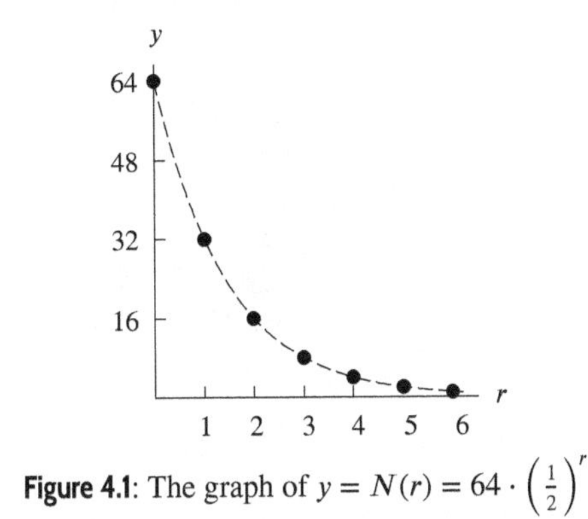

Figure 4.1: The graph of $y = N(r) = 64 \cdot \left(\frac{1}{2}\right)^r$

(b) There is a winner when there is only one team left, Connecticut. So, $N(r) = 1$.

$$\begin{aligned} 64\left(\frac{1}{2}\right)^r &= 1 \\ \left(\frac{1}{2}\right)^r &= \frac{1}{64} \\ \frac{1}{2^r} &= \frac{1}{64} \\ 2^r &= 64 \\ r &= 6. \end{aligned}$$

You can solve $2^r = 64$ either by taking successive powers of 2 until you get to 64 or by substituting values for r until you get the one that works.

Connecticut won after winning 6 rounds.

57. (a) (i) The monthly payment on \$1000 each month at 4% for a loan period of 15 years is \$7.40. For \$60,000, the payment would be $\$7.40 \times 60 = \$444$ per month.

(ii) The monthly payment on \$1000 each month at 4% for a loan period of 30 years is \$4.77. For \$60,000, the payment would be $\$4.77 \times 60 = \$286.20$ per month.

(iii) The monthly payment on \$1000 each month at 6% for a loan period of 15 years is \$8.44. For \$60,000, the payment would be $\$8.44 \times 60 = \$506.40$ per month.

(b) As calculated in part (a)-(i), the monthly payment on a \$60,000 loan at 4% for 15 years would be \$444 per month. In part (a)-(iii) we showed that the the monthly payment on a \$60,000 loan at 6% for 15 years would be \$506.40 per month. So taking the loan out at 4% rather that 6% would save the difference:

$$\text{Amount saved} = \$506.40 - \$444 = \$62.40 \text{ per month}$$

Since there are $15 \times 12 = 180$ months in 15 years,

$$\text{Total amount saved} = \$62.40 \text{ per month} \times 180 \text{ months} = \$11{,}232.$$

(c) In part (a)-(i) we found the monthly payment on an 4% mortgage of \$60,000 for 15 years to be \$444. The total amount paid over 15 years is then

$$\$444 \text{ per month} \times 180 \text{ months} = \$79{,}920.$$

In part (a)-(ii) we found the monthly payment on an 4% mortgage of \$60,000 for 30 years to be \$286.20. The total amount paid over 30 years is then

$$286.20 \text{ per month} \times 360 \text{ months} = \$103{,}032.$$

The amount saved by taking the mortgage over a shorter period of time is the difference:

$$\$103{,}032 - \$79{,}920 = \$23{,}112.$$

61. We have

$$\begin{aligned}
\frac{f(n+2)}{f(n)} &= \frac{1000 \cdot 2^{-\frac{1}{4}-\frac{n+2}{2}}}{1000 \cdot 2^{-\frac{1}{4}-\frac{n}{2}}} \\
&= \frac{1000}{1000} \cdot \frac{2^{-\frac{1}{4}}}{2^{-\frac{1}{4}}} \cdot \frac{2^{-\frac{n+2}{2}}}{2^{-\frac{n}{2}}} \\
&= 2^{-\frac{n+2}{2}} \cdot 2^{\frac{n}{2}} \\
&= 2^{-\frac{n}{2}-\frac{2}{2}+\frac{n}{2}} \\
&= 2^{-1} = 0.5.
\end{aligned}$$

This means a sheet two numbers higher in the series is half as wide. For instance, a sheet of $A3$ is half as wide as a sheet of $A1$.

65. The graph $a_0(b_0)^t$ climbs faster than that of $a_1(b_1)^t$, so $b_0 > b_1$.

Solutions for Section 4.2

Skill Refresher

S1. We have $b^4 \cdot b^6 = b^{4+6} = b^{10}$.

S5. We have $f(0) = 5.6(1.043)^0 = 5.6$ and $f(3) = 5.6(1.043)^3 = 6.354$.

S9. We have

$$\begin{aligned}
\frac{4}{3}x^5 &= 7 \\
x^5 &= 5.25 \\
x &= (5.25)^{1/5} = 1.393.
\end{aligned}$$

Exercises

1. The formula $P_A = 200 + 1.3t$ for City A shows that its population is growing linearly. In year $t = 0$, the city has 200,000 people and the population grows by 1.3 thousand people, or 1,300 people, each year.

The formulas for cities B, C, and D show that these populations are changing exponentially. Since $P_B = 270(1.021)^t$, City B starts with 270,000 people and grows at an annual rate of 2.1%. Similarly, City C starts with 150,000 people and grows at 4.5% annually.

Since $P_D = 600(0.978)^t$, City D starts with 600,000 people, but its population decreases at a rate of 2.2% per year. We find the annual percent rate by taking $b = 0.978 = 1 + r$, which gives $r = -0.022 = -2.2\%$. So City D starts out with more people than the other three but is shrinking.

Figure 4.2 gives the graphs of the three exponential populations. Notice that the P-intercepts of the graphs correspond to the initial populations (when $t = 0$) of the towns. Although the graph of P_C starts below the graph of P_B, it eventually catches up and rises above the graph of P_B, because City C is growing faster than City B.

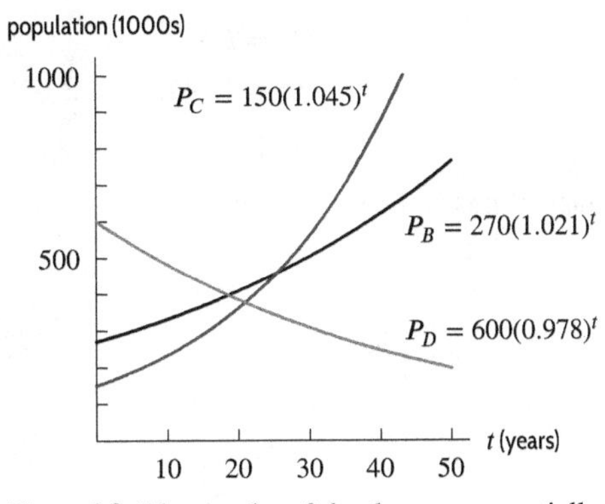

Figure 4.2: The graphs of the three exponentially changing populations

5. **(a)** See Table 4.1.

Table 4.1

t	0	1	2	3	4	5
$f(t)$	1000	1200	1440	1728	2073.6	2488.32

(b) We have

$$\frac{f(1)}{f(0)} = \frac{1200}{1000} = 1.2$$
$$\frac{f(2)}{f(1)} = \frac{1440}{1200} = 1.2$$
$$\frac{f(3)}{f(2)} = \frac{1728}{1440} = 1.2$$
$$\frac{f(4)}{f(3)} = \frac{2073.6}{1728} = 1.2$$
$$\frac{f(5)}{f(4)} = \frac{2488.32}{2073.6} = 1.2.$$

(c) All of the ratios of successive terms are 1.2. This makes sense because we have

$$f(0) = 1000$$
$$f(1) = 1000 \cdot 1.2$$
$$f(2) = 1000 \cdot 1.2 \cdot 1.2$$
$$f(3) = 1000 \cdot 1.2 \cdot 1.2 \cdot 1.2$$

and so on. Each term is the previous term multiplied by 1.2. It follows that the ratio of successive terms will always be the growth factor, which is 1.2 in this case.

9. **(a)** If a function is linear, then the differences in successive function values will be constant. If a function is exponential, the ratios of successive function values will remain constant. Now

$$i(1) - i(0) = 14 - 18 = -4$$

and

$$i(2) - i(1) = 10 - 14 = -4.$$

Checking the rest of the data, we see that the differences remain constant, so $i(x)$ is linear.

(b) We know that $i(x)$ is linear, so it must be of the form

$$i(x) = b + mx,$$

where m is the slope and b is the y-intercept. Since at $x = 0$, $i(0) = 18$, we know that the y-intercept is 18, so $b = 18$. Also, we know that at $x = 1$, $i(1) = 14$, we have

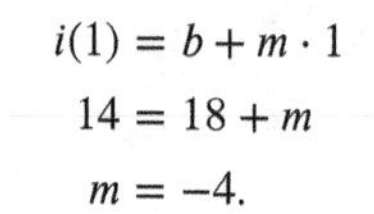

$$\begin{aligned} i(1) &= b + m \cdot 1 \\ 14 &= 18 + m \\ m &= -4. \end{aligned}$$

Thus, $i(x) = 18 - 4x$. The graph of $i(x)$ is shown in Figure 4.3.

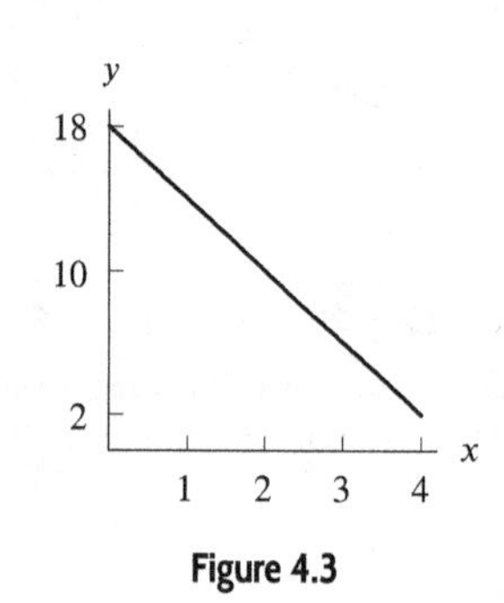

Figure 4.3

13. We know that $f(x) = ab^x$. Taking the ratio of $f(2)$ to $f(-1)$ we have

$$\begin{aligned} \frac{f(2)}{f(1)} &= \frac{1/27}{27} = \frac{ab^2}{ab^{-1}} \\ \frac{1}{(27)^2} &= b^3 \\ b^3 &= \frac{1}{27^2} \\ b &= \left(\frac{1}{27^2}\right)^{\frac{1}{3}}. \end{aligned}$$

Thus, $b = \frac{1}{9}$. Therefore, $f(x) = a(\frac{1}{9})^x$.

Using the fact that $f(-1) = 27$, we have

$$f(-1) = a\left(\frac{1}{9}\right)^{-1} = a \cdot 9 = 27,$$

which means $a = 3$. Thus,

$$f(x) = 3\left(\frac{1}{9}\right)^x.$$

17. Since the function is exponential, we know $y = ab^x$. We also know that $(0, 1/2)$ and $(3, 1/54)$ are on the graph of this function, so $1/2 = ab^0$ and $1/54 = ab^3$. The first equation implies that $a = 1/2$. Substituting this value in the second equation gives $1/54 = (1/2)b^3$ or $b^3 = 1/27$, or $b = 1/3$. Thus, $y = \frac{1}{2}\left(\frac{1}{3}\right)^x$.

21. Let f be the function whose graph is shown. Were f exponential, it would increase by equal factors on equal intervals. However, we see that

$$\frac{f(3)}{f(1)} = \frac{11}{5} = 2.2$$
$$\frac{f(5)}{f(3)} = \frac{30}{11} = 2.7.$$

On these equal intervals, the value of f does not increase by equal factors, so f is not exponential.

25. **(a)** The function h is linear because equal spacing between successive input values ($\Delta x = 3$) results in equal spacing between successive output values ($\Delta h = 1.47$). On the other hand, the function f is exponential because ratios of successive rounded output values equal 1.10.

(b) Since f is exponential, we know that $f(x) = ab^x$, and we must figure out the values of the constants a and b. From our given information, we have $f(0) = 2.23$, which yields

$$a = ab^0 = 2.23.$$

Also, since $f(3) = 2.45$, we have

$$\begin{aligned} 2.23b^3 &= 2.45 \\ b &= \left(\frac{2.45}{2.23}\right)^{1/3} \\ &= 1.10^{1/3} \\ &= 1.03, \end{aligned}$$

so our final answer is $f(x) = 2.23((1.10)^{1/3})^x = 2.23(1.10)^{x/3} = 2.23(1.03)^x$.

Problems

29. Testing the rates of change for $R(t)$, we find that

$$\frac{2.61 - 2.32}{9 - 5} = 0.0725$$

and

$$\frac{3.12 - 2.61}{15 - 9} = 0.085,$$

so we know that $R(t)$ is not linear. If $R(t)$ is exponential, then $R(t) = ab^t$, and

$$R(5) = a(b)^5 = 2.32$$

and

$$R(9) = a(b)^9 = 2.61.$$

So

$$\begin{aligned} \frac{R(9)}{R(5)} &= \frac{ab^9}{ab^5} = \frac{2.61}{2.32} \\ \frac{b^9}{b^5} &= \frac{2.61}{2.32} \\ b^4 &= \frac{2.61}{2.32} \\ b &= (\frac{2.61}{2.32})^{\frac{1}{4}} \approx 1.030. \end{aligned}$$

Since

$$\begin{aligned} R(15) &= a(b)^{15} = 3.12 \\ \frac{R(15)}{R(9)} &= \frac{ab^{15}}{ab^9} = \frac{3.12}{2.61} \\ b^6 &= \frac{3.12}{2.61} \\ b &= \left(\frac{3.12}{2.61}\right)^{\frac{1}{6}} \approx 1.030. \end{aligned}$$

Since the growth factor, b, is constant, we know that $R(t)$ could be an exponential function and that $R(t) = ab^t$. Taking the ratios of $R(5)$ and $R(9)$, we have

$$\begin{aligned}\frac{R(9)}{R(5)} = \frac{ab^9}{ab^5} &= \frac{2.61}{2.32}\\ b^4 &= 1.125\\ b &= 1.030.\end{aligned}$$

So $R(t) = a(1.030)^t$. We now solve for a by using $R(5) = 2.32$,

$$\begin{aligned}R(5) &= a(1.030)^5\\ 2.32 &= a(1.030)^5\\ a &= \frac{2.32}{1.030^5} \approx 2.001.\end{aligned}$$

Thus, $R(t) = 2.001(1.030)^t$.

33. One approach is to graph both functions and to see where the graph of $p(x)$ is below the graph of $q(x)$. From Figure 4.4, we see that $p(x)$ intersects $q(x)$ in two places: namely, at $x \approx -1.69$ and $x = 2$. We notice that $p(x)$ is above $q(x)$ between these two points and below $q(x)$ outside the segment defined by these two points. Hence $p(x) < q(x)$ for $x < -1.69$ and for $x > 2$.

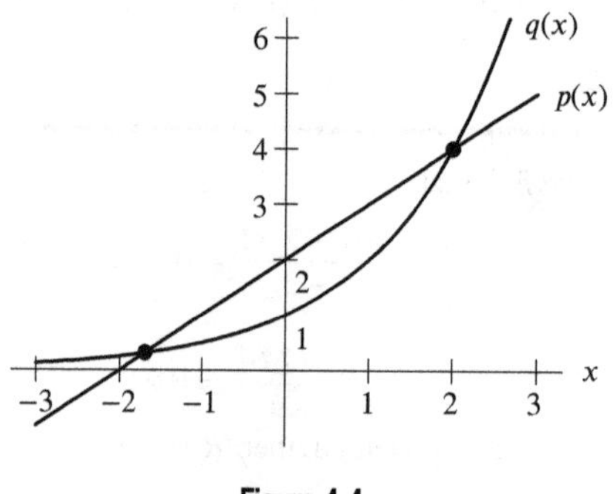

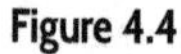
Figure 4.4

37. **(a)** Since, for $t < 0$, we know that the voltage is a constant 80 volts, $V(t) = 80$ on that interval.

For $t \geq 0$, we know that $v(t)$ is an exponential function, so $V(t) = ab^t$. According to this formula, $V(0) = ab^0 = a(1) = a$. According to the graph, $V(0) = 80$. From these two facts, we know that $a = 80$, so $V(t) = 80b^t$. If $V(10) = 80b^{10}$ and $V(10) = 15$ (from the graph), then

$$\begin{aligned}80b^{10} &= 15\\ b^{10} &= \frac{15}{80}\\ (b^{10})^{\frac{1}{10}} &= (\frac{15}{80})^{\frac{1}{10}}\\ b &\approx 0.8459\end{aligned}$$

so that $V(t) = 80(0.8459)^t$ on this interval. Combining the two pieces, we have

$$V(t) = \begin{cases} 80 & \text{for } t < 0\\ 80(0.8459)^t & \text{for } t \geq 0.\end{cases}$$

(b) Using a computer or graphing calculator, we can find the intersection of the line $y = 0.1$ with $y = 80(0.8459)^t$. We find $t \approx 39.933$ seconds.

41. We let W represent the winning time and t represent the number of years since 1994.

(a) To find the linear function, we first find the slope:

$$\text{Slope} = \frac{\Delta W}{\Delta t} = \frac{40.981 - 43.45}{16 - 0} = -0.154.$$

The vertical intercept is 43.45 so the linear function is $W = 43.45 - 0.154t$. The predicted winning time in 2018 is $W = 43.45 - 0.154(24) = 39.754$ seconds.

(b) The time at $t = 0$ is 43.45, so the exponential function is $W = 43.45(a)^t$. We use the fact that $W = 40.981$ when $t = 16$ to find a:

$$\begin{aligned} 40.981 &= 43.45(a)^{16} \\ 0.943176 &= a^{16} \\ a &= (0.943176)^{1/16} = 0.99635. \end{aligned}$$

The exponential function is $W = 43.45(0.99635)^t$. The predicted winning time in 2018 is $W = 43.45(0.99635)^{24} = 39.799$ seconds.

45. (a) Since the rate of change is constant, the increase is linear.

(b) Life expectancy is increasing at a constant rate of 3 months, or 0.25 years, each year. The slope is 0.25. When $t = 11$ we have $L = 78.7$. We use the point-slope form to find the linear function:

$$\begin{aligned} L - 78.7 &= 0.25(t - 11) \\ L - 78.7 &= 0.25t - 2.75 \\ L &= 0.25t + 75.95. \end{aligned}$$

(c) When $t = 50$, we have $L = 0.25(50) + 75.95 = 88.35$. It the rate of increase continues, babies born in 2050 will have a life expectancy of 88.45 years.

49. (a) Assuming linear growth at 250 per year, the population in 2013 would be

$$18{,}500 + 250 \cdot 10 = 21{,}000.$$

Using the population after one year, we find that the percent rate would be $250/18{,}500 \approx 0.013514 = 1.351\%$ per year, so after 10 years the population would be

$$18{,}500(1.013514)^{10} \approx 21{,}158.$$

The town's growth is poorly modeled by both linear and exponential functions.

(b) We do not have enough information to make even an educated guess about a formula.

Solutions for Section 4.3

Exercises

1. The function could not be exponential because it first decreases and then increases.

5. (a) See Table 4.2.

Table 4.2

x	−3	−2	−1	0	1	2	3
$f(x)$	1/8	1/4	1/2	1	2	4	8

(b) For large negative values of x, $f(x)$ is close to the x-axis. But for large positive values of x, $f(x)$ climbs rapidly away from the x-axis. As x gets larger, y grows more and more rapidly. See Figure 4.5.

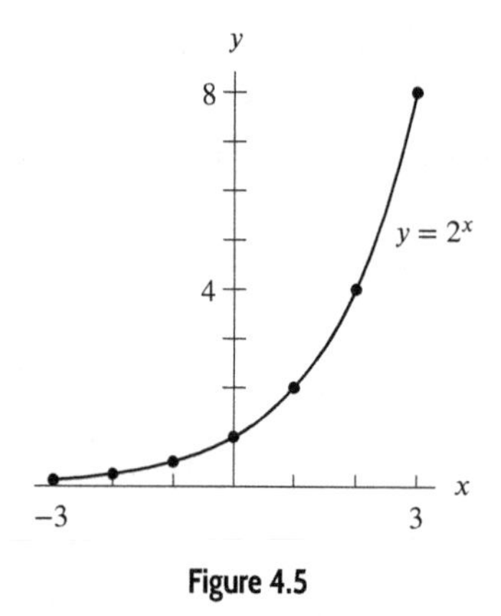

Figure 4.5

9. Yes, the graphs will cross. The graph of $g(x)$ has a smaller y-intercept but increases faster and will eventually overtake the graph of $f(x)$.

13. No, the graphs will not cross. Both functions are decreasing but the graph of $f(x)$ has a larger y-intercept and is decreasing at a slower rate than $g(x)$, so it will always be above the graph of $g(x)$.

17. The graph of $g(x)$ is shifted four units to the left of $f(x)$, and the graph of $h(x)$ is shifted two units to the right of $f(x)$.

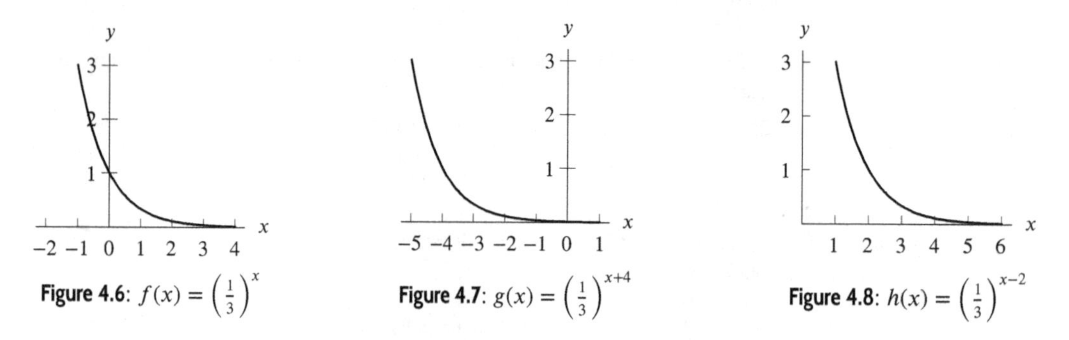

Figure 4.6: $f(x) = \left(\frac{1}{3}\right)^x$ Figure 4.7: $g(x) = \left(\frac{1}{3}\right)^{x+4}$ Figure 4.8: $h(x) = \left(\frac{1}{3}\right)^{x-2}$

21. Solve for P to obtain $P = 7(0.6)^t$. Graphing $P = 7(0.6)^t$ and tracing along the graph on a calculator gives us an answer of $t = 2.452$. See Figure 4.9.

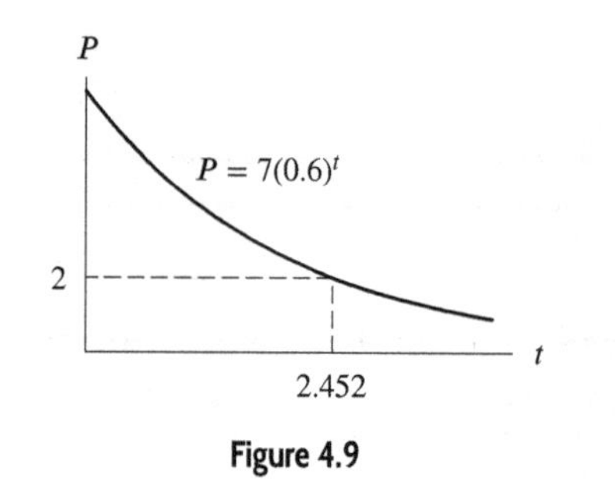

Figure 4.9

Problems

25. Since $y = a$ when $t = 0$ in $y = ab^t$, a is the y-intercept. Thus, the function with the greatest y-intercept, D, has the largest a.

29. As t approaches $-\infty$, the value of ab^t approaches zero for any a, so the horizontal asymptote is $y = 0$ (the t-axis).

33. A possible graph is shown in Figure 4.10. There are many possible answers.

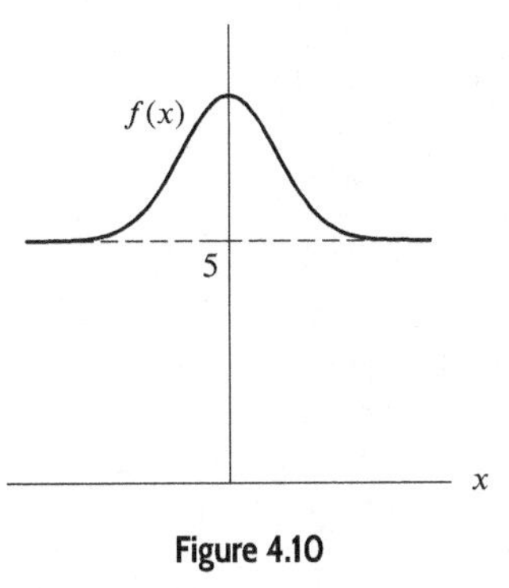

Figure 4.10

37. As r increases, the graph of $y = a(1+r)^t$ rises more steeply, so the point of intersection moves to the left and down. However, no matter how steep the graph becomes, the point of intersection remains above and to the right of the y-intercept of the second curve, or the point $(0, b)$. Thus, the value of y_0 decreases but does not reach b.

41. The function, when entered as $y = 1.04\hat{}5x$, is interpreted as $y = (1.04^5)x = 1.217x$. This function's graph is a straight line in all windows. Parentheses must be used to ensure that x is in the exponent.

45. **(a)** See Figure 4.11.

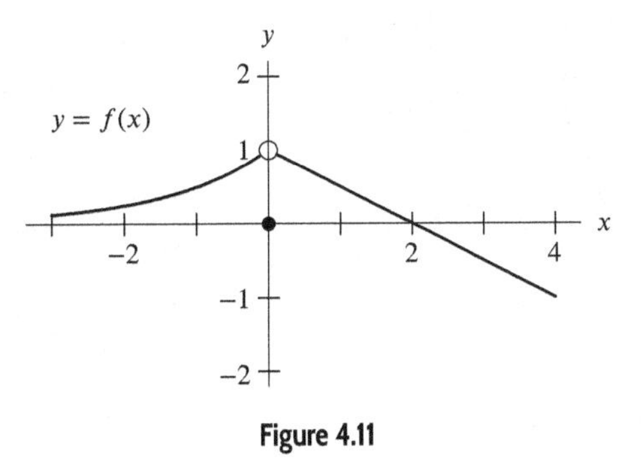

Figure 4.11

(b) The range of this function is all real numbers less than one — i.e. $f(x) < 1$.

(c) The y-intercept occurs at $(0, 0)$. This point is also an x-intercept. To solve for other x-intercepts we must attempt to solve $f(x) = 0$ for each of the two remaining parts of f. In the first case, we know that the function $f(x) = 2^x$ has no x-intercepts, as there is no value of x for which 2^x is equal to zero. In the last case, for $x > 0$, we set $f(x) = 0$ and solve for x:

$$\begin{aligned} 0 &= 1 - \frac{1}{2}x \\ \frac{1}{2}x &= 1 \\ x &= 2. \end{aligned}$$

Hence $x = 2$ is another x-intercept of f.

(d) As x gets large, the function is defined by $f(x) = 1 - 1/2x$. To determine what happens to f as $x \to +\infty$, find values of f for very large values of x. For example,

$$f(100) = 1 - \frac{1}{2}(100) = -49, \quad f(10000) = 1 - \frac{1}{2}(10000) = -4999$$

$$\text{and} \quad f(1{,}000{,}000) = 1 - \frac{1}{2}(1{,}000{,}000) = -499{,}999.$$

As x becomes larger, $f(x)$ becomes more and more negative. A way to write this is:

$$\text{As } x \to +\infty, \; f(x) \to -\infty.$$

As x gets very negative, the function is defined by $f(x) = 2^x$.

Choosing very negative values of x, we get $f(-100) = 2^{-100} = 1/2^{100}$, and $f(-1000) = 2^{-1000} = 1/2^{1000}$. As x becomes more negative the function values get closer to zero. We write

$$\text{As } x \to -\infty, \; f(x) \to 0.$$

(e) Increasing for $x < 0$, decreasing for $x > 0$.

49. **(a)** Figure 4.12 shows the three populations. From this graph, the three models seem to be in good agreement. Models 1 and 3 are indistinguishable; model 2 appears to rise a little faster. However, notice that we cannot see the behavior beyond 50 months because our function values go beyond the top of the viewing window.

(b) Figure 4.13 shows the population differences. The graph of $y = f_2(x) - f_1(x) = 3(1.21)^x - 3(1.2)^x$ grows very rapidly, especially after 40 months. The graph of $y = f_3(x) - f_1(x) = 3.01(1.2)^x - 3(1.2)^x$ is hardly visible on this scale.

(c) Models 1 and 3 are in good agreement, but model 2 predicts a much larger mussel population than does model 1 after only 50 months. We can come to at least two conclusions. First, even small differences in the base of an exponential function can be highly significant, while differences in initial values are not as significant. Second, although two exponential curves can look very similar, they can actually be making very different predictions as time increases.

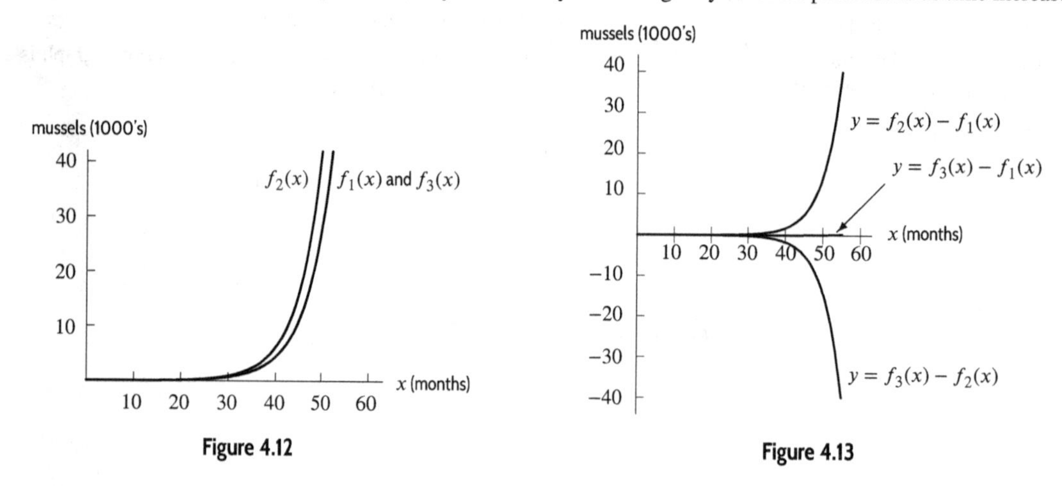

Figure 4.12

Figure 4.13

53. It appears in the graph that

(a) $\lim_{x\to-\infty} f(x) = 5$

(b) $\lim_{x\to\infty} f(x) = -3$.

Of course, we need to be sure that we are seeing all the important features of the graph in order to have confidence in these estimates.

Solutions for Section 4.4

Exercises

1. Let b be the effective annual growth factor. Since the amount in the account at time t is given by $1000b^t$, we set $1000b^{15}$ equal to 3500 and solve for b:

$$\begin{aligned} 1000b^{15} &= 3500 \\ b^{15} &= 3.5 \\ b &= (3.5)^{1/15} = 1.0871. \end{aligned}$$

The effective annual yield over the 15-year period was 8.71% per year.

5. If P is the initial amount, the amount after 8 years is $0.5P$. To find the effective annual yield, we set Pb^8 equal to $0.5P$ and solve for b:

$$b^8 = 0.5$$
$$b = (0.5)^{1/8} = 0.917.$$

Since $0.917 = 1 - 0.083$, the investment has decreased by an effective annual rate of -8.3% per year.

9. **(a)** If the interest is compounded annually, there will be $\$500 \cdot 1.05 = \$525$ after one year.
(b) If the interest is compounded weekly, there will be $500 \cdot (1 + 0.05/52)^{52} = \$525.62$ after one year.
(c) If the interest is compounded every minute, there will be $500 \cdot (1 + 0.05/525{,}600)^{525600} = \$525.64$ after one year.

13. **(a)** The nominal rate is the stated annual interest without compounding, thus 3%.
The effective annual rate for an account paying 1% compounded annually is 3%.
(b) The nominal rate is the stated annual interest without compounding, thus 3%.
With quarterly compounding, there are four interest payments per year, each of which is $3/4 = 0.75\%$. Over the course of the year, this occurs four times, giving an effective annual rate of $1.0075^4 = 1.03034$, which is 3.034%.
(c) The nominal rate is the stated annual interest without compounding, thus 3%.
With daily compounding, there are 365 interest payments per year, each of which is $(3/365)\%$. Over the course of the year, this occurs 365 times, giving an effective annual rate of $(1 + 0.03/365)^{365} = 1.03045$, which is 3.045%.

Problems

17. If the investment is growing by 3% per year, we know that, at the end of one year, the investment will be worth 103% of what it had been the previous year. At the end of two years, it will be 103% of 103% $= (1.03)^2$ as large. At the end of 10 years, it will have grown by a factor of $(1.03)^{10}$, or 1.34392. The investment will be 134.392% of what it had been, so we know that it will have increased by 34.392%. Since $(1.03)^{10} \approx 1.34392$, it increases by 34.392%.

21. (i) Equation (b). Since the growth factor is 1.12, or 112%, the annual interest rate is 12%.
(ii) Equation (a). An account earning at least 1% monthly will have a monthly growth factor of at least 1.01, which means that the annual (12-month) growth factor will be at least

$$(1.01)^{12} = 1.1268.$$

Thus, an account earning at least 1% monthly will earn at least 12.68% yearly. The only account that earns this much interest is account (a).
(iii) Equation (c). An account earning 12% annually compounded semi-annually will earn 6% twice yearly. In t years, there are $2t$ half-years.
(iv) Equations (b), (c) and (d). An account that earns 3% each quarter ends up with a yearly growth factor of $(1.03)^4 = 1.1255$. This corresponds to an annual percentage rate of 12.55%. Accounts (b), (c) and (d) earn less than this. Check this by determining the growth factor in each case.
(v) Equations (a) and (e). An account that earns 6% every 6 months will have a growth factor, after 1 year, of $(1+0.06)^2 = 1.1236$, which is equivalent to a 12.36% annual interest rate, compounded annually. Account (a), earning 20% each year, earns more than 6% twice each year, or 12.36% annually. Account (e), which earns 3% each quarter, earns $(1.03)^2 = 1.0609$, or 6.09% every 6 months, which is greater than 6%.

Solutions for Section 4.5

Skill Refresher

S1. We have $e^{0.07} = 1.073$.

S5. We have $f(0) = 2.3e^{0.3(0)} = 2.3$ and $f(4) = 2.3e^{0.3(4)} = 7.636$.

S9. Writing the function as

$$f(t) = \left(3e^{0.04t}\right)^3 = 3^3 e^{0.04t\cdot 3} = 27e^{0.12t},$$

we have $a = 27$ and $k = 0.12$.

S13. Writing the function as

$$m(x) = \frac{7e^{0.2x}}{\sqrt{3e^x}} = \frac{7}{\sqrt{3}}e^{0.2x} \cdot e^{-0.5x} = \frac{7}{\sqrt{3}}e^{-0.3x},$$

we have $a = \frac{7}{\sqrt{3}}$ and $k = -0.3$.

Exercises

1. We know that $e \approx 2.71828$, so $2 < e < 3$. Since e lies between 2 and 3, the graph of $y = e^x$ lies between the graphs of $y = 2^x$ and $y = 3^x$. Since 3^x increases faster than 2^x, the correct matching is shown in Figure 4.14.

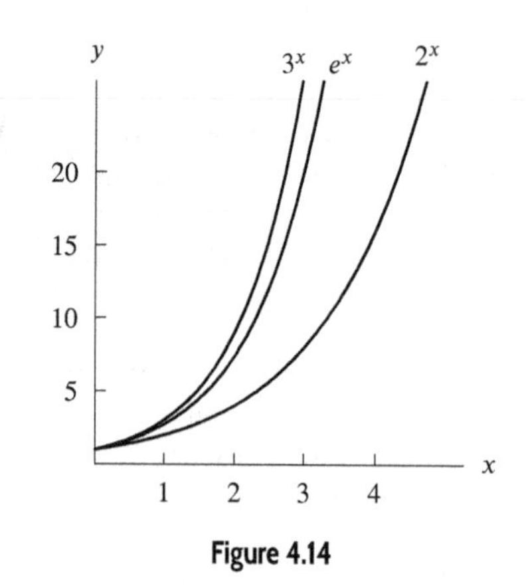

Figure 4.14

5. Calculating the equivalent continuous rates, we find $e^{0.45} = 1.568$, $e^{0.47} = 1.600$, $e^{0.5} = 1.649$. Thus the functions to be matched are

(a) $(1.5)^x$ **(b)** $(1.568)^x$ **(c)** $(1.6)^x$ **(d)** $(1.649)^x$

So (a) is (IV), (b) is (III), (c) is (II), (d) is (I).

9. **(a)** We see that $Q_0 = 2.7$
(b) Since the base is less than one, the quantity is decreasing.
(c) Since the base is $0.12 = 1 - 0.88$, the decay rate is 88% per unit time.
(d) No, the growth rate is not continuous.

13. **(a)** We see that $Q_0 = 1$
(b) Since the base is greater than one, the quantity is increasing.
(c) Since the base is $2 = 1 + 1$, the growth rate is 100% per unit time. The quantity is doubling every time unit.
(d) No, the growth rate is not continuous.

Problems

17. **(a)** Since the decay rate is not continuous, we have $Q = 500(0.93)^t$. At $t = 10$ we have $Q = 500(0.93)^{10} = 241.991$.
(b) Since the decay rate is continuous, we have $Q = 500e^{-0.07t}$. At $t = 10$ we have $Q = 500e^{-0.07(10)} = 248.293$. As we expect, the results are similar for continuous and not-continuous assumptions, but slightly larger if we assume a continuous decay rate.

21. **(a)** Using $P = P_0e^{kt}$ where $P_0 = 25{,}000$ and $k = 7.5\%$, we have

$$P(t) = 25{,}000e^{0.075t}.$$

(b) We first need to find the growth factor so will rewrite

$$P = 25{,}000e^{0.075t} = 25{,}000(e^{0.075})^t \approx 25{,}000(1.07788)^t.$$

At the end of a year, the population is 107.788% of what it had been at the end of the previous year. This corresponds to an increase of approximately 7.788%. This is greater than 7.5% because the rate of 7.5% per year is being applied to larger and larger amounts. In one instant, the population is growing at a rate of 7.5% per year. In the next instant, it grows again at a rate of 7.5% a year, but 7.5% of a slightly larger number. The fact that the population is increasing in tiny increments continuously results in an actual increase greater than the 7.5% increase that would result from one single jump of 7.5% at the end of the year.

25. **(a)** First, we note that after t hours, the population, P, of the colony is given by $P = 100e^{0.25t}$. Substituting 4 for t in this equation, we obtain

$$P = 100e^{0.25(4)} = 271.828.$$

Therefore, there are about 272 bacteria in the colony after 4 hours.

(b) Since

$$P = 100e^{0.25t} = 100(e^{0.25})^t = 100(1.2840)^t,$$

the hourly growth rate of the colony is $1.2840 - 1 = 0.2840$. Therefore, the colony grows by 28.4% each hour.

29. **(a)** For an annual interest rate of 2%, the balance B after 10 years is

$$B = 5000(1.02)^{10} = 6094.97 \text{ dollars.}$$

(b) For a continuous interest rate of 2% per year, the balance B after 10 years is

$$B = 5000e^{0.02 \cdot 10} = 6107.01 \text{ dollars.}$$

33. With continuous compounding, the interest earns interest during the year, so the balance grows faster with continuous compounding than with annual compounding. Curve A corresponds to continuous compounding and curve B corresponds to annual compounding. The initial amount in both cases is the vertical intercept, $500.

37. Since $e^{0.053} = 1.0544$, the effective annual yield of the account paying 5.3% interest compounded continuously is 5.44%. Since this is less than the effective annual yield of 5.5% from the 5.5% compounded annually, we see that the account paying 5.5% interest compounded annually is slightly better.

41. The balance in the first bank is $10{,}000(1.05)^8 = \$14{,}774.55$. The balance in the second bank is $10{,}000e^{0.05(8)} = \$14{,}918.25$. The bank with continuously compounded interest has a balance $143.70 higher.

45. **(a)** Since poultry production is increasing at a constant continuous percent rate, we use the exponential formula $P = ae^{kt}$. Since $P = 106$ when $t = 0$, we have $a = 106$. Since $k = 0.018$, we have

$$P = 106e^{0.018t}.$$

(b) When $t = 5$, we have $P = 106e^{0.018(5)} = 115.98$. In the year 2018, the formula predicts that world poultry production will be about 116 million tons.

(c) A graph of $P = 106e^{0.018t}$ is given in Figure 4.15. We see that when $P = 120$ we have $t = 6.9$. We expect production to be 120 million tons late in the year 2019.

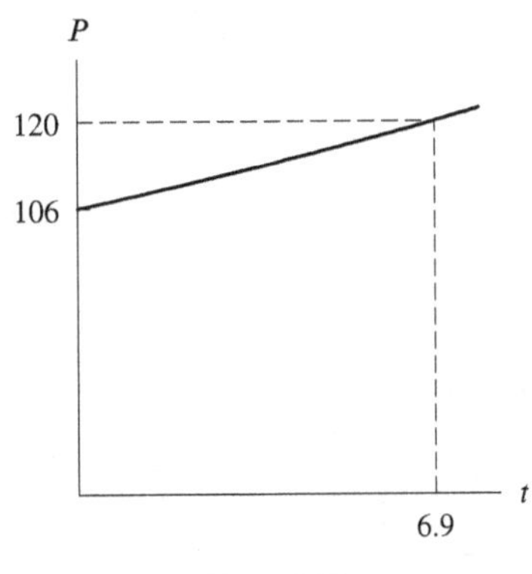

Figure 4.15

49. **(a)** Let p_0 be the price of an item at the beginning of 2000. At the beginning of 2001, its price will be 103.4% of that initial price or $1.034p_0$. At the beginning of 2002, its price will be 102.8% of the price from the year before, that is:

$$\text{Price beginning 2002} = (1.028)(1.034p_0).$$

By the beginning of 2003, the price will be 101.6% of its price the previous year.

$$\begin{aligned}\text{Price beginning 2003} &= 1.016(\text{price beginning 2002})\\ &= 1.016(1.028)(1.034p_0).\end{aligned}$$

Continuing this process,

$$\begin{aligned}(\text{Price beginning 2005}) &= (1.027)(1.023)(1.016)(1.028)(1.034)p_0\\ &\approx 1.135p_0.\end{aligned}$$

So, the cost at the beginning of 2005 is 113.5% of the cost at the beginning of 2000 and the total percent increase is 13.5%.

(b) If r is the average inflation rate for this time period, then $b = 1 + r$ is the factor by which the population on the average grows each year. Using this average growth factor, if the price of an item is initially p_0, at the end of a year its value would be p_0b, at the end of two years it would be $(p_0b)b = p_0b^2$, and at the end of five years p_0b^5. According to the answer in part (a), the price at the end of five years is $1.135p_0$. So

$$\begin{aligned}p_0b^5 &= 1.135p_0\\ b^5 &= 1.135\\ b &= (1.135)^{1/5} \approx 1.026.\end{aligned}$$

If $b = 1.026$, then $r = 0.026$ or 2.6%, the average annual inflation rate.

(c) We assume that the average rate of 2.6% inflation for 2000 through 2004 holds through the beginning of 2010. So, on average, the price of the shower curtain is 102.6% of what it was the previous year for ten years. Then the price of the shower curtain would be $20(1.026)^{10} \approx \$25.85$.

53. $\lim_{x\to\infty} e^{-3x} = 0.$

57. The values of a and k are both positive.

Solutions for Chapter 4 Review

Exercises

1. For a 10% increase, we multiply by 1.10 to obtain $500 \cdot 1.10 = 550$.

5. For a 42% increase, we multiply by 1.42 to obtain $500 \cdot 1.42 = 710$. For a 42% decrease, we multiply by $1 - 0.42 = 0.58$ to obtain $710 \cdot 0.58 = 411.8$.

9. The percent of change is given by

$$\text{Percent of change} = \frac{\text{Amount of change}}{\text{Old amount}} \cdot 100\%.$$

So in these two cases,

$$\begin{aligned}\text{Percent of change from 10 to 12} &= \frac{12-10}{10} \cdot 100\% = 20\%\\ \text{Percent of change from 100 to 102} &= \frac{102-100}{100} \cdot 100\% = 2\%\end{aligned}$$

13. This cannot be linear, since $\Delta f(x)/\Delta x$ is not constant, nor can it be exponential, since between $x = 15$ and $x = 12$, we see that $f(x)$ doubles while $\Delta x = 3$. Between $x = 15$ and $x = 16$, we see that $f(x)$ doubles while $\Delta x = 1$, so the percentage increase is not constant. Thus, the function is neither.

Problems

17. **(a)** $B = B_0(1.042)^1 = B_0(1.042)$, so the effective annual rate is 4.2%.

(b) $B = B_0\left(1 + \dfrac{.042}{12}\right)^{12} \approx B_0(1.0428)$, so the effective annual rate is approximately 4.28%.

(c) $B = B_0 e^{0.042(1)} \approx B_0(1.0429)$, so the effective annual rate is approximately 4.29%.

21. Since $g(x) = ab^x$, we can say that $g(\frac{1}{2}) = ab^{1/2}$ and $g(\frac{1}{4}) = ab^{1/4}$. Since we know that $g(\frac{1}{2}) = 4$ and $g(\frac{1}{4}) = 2\sqrt{2}$, we can conclude that

$$ab^{1/2} = 4 = 2^2$$

and

$$ab^{1/4} = 2\sqrt{2} = 2 \cdot 2^{1/2} = 2^{3/2}.$$

Forming ratios, we have

$$\begin{aligned} \frac{ab^{1/2}}{ab^{1/4}} &= \frac{2^2}{2^{3/2}} \\ b^{1/4} &= 2^{1/2} \\ (b^{1/4})^4 &= (2^{1/2})^4 \\ b &= 2^2 = 4. \end{aligned}$$

Now we know that $g(x) = a(4)^x$, so $g(\frac{1}{2}) = a(4)^{1/2} = 2a$. Since we also know that $g(\frac{1}{2}) = 4$, we can say

$$\begin{aligned} 2a &= 4 \\ a &= 2. \end{aligned}$$

Therefore $g(x) = 2(4)^x$.

25. **(a)** If f is linear, then $f(x) = b + mx$, where m, the slope, is given by:

$$m = \frac{\Delta y}{\Delta x} = \frac{f(2) - f(-3)}{(2) - (-3)} = \frac{20 - \frac{5}{8}}{5} = \frac{\frac{155}{8}}{5} = \frac{31}{8}.$$

Using the fact that $f(2) = 20$ and substituting the known values for m, we write

$$\begin{aligned} 20 &= b + m(2) \\ 20 &= b + \left(\frac{31}{8}\right)(2) \\ 20 &= b + \frac{31}{4} \end{aligned}$$

which gives

$$b = 20 - \frac{31}{4} = \frac{49}{4}.$$

So, $f(x) = \dfrac{31}{8}x + \dfrac{49}{4}$.

(b) If f is exponential, then $f(x) = ab^x$. We know that $f(2) = ab^2$ and $f(2) = 20$. We also know that $f(-3) = ab^{-3}$ and $f(-3) = \frac{5}{8}$. So

$$\begin{aligned} \frac{f(2)}{f(-3)} &= \frac{ab^2}{ab^{-3}} = \frac{20}{\frac{5}{8}} \\ b^5 &= 20 \times \frac{8}{5} = 32 \\ b &= 2. \end{aligned}$$

Thus, $f(x) = a(2)^x$. Solve for a by using $f(2) = 20$ and (with $b = 2$), $f(2) = a(2)^2$.

$$\begin{aligned} 20 &= a(2)^2 \\ 20 &= 4a \\ a &= 5. \end{aligned}$$

Thus, $f(x) = 5(2)^x$.

29. The formula is of the form $y = ab^x$. Since the points $(-1, 1/15)$ and $(2, 9/5)$ are on the graph,

$$\frac{1}{15} = ab^{-1}$$
$$\frac{9}{5} = ab^2.$$

Taking the ratio of the second equation to the first, we obtain

$$\frac{9/5}{1/15} = \frac{ab^2}{ab^{-1}}$$
$$27 = b^3$$
$$b = 3.$$

Substituting this value of b into $\frac{1}{15} = ab^{-1}$ gives

$$\frac{1}{15} = a(3)^{-1}$$
$$\frac{1}{15} = \frac{1}{3}a$$
$$a = \frac{1}{15} \cdot 3$$
$$a = \frac{1}{5}.$$

Therefore $y = \frac{1}{5}(3)^x$ is a possible formula for this function.

33. **(a)** If P is linear, then $P(t) = b + mt$ and

$$m = \frac{\Delta P}{\Delta t} = \frac{P(13) - P(7)}{13 - 7} = \frac{3.75 - 3.21}{13 - 7} = \frac{0.54}{6} = 0.09.$$

So $P(t) = b + 0.09t$ and $P(7) = b + 0.09(7)$. We can use this and the fact that $P(7) = 3.21$ to say that

$$3.21 = b + 0.09(7)$$
$$3.21 = b + 0.63$$
$$2.58 = b.$$

So $P(t) = 2.58 + 0.09t$. The slope is 0.09 million people per year. This tells us that, if its growth is linear, the country grows by 0.09(1,000,000) = 90,000 people every year.

(b) If P is exponential, $P(t) = ab^t$. So

$$P(7) = ab^7 = 3.21$$

and

$$P(13) = ab^{13} = 3.75.$$

We can say that

$$\frac{P(13)}{P(7)} = \frac{ab^{13}}{ab^7} = \frac{3.75}{3.21}$$
$$b^6 = \frac{3.75}{3.21}$$
$$(b^6)^{1/6} = \left(\frac{3.75}{3.21}\right)^{1/6}$$
$$b = 1.026.$$

Thus, $P(t) = a(1.026)^t$. To find a, note that

$$P(7) = a(1.026)^7 = 3.21$$
$$a = \frac{3.21}{(1.026)^7} = 2.68.$$

We have $P(t) = 2.68(1.026)^t$. Since $b = 1.026$ is the growth factor, the country's population grows by about 2.6% per year, assuming exponential growth.

37. **(a)** To see if an exponential function fits the data well, we can look at ratios of successive terms. Giving each ratio to two decimal places, we have

$$\frac{128.4}{109.5} = 1.17, \quad \frac{140.8}{128.4} = 1.10, \quad \frac{158.7}{140.8} = 1.13, \quad \frac{182.1}{158.7} = 1.15, \quad \frac{207.9}{182.1} = 1.14$$

Since the ratios are all similar, an exponential function approximates this data using an average growth factor (or base) of 1.138. Since $S = 109.5$ when $t = 0$, an exponential function to model these data is $S = 109.5(1.138)^t$.

(b) The number of cell phone subscribers in the United States was growing at a rate of approximately 13.8% per year during this period.

(c) Using the model $S = 109.5(1.138)^t$ with $t = 6$ for 2006, we find $S \approx 237.8$. The model is a reasonable fit with the 2006 data.

(d) Using the model $S = 109.5(1.138)^t$ with $t = 10$ for 2010, we find $S \approx 398.9$. which is greater than the population in the United States. The model does not fit the 2010 data; growth has slowed.

41. According to Figure 4.16, f seems to approach its horizontal asymptote, $y = 0$, faster. To convince yourself, compare values of f and g for very large values of x.

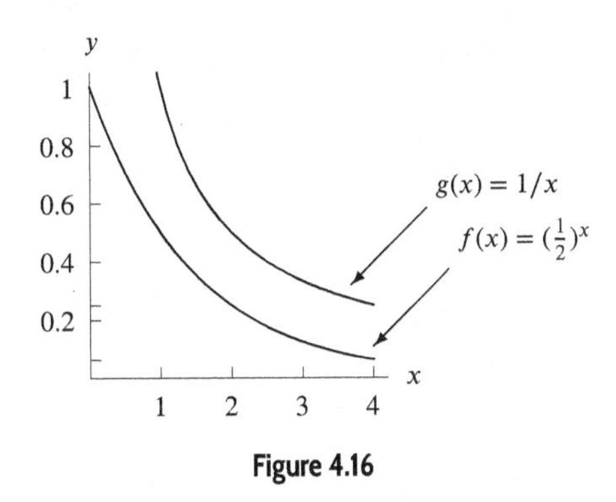

Figure 4.16

45. **(a)** This population initially numbers 5200, and it grows in size by 11.8% every year.

(b) This population initially numbers 4600. There are $12t$ months in t years, which means the population grows by a factor 1.01 twelve times each year, or once every month. In other words, this population grows by 1% every month.

(c) This population initially numbers 3800. It decreases by one-half (50%) every twelve years.

(d) This population initially numbers 8000. It grows at a continuous annual rate of 7.78%.

(e) Note that, unlike the other functions, this is a linear function in point-slope form. It tells us that this population numbers 1675 in year $t = 30$ and that it falls by 25 members every year.

49. Writing the function as

$$p(x) = \frac{7e^{6x} \cdot \sqrt{e} \cdot (2e^x)^{-1}}{10e^{4x}} = \frac{7 \cdot \sqrt{e} \cdot 2^{-1}}{10} e^{(6x-x-4x)} = \frac{7\sqrt{e}}{20} e^x,$$

we have $a = \frac{7\sqrt{e}}{20}$ and $k = 1$.

53. We have $g(n) = \sqrt{f(n) \cdot f(n-1)}$ where

$$\begin{aligned}
f(n) &= 1000 \cdot 2^{-\frac{1}{4}-\frac{n}{2}} \\
&= 1000 \cdot 2^{-\frac{1}{4}} \cdot 2^{-\frac{n}{2}} \\
\text{and} \quad f(n-1) &= 1000 \cdot 2^{-\frac{1}{4}-\frac{n-1}{2}} \\
&= 1000 \cdot 2^{-\frac{1}{4}} \cdot 2^{-\frac{n-1}{2}} \\
&= 1000 \cdot 2^{-\frac{1}{4}} \cdot 2^{\frac{1-n}{2}} \\
&= 1000 \cdot 2^{-\frac{1}{4}} \cdot 2^{\frac{1}{2}-\frac{n}{2}} \\
&= 1000 \cdot 2^{-\frac{1}{4}} \cdot 2^{\frac{1}{2}} \cdot 2^{-\frac{n}{2}} \\
&= 1000 \cdot 2^{-\frac{1}{4}+\frac{1}{2}} \cdot 2^{-\frac{n}{2}} \\
&= 1000 \cdot 2^{\frac{1}{4}} \cdot 2^{-\frac{n}{2}}
\end{aligned}$$

$$\begin{aligned}
\text{so} \quad f(n)\cdot f(n-1) &= \left(1000\cdot 2^{-\frac{1}{4}}2^{-\frac{n}{2}}\right)\left(1000\cdot 2^{\frac{1}{4}}\cdot 2^{-\frac{n}{2}}\right)\\
&= 1000\cdot 1000\cdot 2^{-\frac{1}{4}}\cdot 2^{\frac{1}{4}}\cdot 2^{-\frac{n}{2}}\cdot 2^{-\frac{n}{2}}\\
&= 1{,}000{,}000\cdot 2^{-n}.
\end{aligned}$$

This means

$$\begin{aligned}
g(n) &= \sqrt{f(n)\cdot f(n-1)}\\
&= \sqrt{1{,}000{,}000\cdot 2^{-n}}\\
&= \sqrt{1{,}000{,}000}\sqrt{\cdot 2^{-n}}\\
&= 1000\,(2^{-n})^{0.5}\\
&= 1000\cdot 2^{-0.5n}\\
&= 1000\left(2^{-0.5}\right)^{n}\\
&= 1000(0.7071)^{n}.
\end{aligned}$$

57. In (a), we see that the graph of g starts out below the graph of f. In (c), we see that at some point, the lower graph rises to intersect the higher graph, which tells us that g grows faster than f. This means that $d > b$.

61.  **(a)**

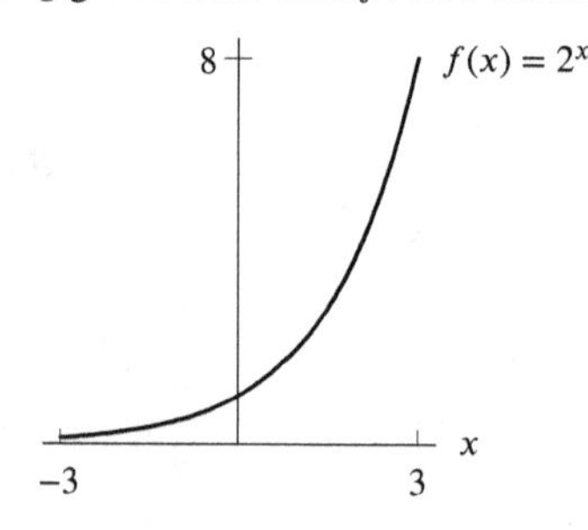

(b) The point $(0, 1)$ is on the graph. So is $(0.01, 1.00696)$. Taking $\frac{y_2 - y_1}{x_2 - x_1}$, we get an estimate for the slope of 0.696. We may zoom in still further to find that $(0.001, 1.000693)$ is on the graph. Using this and the point $(0, 1)$ we would get a slope of 0.693. Zooming in still further we find that the slope stabilizes at around 0.693; so, to two digits of accuracy, the slope is 0.69.

(c) Using the same method as in part (b), we find that the slope is ≈ 1.10.

(d) We might suppose that the slope of the tangent line at $x = 0$ increases as b increases. Trying a few values, we see that this is the case. Then we can find the correct b by trial and error: $b = 2.5$ has slope around 0.916, $b = 3$ has slope around 1.1, so $2.5 < b < 3$. Trying $b = 2.75$ we get a slope of 1.011, just a little too high. $b = 2.7$ gives a slope of 0.993, just a little too low. $b = 2.72$ gives a slope of 1.0006, which is as good as we can do by giving b to two decimal places. Thus $b \approx 2.72$.

In fact, the slope is exactly 1 when $b = e = 2.718\ldots$.

65. At $x = 50$,

$$y = 5000e^{-50/40} = 1432.5240.$$

At $x = 150$,

$$y = 5000e^{-150/40} = 117.5887.$$

We have $q(50) = 1432.524$ and $q(150) = 117.5887$. This gives $y = b + mx$ where

$$m = \frac{q(150) - q(50)}{150 - 50} = \frac{117.5887 - 1432.524}{100} = -13.1.$$

Solving for b, we have

$$\begin{aligned}
q(50) &= b - 13.1(50)\\
b &= q(50) + 13.1(50)\\
&= 1432.524 + 13.1(50)\\
&= 2090,
\end{aligned}$$

so $y = -13.1x + 2090$.

69. **(a)** See Figure 4.17.

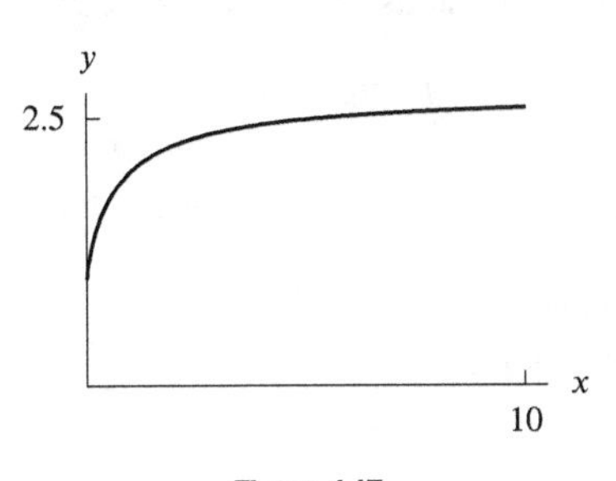

Figure 4.17

(b) The y values are increasing.

(c) The values of $(1 + 1/x)^x$ appear to approach a limiting value as x gets larger.

(d) Figures 4.18 and 4.19 show $y = (1 + 1/x)^x$ for $1 \leq x \leq 100$ and $1 \leq x \leq 1000$, respectively. We see y appears to approach a limiting value of slightly above 2.5.

Figure 4.18 **Figure 4.19**

(e) The graphs of $y = (1 + 1/x)^x$ and $y = e$ are indistinguishable in Figure 4.20, suggesting that $(1 + 1/x)^x$ approaches e as x gets larger. However, a graph cannot tell us that $(1 + 1/x)^x$ approaches e exactly as x gets larger—only that $(1 + 1/x)^x$ gets very close to e.

(f) Table 4.3 shows that the value of $(1 + 1/x)^x$ agrees with $e = 2.718281828 \approx 2.7183$ for $x = 50{,}000$ and above.

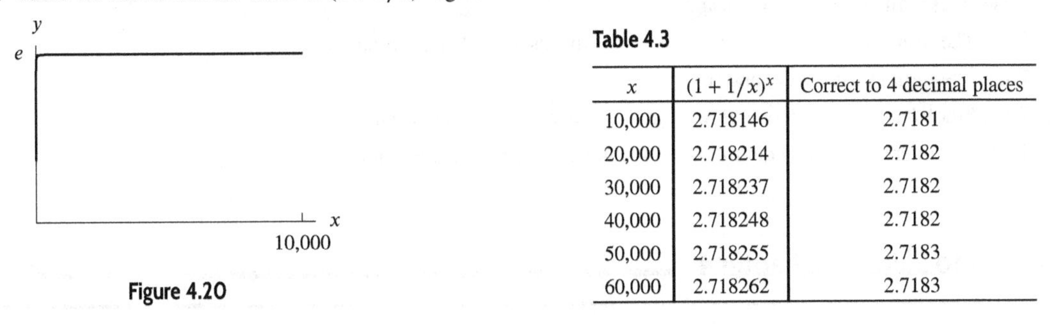

Figure 4.20

Table 4.3

x	$(1+1/x)^x$	Correct to 4 decimal places
10,000	2.718146	2.7181
20,000	2.718214	2.7182
30,000	2.718237	2.7182
40,000	2.718248	2.7182
50,000	2.718255	2.7183
60,000	2.718262	2.7183

73. Figure 4.21 shows three different values of r, labeled r_1, r_2, r_3, and the corresponding values of t, labeled t_1, t_2, t_3. As you can see from the figure, as r is increased, the point of intersection shifts to the left, so the t-coordinate decreases.

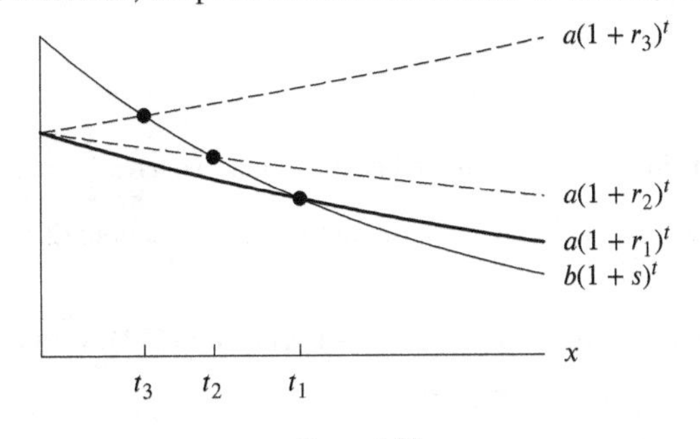

Figure 4.21

77. The wavelength of blue light is $\lambda = 475$ nm, and its absorption coefficient is $\mu(475) = 0.000114$. We have

$$\begin{aligned} T_\lambda(200) &= e^{-\mu(475)\cdot 200} \\ &= e^{-0.000114(200)} \\ &= 0.97746, \end{aligned}$$

so about 97.7% of blue light is transmitted.

81. We have $\lim_{t\to-\infty} (21(1.2)^t + 5.1) = 21(0) + 5.1 = 5.1$.

85. A possible graph is shown in Figure 4.22.

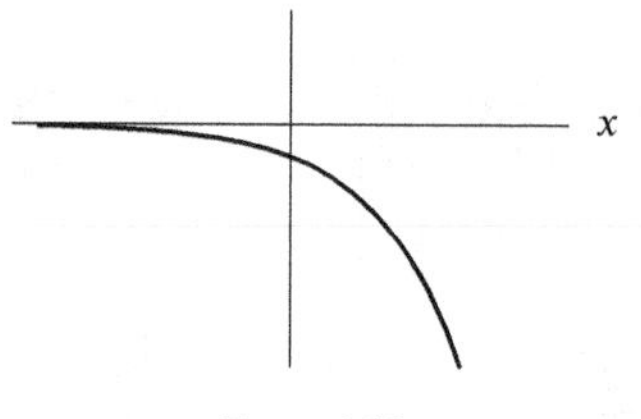

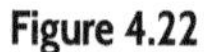

Figure 4.22

STRENGTHEN YOUR UNDERSTANDING

1. True. If the constant rate is r then the formula is $f(t) = a \cdot (1+r)^t$. The function decreases when $0 < 1 + r < 1$ and increases when $1 + r > 1$.

5. False. The annual growth factor would be 1.04, so $S = S_0(1.04)^t$.

9. True. The initial value means the value of Q when $t = 0$, so $Q = f(0) = a \cdot b^0 = a \cdot 1 = a$.

13. False. This is the formula of a linear function.

17. True. The irrational number $e = 2.71828\cdots$ has this as a good approximation.

21. True. The initial value is 200 and the growth factor is 1.04.

25. True. Since k is the continuous growth rate and negative, Q is decreasing.

29. True. The interest from any quarter is compounded in subsequent quarters.

Solutions to Skills for Chapter 4

1. $(-5)^2 = (-5)(-5) = 25$

5. $\frac{5^3}{5^2} = 5^{3-2} = 5^1 = 5$

9. $\sqrt{4^2} = 4$

13. Since the base of 2 is the same in both numerator and denominator, we have $\dfrac{2^7}{2^3} = 2^{7-3} = 2^4$ or $2 \cdot 2 \cdot 2 \cdot 2$ or 16.

17. The order of operations tells us to find 10^3 and then multiply by 2.1. Therefore $(2.1)\left(10^3\right) = (2.1)(1{,}000) = 2{,}100$.

21. $16^{5/4} = (2^4)^{5/4} = 2^5 = 32$

25. Exponentiation is done first, with the result that $(-1)^3 = -1$. Therefore $(-1)^3\sqrt{36} = (-1)\sqrt{36} = (-1)(6) = -6$.

29. $3^{-3/2} = \frac{1}{3^{3/2}} = \frac{1}{(3^3)^{1/2}} = \frac{1}{(27)^{1/2}} = \frac{1}{(9\cdot3)^{1/2}} = \frac{1}{9^{1/2}\cdot3^{1/2}} = \frac{1}{3\sqrt{3}}$

33. The cube root of 0.125 is 0.5. Therefore $(0.125)^{1/3} = \sqrt[3]{0.125} = 0.5$.

37. $\sqrt{x^5y^4} = (x^5 \cdot y^4)^{1/2} = x^{5/2} \cdot y^{4/2} = x^{5/2}y^2$

41. $\sqrt{r^3} = (r^3)^{1/2} = r^{3/2}$

45.

$$\begin{aligned}\sqrt{48u^{10}v^{12}y^5} &= (48)^{1/2} \cdot (u^{10})^{1/2} \cdot (v^{12})^{1/2} \cdot (y^5)^{1/2} \\ &= (16 \cdot 3)^{1/2}u^5v^6y^{5/2} \\ &= 16^{1/2} \cdot 3^{1/2} \cdot u^5v^6y^{5/2} \\ &= 4\sqrt{3}u^5v^6y^{5/2}\end{aligned}$$

49.

$$(3AB)^{-1}\left(A^2B^{-1}\right)^2 = \left(3^{-1} \cdot A^{-1} \cdot B^{-1}\right)\left(A^4 \cdot B^{-2}\right) = \frac{A^4}{3^1 \cdot A^1 \cdot B^1 \cdot B^2} = \frac{A^3}{3B^3}.$$

53. $\dfrac{a^{n+1}3^{n+1}}{a^n3^n} = a^{n+1-n}3^{n+1-n} = a^1 \cdot 3^1 = 3a$

57. $-32^{3/5} = -(\sqrt[5]{32})^3 = -(2)^3 = -8$

61. $64^{-3/2} = (\sqrt{64})^{-3} = (8)^{-3} = \left(\dfrac{1}{8}\right)^3 = \dfrac{1}{512}$

65. We have

$$\begin{aligned}7x^4 &= 20x^2 \\ \frac{x^4}{x^2} &= \frac{20}{7} \\ x^2 &= 20/7 \\ x &= \pm(20/7)^{1/2} = \pm 1.690.\end{aligned}$$

69. False

73. True

77. We have

$$\begin{aligned}5^x &= 32 \\ (2^a)^x &= 2^5 \\ 2^{ax} &= 2^5 \\ ax &= 5 \\ x &= \frac{5}{a}.\end{aligned}$$

81. We have

$$\begin{aligned}5^x &= 7 \\ (2^a)^x &= 2^b \\ 2^{ax} &= 2^b \\ ax &= b \\ x &= \frac{b}{a}.\end{aligned}$$

CHAPTER FIVE

Solutions for Section 5.1

Skill Refresher

S1. Since $1{,}000{,}000 = 10^6$, we have $x = 6$.

S5. Since e^w can never equal zero, the equation $e^w = 0$ has no solution. It similarly follows for any constant b, b^x can never equal zero.

S9. Since $\sqrt[4]{0.1} = 0.1^{1/4} = \left(10^{-1}\right)^{1/4} = 10^{-1/4}$, we have $2t = -\frac{1}{4}$. Solving for t, we therefore have $t = -\frac{1}{8}$.

Exercises

1. The statement is equivalent to $19 = 10^{1.279}$.

5. The statement is equivalent to $P = 10^t$.

9. The statement is equivalent to $v = \log \alpha$.

13. We are looking for a power of 10, and log 1 is asking for the power of 10 which gives 1. Since $1 = 10^0$, $\log 1 = 0$.

17. Using the identity $\log 10^N = N$, we have $\log 10^5 = 5$.

21. Using the identity $10^{\log N} = N$, we have $10^{\log 0.01} = 0.01$.

25. Recall that $\sqrt{e} = e^{1/2}$. Using the identity $\ln e^N = N$, we get $\ln \sqrt{e} = \ln e^{1/2} = \frac{1}{2}$.

29. We are solving for an exponent, so we use logarithms. We can use either the common logarithm or the natural logarithm. Since $2^3 = 8$ and $2^4 = 16$, we know that x must be between 3 and 4. Using the log rules, we have

$$\begin{aligned} 2^x &= 11 \\ \log(2^x) &= \log(11) \\ x\log(2) &= \log(11) \\ x &= \frac{\log(11)}{\log(2)} = 3.459. \end{aligned}$$

If we had used the natural logarithm, we would have

$$x = \frac{\ln(11)}{\ln(2)} = 3.459.$$

33. We begin by dividing both sides by 17 to isolate the exponent:

$$\frac{48}{17} = (2.3)^w.$$

We then take the log of both sides and use the rules of logs to solve for w:

$$\begin{aligned} \log\frac{48}{17} &= \log(2.3)^w \\ \log\frac{48}{17} &= w\log(2.3) \\ w &= \frac{\log\frac{48}{17}}{\log(2.3)} = 1.246. \end{aligned}$$

Problems

37. **(a)** $\log(10 \cdot 100) = \log 1000 = 3$

$\log 10 + \log 100 = 1 + 2 = 3$

(b) $\log(100 \cdot 1000) = \log 100{,}000 = 5$

$\log 100 + \log 1000 = 2 + 3 = 5$

(c) $\log \frac{10}{100} = \log \frac{1}{10} = \log 10^{-1} = -1$

$\log 10 - \log 100 = 1 - 2 = -1$

(d) $\log \frac{100}{1000} = \log \frac{1}{10} = \log 10^{-1} = -1$

$\log 100 - \log 1000 = 2 - 3 = -1$

(e) $\log 10^2 = 2$

$2 \log 10 = 2(1) = 2$

(f) $\log 10^3 = 3$

$3 \log 10 = 3(1) = 3$

In each case, both answers are equal. This reflects the properties of logarithms.

41. False. $\log A \log B = \log A \cdot \log B$, not $\log A + \log B$.

45. Using properties of logs, we have

$$\begin{aligned} \log(3 \cdot 2^x) &= 8 \\ \log 3 + x \log 2 &= 8 \\ x \log 2 &= 8 - \log 3 \\ x &= \frac{8 - \log 3}{\log 2} = 24.990. \end{aligned}$$

49. Using properties of logs, we have

$$\begin{aligned} \ln(3x^2) &= 8 \\ \ln 3 + 2 \ln x &= 8 \\ 2 \ln x &= 8 - \ln 3 \\ \ln x &= \frac{8 - \ln 3}{2} \\ x &= e^{(8 - \ln 3)/2} = 31.522. \end{aligned}$$

Notice that to solve for x, we had to convert from an equation involving logs to an equation involving exponents in the last step.

An alternate way to solve the original equation is to begin by converting from an equation involving logs to an equation involving exponents:

$$\begin{aligned} \ln(3x^2) &= 8 \\ 3x^2 &= e^8 \\ x^2 &= \frac{e^8}{3} \\ x &= \sqrt{e^8/3} = 31.522. \end{aligned}$$

Of course, we get the same answer with both methods.

53. **(a)** The initial value of P is 25. The growth rate is 7.5% per time unit.

(b) We see on the graph that $P = 100$ at approximately $t = 19$. We can use graphing technology to estimate t as accurately as we like.

(c) We substitute $P = 100$ and use logs to solve for t:

$$\begin{aligned} P &= 25(1.075)^t \\ 100 &= 25(1.075)^t \\ 4 &= (1.075)^t \\ \log(4) &= \log(1.075^t) \\ t\log(1.075) &= \log(4) \\ t &= \frac{\log(4)}{\log(1.075)} = 19.169. \end{aligned}$$

57. Since the goal is to get t by itself as much as possible, first divide both sides by 3, and then use logs.

$$\begin{aligned} 3(1.081)^t &= 14 \\ 1.081^t &= \frac{14}{3} \\ \log(1.081)^t &= \log(\frac{14}{3}) \\ t\log 1.081 &= \log(\frac{14}{3}) \\ t &= \frac{\log(\frac{14}{3})}{\log 1.081}. \end{aligned}$$

61. We have

$$\begin{aligned} 6000\left(\frac{1}{2}\right)^{t/15} &= 1000 \\ \left(\frac{1}{2}\right)^{t/15} &= \frac{1}{6} \\ \ln\left(\frac{1}{2}\right)^{t/15} &= \ln\left(\frac{1}{6}\right) \\ \frac{t}{15}\cdot\ln\left(\frac{1}{2}\right) &= -\ln 6 \\ t(-\ln 2) &= -15\ln 6 \\ t &= \frac{15\ln 6}{\ln 2} \end{aligned}$$

65. Taking natural logs, we get

$$\begin{aligned} 3^{x+4} &= 10 \\ \ln 3^{x+4} &= \ln 10 \\ (x+4)\ln 3 &= \ln 10 \\ x\ln 3 + 4\ln 3 &= \ln 10 \\ x\ln 3 &= \ln 10 - 4\ln 3 \\ x &= \frac{\ln 10 - 4\ln 3}{\ln 3} \end{aligned}$$

69.

$$\begin{aligned} 58e^{4t+1} &= 30 \\ e^{4t+1} &= \frac{30}{58} \\ \ln e^{4t+1} &= \ln(\frac{30}{58}) \\ 4t+1 &= \ln(\frac{30}{58}) \\ t &= \frac{1}{4}\left(\ln(\frac{30}{58}) - 1\right). \end{aligned}$$

73. We have

$$\log(2x+5)\cdot\log(9x^2)=0.$$

In order for this product to equal zero, we know that one or both factors must be equal to zero. Thus, we will set each of the factors equal to zero to determine the values of x for which the factors will equal zero. We have

$$\begin{aligned}\log(2x+5)&=0\\2x+5&=1\\2x&=-4\\x&=-2\end{aligned}\qquad\text{or}\qquad\begin{aligned}\log(9x^2)&=0\\9x^2&=1\\x^2&=\frac{1}{9}\\x&=\frac{1}{3}\text{ or }x=-\frac{1}{3}.\end{aligned}$$

Thus our solutions are $x=-2,\frac{1}{3}$, or $-\frac{1}{3}$.

77. **(a)** We combine like terms and then use properties of logs.

$$\begin{aligned}e^{2x}+e^{2x}&=1\\2e^{2x}&=1\\e^{2x}&=0.5\\2x&=\ln(0.5)\\x&=\frac{\ln(0.5)}{2}.\end{aligned}$$

(b) We combine like terms and then use properties of logs.

$$\begin{aligned}2e^{3x}+e^{3x}&=b\\3e^{3x}&=b\\e^{3x}&=\frac{b}{3}\\3x&=\ln(b/3)\\x&=\frac{\ln(b/3)}{3}.\end{aligned}$$

81. Since $Q=r\cdot s^t$ and $q=\ln Q$, we see that

$$\begin{aligned}q&=\ln\underbrace{\left(r\cdot s^t\right)}_{Q}\\&=\ln r+\ln\left(s^t\right)\\&=\underbrace{\ln r}_{b}+\underbrace{(\ln s)}_{m}\,t,\end{aligned}$$

so $b=\ln r$ and $m=\ln s$.

Solutions for Section 5.2

Skill Refresher

S1. Rewrite as $10^{-\log 5x}=10^{\log(5x)^{-1}}=(5x)^{-1}$.

S5. Taking logs of both sides, we get

$$\log(4^x)=\log 9.$$

This gives

$$x\log 4=\log 9$$

or in other words

$$x=\frac{\log 9}{\log 4}=1.585.$$

S9. We begin by converting to exponential form:

$$
\begin{aligned}
\log(2x+7) &= 2 \\
10^{\log(2x+7)} &= 10^2 \\
2x+7 &= 100 \\
2x &= 93 \\
x &= \frac{93}{2}.
\end{aligned}
$$

Exercises

1. The continuous percent growth rate is the value of k in the equation $Q = ae^{kt}$, which is 7.

To convert to the form $Q = ab^t$, we first say that the right sides of the two equations equal each other (since each equals Q), and then we solve for a and b. Thus, we have $ab^t = 4e^{7t}$. At $t = 0$, we can solve for a:

$$
\begin{aligned}
ab^0 &= 4e^{7\cdot 0} \\
a\cdot 1 &= 4\cdot 1 \\
a &= 4.
\end{aligned}
$$

Thus, we have $4b^t = 4e^{7t}$, and we solve for b:

$$
\begin{aligned}
4b^t &= 4e^{7t} \\
b^t &= e^{7t} \\
b^t &= \left(e^7\right)^t \\
b &= e^7 \approx 1096.633.
\end{aligned}
$$

Therefore, the equation is $Q = 4\cdot 1096.633^t$.

5. We want $25e^{0.053t} = 25(e^{0.053})^t = ab^t$, so we choose $a = 25$ and $b = e^{0.053} = 1.0544$. The given exponential function is equivalent to the exponential function $y = 25(1.0544)^t$. The annual percent growth rate is 5.44% and the continuous percent growth rate per year is 5.3% per year.

9. To convert to the form $Q = ae^{kt}$, we first say that the right sides of the two equations equal each other (since each equals Q), and then we solve for a and k. Thus, we have $ae^{kt} = 14(0.862)^{1.4t}$. At $t = 0$, we can solve for a:

$$
\begin{aligned}
ae^{k\cdot 0} &= 14(0.862)^0 \\
a\cdot 1 &= 14\cdot 1 \\
a &= 14.
\end{aligned}
$$

Thus, we have $14e^{kt} = 14(0.862)^{1.4t}$, and we solve for k:

$$
\begin{aligned}
14e^{kt} &= 14(0.862)^{1.4t} \\
e^{kt} &= \left(0.862^{1.4}\right)^t \\
\left(e^k\right)^t &= (0.812)^t \\
e^k &= 0.812 \\
\ln e^k &= \ln 0.812 \\
k &= -0.208.
\end{aligned}
$$

Therefore, the equation is $Q = 14e^{-0.208t}$.

13. We have $a = 230$, $b = 1.182$, $r = b - 1 = 18.2\%$, and $k = \ln b = 0.1672 = 16.72\%$.

17. Writing this as $Q = 12.1(10^{-0.11})^t$, we have $a = 12.1$, $b = 10^{-0.11} = 0.7762$, $r = b - 1 = -22.38\%$, and $k = \ln b = -25.32\%$.

Problems

21. Since the growth factor is $1.027 = 1 + 0.027$, the formula for the bank account balance, with an initial balance of a and time t in years, is
$$B = a(1.027)^t.$$
The balance doubles for the first time when $B = 2a$. Thus, we solve for t after putting B equal to $2a$ to give us the doubling time:
$$\begin{aligned} 2a &= a(1.027)^t \\ 2 &= (1.027)^t \\ \log 2 &= \log(1.027)^t \\ \log 2 &= t\log(1.027) \\ t &= \frac{\log 2}{\log(1.027)} = 26.017. \end{aligned}$$
So the doubling time is about 26 years.

25. The growth factor for Einsteinium-253 should be $1-0.03406 = 0.96594$, since it is decaying by 3.406% per day. Therefore, the decay equation starting with a quantity of a should be:
$$Q = a(0.96594)^t,$$
where Q is quantity remaining and t is time in days. The half-life will be the value of t for which Q is $a/2$, or half of the initial quantity a. Thus, we solve the equation for $Q = a/2$:
$$\begin{aligned} \frac{a}{2} &= a(0.96594)^t \\ \frac{1}{2} &= (0.96594)^t \\ \log(1/2) &= \log(0.96594)^t \\ \log(1/2) &= t\log(0.96594) \\ t &= \frac{\log(1/2)}{\log(0.96594)} = 20.002. \end{aligned}$$
So the half-life is about 20 days.

29. **(a)** Let $P(t) = P_0 b^t$ describe our population at the end of t years. Since P_0 is the initial population, and the population doubles every 15 years, we know that, at the end of 15 years, our population will be $2P_0$. But at the end of 15 years, our population is $P(15) = P_0 b^{15}$. Thus
$$\begin{aligned} P_0 b^{15} &= 2P_0 \\ b^{15} &= 2 \\ b &= 2^{\frac{1}{15}} \approx 1.04729 \end{aligned}$$
Since b is our growth factor, the population is, yearly, 104.729% of what it had been the previous year. Thus it is growing by 4.729% per year.

(b) Writing our formula as $P(t) = P_0 e^{kt}$, we have $P(15) = P_0 e^{15k}$. But we already know that $P(15) = 2P_0$. Therefore,
$$\begin{aligned} P_0 e^{15k} &= 2P_0 \\ e^{15k} &= 2 \\ \ln e^{15k} &= \ln 2 \\ 15k \ln e &= \ln 2 \\ 15k &= \ln 2 \\ k &= \frac{\ln 2}{15} \approx 0.04621. \end{aligned}$$
This tells us that we have a continuous annual growth rate of 4.621%.

33.

$$
\begin{aligned}
\text{Value of first investment} &= 9000e^{0.056t} && a = 9000, k = 5.6\% \\
\text{Value of second investment} &= 4000e^{0.083t} && a = 4000, k = 8.3\% \\
\text{so} \quad 4000e^{0.083t} &= 9000e^{0.056t} && \text{set values equal} \\
\frac{e^{0.083t}}{e^{0.056t}} &= \frac{9000}{4000} && \text{divide} \\
e^{0.083t-0.056t} &= 2.25 && \text{simplify} \\
e^{0.027t} &= 2.25 \\
0.027t &= \ln 2.25 && \text{take logs} \\
t &= \frac{\ln 2.25}{0.027} \\
&= 30.034,
\end{aligned}
$$

so it will take almost exactly 30 years. To check our answer, we see that

$$
\begin{aligned}
9000e^{0.056(30.034)} &= 48{,}383.26 \\
4000e^{0.083(30.034)} &= 48{,}383.26. \quad \text{values are equal}
\end{aligned}
$$

37. We know the y-intercept is 0.8 and that the y-value doubles every 12 units. We can make a quick table and then plot points. See Table 5.1 and Figure 5.1.

Table 5.1

t	y
0	0.8
12	1.6
24	3.2
36	6.4
48	12.8
60	25.6

Figure 5.1

41. **(a)** We see that the initial value of the function is 50 and that the value has doubled to 100 at $t = 12$ so the doubling time is 12 days. Notice that 12 days later, at $t = 24$, the value of the function has doubled again, to 200. No matter where we start, the value will double 12 days later.

(b) We use $P = 50e^{kt}$ and the fact that the function increases from 50 to 100 in 12 days. Solving for k, we have:

$$
\begin{aligned}
50e^{k(12)} &= 100 \\
e^{12k} &= 2 \\
12k &= \ln 2 \\
k &= \frac{\ln 2}{12} = 0.0578.
\end{aligned}
$$

The continuous percent growth rate is 5.78% per day. The formula is $P = 50e^{0.0578t}$.

45. **(a)** If prices rise at 3% per year, then each year they are 103% of what they had been the year before. After 5 years, they will be $(103\%)^5 = (1.03)^5 \approx 1.15927$, or 115.927% of what they had been initially. In other words, they have increased by 15.927% during that time.

(b) If it takes t years for prices to rise 25%, then

$$
\begin{aligned}
1.03^t &= 1.25 \\
\log 1.03^t &= \log 1.25 \\
t \log 1.03 &= \log 1.25 \\
t &= \frac{\log 1.25}{\log 1.03} \approx 7.549.
\end{aligned}
$$

With an annual inflation rate of 3%, it takes approximately 7.5 years for prices to increase by 25%.

49. **(a)** For a function of the form $N(t) = ae^{kt}$, the value of a is the population at time $t = 0$ and k is the continuous growth rate. So the continuous growth rate is $0.026 = 2.6\%$.

(b) In year $t = 0$, the population is $N(0) = a = 6.72$ million.

(c) We want to find t such that the population of 6.72 million triples to 20.16 million. So, for what value of t does $N(t) = 6.72e^{0.026t} = 20.16$?

$$\begin{aligned} 6.72e^{0.026t} &= 20.16 \\ e^{0.026t} &= 3 \\ \ln e^{0.026t} &= \ln 3 \\ 0.026t &= \ln 3 \\ t &= \frac{\ln 3}{0.026} \approx 42.2543 \end{aligned}$$

So the population will triple in approximately 42.25 years.

(d) Since $N(t)$ is in millions, we want to find t such that $N(t) = 0.000001$.

$$\begin{aligned} 6.72e^{0.026t} &= 0.000001 \\ e^{0.026t} &= \frac{0.000001}{6.72} \approx 0.000000149 \\ \ln e^{0.026t} &\approx \ln(0.000000149) \\ 0.026t &\approx \ln(0.000000149) \\ t &\approx \frac{\ln(0.000000149)}{0.026} \approx -604.638 \end{aligned}$$

According to this model, the population of Washington State was 1 person approximately 605 years ago. It is unreasonable to suppose the formula extends so far into the past.

53. **(a)** Since $f(x)$ is exponential, its formula will be $f(x) = ab^x$. Since $f(0) = 0.5$,

$$f(0) = ab^0 = 0.5.$$

But $b^0 = 1$, so

$$\begin{aligned} a(1) &= 0.5 \\ a &= 0.5. \end{aligned}$$

We now know that $f(x) = 0.5b^x$. Since $f(1) = 2$, we have

$$\begin{aligned} f(1) = 0.5b^1 &= 2 \\ 0.5b &= 2 \\ b &= 4 \end{aligned}$$

So $f(x) = 0.5(4)^x$.

We will find a formula for $g(x)$ the same way.

$$g(x) = ab^x.$$

Since $g(0) = 4$,

$$\begin{aligned} g(0) = ab^0 &= 4 \\ a &= 4. \end{aligned}$$

Therefore,

$$g(x) = 4b^x.$$

We'll use $g(2) = \frac{4}{9}$ to get

$$\begin{aligned} g(2) = 4b^2 &= \frac{4}{9} \\ b^2 &= \frac{1}{9} \\ b &= \pm\frac{1}{3}. \end{aligned}$$

Since $b > 0$,

$$g(x) = 4\left(\frac{1}{3}\right)^x.$$

Since $h(x)$ is linear, its formula will be

$$h(x) = b + mx.$$

We know that b is the y-intercept, which is 2, according to the graph. Since the points $(a, a+2)$ and $(0, 2)$ lie on the graph, we know that the slope, m, is

$$\frac{(a+2)-2}{a-0} = \frac{a}{a} = 1,$$

so the formula is

$$h(x) = 2 + x.$$

(b) We begin with

$$\begin{aligned} f(x) &= g(x) \\ \frac{1}{2}(4)^x &= 4\left(\frac{1}{3}\right)^x. \end{aligned}$$

Since the variable is an exponent, we need to use logs, so

$$\begin{aligned} \log\left(\frac{1}{2}\cdot 4^x\right) &= \log\left(4\cdot\left(\frac{1}{3}\right)^x\right) \\ \log\frac{1}{2} + \log(4)^x &= \log 4 + \log\left(\frac{1}{3}\right)^x \\ \log\frac{1}{2} + x\log 4 &= \log 4 + x\log\frac{1}{3}. \end{aligned}$$

Now we will move all expressions containing the variable to one side of the equation:

$$x\log 4 - x\log\frac{1}{3} = \log 4 - \log\frac{1}{2}.$$

Factoring out x, we get

$$\begin{aligned} x(\log 4 - \log\frac{1}{3}) &= \log 4 - \log\frac{1}{2} \\ x\log\left(\frac{4}{1/3}\right) &= \log\left(\frac{4}{1/2}\right) \\ x\log 12 &= \log 8 \\ x &= \frac{\log 8}{\log 12}. \end{aligned}$$

This is the exact value of x. Note that $\frac{\log 8}{\log 12} \approx 0.837$, so $f(x) = g(x)$ when x is exactly $\frac{\log 8}{\log 12}$ or about 0.837.

(c) Since $f(x) = h(x)$, we want to solve

$$\frac{1}{2}(4)^x = x + 2.$$

The variable does not occur only as an exponent, so logs cannot help us solve this equation. Instead, we need to graph the two functions and note where they intersect. The points occur when $x \approx 1.378$ or $x \approx -1.967$.

57. Let $Q = ae^{kt}$ be an increasing exponential function, so that k is positive. To find the doubling time, we find how long it takes Q to double from its initial value a to the value $2a$:

$$\begin{aligned} ae^{kt} &= 2a \\ e^{kt} &= 2 \qquad \text{(dividing by } a\text{)} \\ \ln e^{kt} &= \ln 2 \\ kt &= \ln 2 \qquad \text{(because } \ln e^x = x \text{ for all } x\text{)} \\ t &= \frac{\ln 2}{k}. \end{aligned}$$

Using a calculator, we find $\ln 2 = 0.693 \approx 0.70$. This is where the 70 comes from.
If, for example, the continuous growth rate is $k = 0.07 = 7\%$, then

$$\text{Doubling time} = \frac{\ln 2}{0.07} = \frac{0.693}{0.07} \approx \frac{0.70}{0.07} = \frac{70}{7} = 10.$$

If the growth rate is $r\%$, then $k = r/100$. Therefore

$$\text{Doubling time} = \frac{\ln 2}{k} = \frac{0.693}{k} \approx \frac{0.70}{r/100} = \frac{70}{r}.$$

61. We have

$$\begin{aligned} 6e^{-0.5e^{-0.1t}} &= 3 \\ e^{-0.5e^{-0.1t}} &= 0.5 && \text{divide} \\ -0.5e^{-0.1t} &= \ln 0.5 && \text{take logs} \\ e^{-0.1t} &= -2\ln 0.5 && \text{multiply} \\ -0.1t &= \ln(-2\ln 0.5) && \text{take logs} \\ t &= -10\ln(-2\ln 0.5) && \text{multiply} \\ &= -3.266. \end{aligned}$$

Checking our answer, we see that

$$\begin{aligned} f(-3.266) &= 6e^{-0.5e^{-0.1(-3.266)}} \\ &= 6e^{-0.5e^{0.3266}} \\ &= 6e^{-0.5(1.3862)} \\ &= 6e^{-0.6931} = 3, \end{aligned}$$

as required.

Solutions for Section 5.3

Skill Refresher

S1. $\log 0.0001 = \log 10^{-4} = -4\log 10 = -4$.

S5. The equation $-\ln x = 12$ can be expressed as $\ln x = -12$, which is equivalent to $x = e^{-12}$.

S9. Rewrite the sum as $\ln x^3 + \ln x^2 = \ln(x^3 \cdot x^2) = \ln x^5$.

Exercises

1. For $y = \ln x$, we have

$$\text{Domain is } x > 0$$
$$\text{Range is all } y.$$

The graph of $y = \ln(x-3)$ is the graph of $y = \ln x$ shifted right by 3 units. Thus for $y = \ln(x-3)$, we have

$$\text{Domain is } x > 3$$
$$\text{Range is all } y.$$

5. The rate of change is negative and decreases as we move right. The graph is concave down.

9. The graphs of $y = 10^x$ and $y = 2^x$ both have horizontal asymptotes, $y = 0$. The graph of $y = \log x$ has a vertical asymptote, $x = 0$.

13. See Figure 5.2. The graph of $y = \log(x - 4)$ is the graph of $y = \log x$ shifted to the right 4 units.

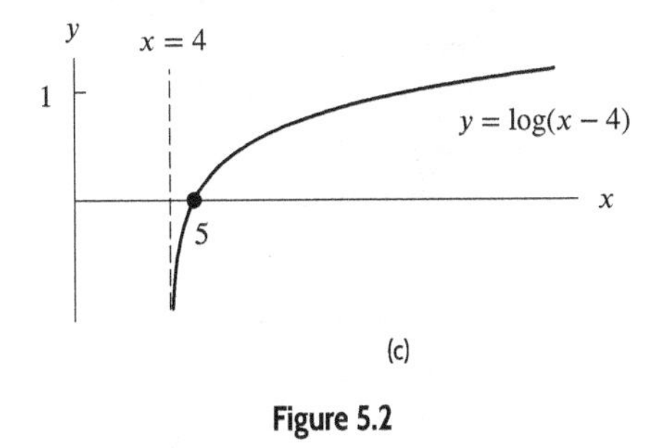

Figure 5.2

17. We know, by the definition of pH, that $0 = -\log[H^+]$. Therefore, $-0 = \log[H^+]$, and $10^{-0} = [H^+]$. Thus, the hydrogen ion concentration is $10^{-0} = 10^0 = 1$ mole per liter.

21. We know, by the definition of pH, that $4.5 = -\log[H^+]$. Therefore, $-4.5 = \log[H^+]$, and $10^{-4.5} = [H^+]$. Thus, the hydrogen ion concentration is $10^{-4.5} = 3.162 \times 10^{-5}$ moles per liter.

Problems

25. **(a)** Let the functions graphed in (a), (b), and (c) be called $f(x)$, $g(x)$, and $h(x)$ respectively. Looking at the graph of $f(x)$, we see that $f(10) = 3$. In the table for $r(x)$ we note that $r(10) = 1.699$ so $f(x) \neq r(x)$. Similarly, $s(10) = 0.699$, so $f(x) \neq s(x)$. The values describing $t(x)$ do seem to satisfy the graph of $f(x)$, however. In the graph, we note that when $0 < x < 1$, then y must be negative. The data point $(0.1, -3)$ satisfies this. When $1 < x < 10$, then $0 < y < 3$. In the table for $t(x)$, we see that the point $(2, 0.903)$ satisfies this condition. Finally, when $x > 10$ we see that $y > 3$. The values $(100, 6)$ satisfy this. Therefore, $f(x)$ and $t(x)$ could represent the same function.

(b) For $g(x)$, we note that

$$\begin{cases} \text{when } 0 < x < 0.2, & \text{then } y < 0; \\ \text{when } 0.2 < x < 1, & \text{then } 0 < y < 0.699; \\ \text{when } x > 1, & \text{then } y > 0.699. \end{cases}$$

All the values of x in the table for $r(x)$ are greater than 1 and all the corresponding values of y are greater than 0.699, so $g(x)$ could equal $r(x)$. We see that, in $s(x)$, the values $(0.5, -0.060)$ do not satisfy the second condition so $g(x) \neq s(x)$. Since we already know that $t(x)$ corresponds to $f(x)$, we conclude that $g(x)$ and $r(x)$ correspond.

(c) By elimination, $h(x)$ must correspond to $s(x)$. We see that in $h(x)$,

$$\begin{cases} \text{when } x < 2, & \text{then } y < 0; \\ \text{when } 2 < x < 20, & \text{then } 0 < y < 1; \\ \text{when } x > 20, & \text{then } y > 1. \end{cases}$$

Since the values in $s(x)$ satisfy these conditions, it is reasonable to say that $h(x)$ and $s(x)$ correspond.

29. A possible formula is $y = \ln x$.

33. **(a)** The graph in (II) is the only graph with two horizontal asymptotes.

(b) The graph in (I) has a vertical asymptote at $x = 0$ and a horizontal asymptote coinciding with the x-axis.

(c) The graph in (III) tends to infinity both as x tends to 0 and as x gets larger and larger.

(d) The graphs in each of (I), (III) and (IV) have a vertical asymptote at $x = 0$.

37. **(a)** (i) $\text{pH} = -\log x = 2$ so $\log x = -2$ so $x = 10^{-2}$

(ii) $\text{pH} = -\log x = 4$ so $\log x = -4$ so $x = 10^{-4}$

(iii) $\text{pH} = -\log x = 7$ so $\log x = -7$ so $x = 10^{-7}$

(b) Solutions with high pHs have low concentrations and so are less acidic.

41. **(a)** The pH is 2.3, which, according to our formula for pH, means that

$$-\log\left[\text{H}^+\right] = 2.3.$$

This means that

$$\log\left[\text{H}^+\right] = -2.3.$$

This tells us that the exponent of 10 that gives $[H^+]$ is -2.3, so

$$\begin{aligned} [H^+] &= 10^{-2.3} && \text{because } -2.3 \text{ is exponent of 10} \\ &= 0.005 \text{ moles/liter.} \end{aligned}$$

(b) From part (a) we know that 1 liter of lime juice contains 0.005 moles of H^+ ions. To find out how many H^+ ions our lime juice has, we need to convert ounces of juice to liters of juice and moles of ions to numbers of ions. We have

$$2 \text{ oz} \times \frac{1 \text{ liter}}{30.3 \text{ oz}} = 0.066 \text{ liters.}$$

We see that

$$0.066 \text{ liters juice} \times \frac{0.005 \text{ moles } H^+ \text{ ions}}{1 \text{ liter}} = 3.3 \times 10^{-4} \text{ moles } H^+ \text{ ions.}$$

There are 6.02×10^{23} ions in one mole, and so

$$3.3 \times 10^{-4} \text{ moles} \times \frac{6.02 \times 10^{23} \text{ ions}}{\text{mole}} = 1.987 \times 10^{20} \text{ ions.}$$

45. If I_T is the sound intensity of Turkish fans' roar, then

$$10 \log \left(\frac{I_T}{I_0} \right) = 131.76 \text{ dB.}$$

Similarly, if I_B is the sound intensity of the Broncos fans, then

$$10 \log \left(\frac{I_B}{I_0} \right) = 128.7 \text{ dB.}$$

Computing the difference of the decibel ratings gives

$$10 \log \left(\frac{I_T}{I_0} \right) - 10 \log \left(\frac{I_B}{I_0} \right) = 131.76 - 128.7 = 3.06.$$

Dividing by 10 gives

$$\begin{aligned} \log \left(\frac{I_T}{I_0} \right) - \log \left(\frac{I_B}{I_0} \right) &= 0.306 \\ \log \left(\frac{I_T/I_0}{I_B/I_0} \right) &= 0.306 && \text{Using the property } \log b - \log a = \log(b/a) \\ \log \left(\frac{I_T}{I_B} \right) &= 0.3063 && \text{Canceling } I_0 \\ \frac{I_T}{I_B} &= 10^{0.306} && \text{Raising 10 to the power of both sides} \end{aligned}$$

So $I_T = 10^{0.306} I_B$, which means that crowd of Turkish fans was $10^{0.306} \approx 2$ times as intense as the Broncos fans.

Notice that although the difference in decibels between the Turkish fans' roar and the Broncos fans' roar is only 3.06 dB, the sound intensity is twice as large.

49. We set $M = 3.5$ and solve for the ratio W/W_0.

$$\begin{aligned} 3.5 &= \log \left(\frac{W}{W_0} \right) \\ 10^{3.5} &= \frac{W}{W_0} \\ &= 3{,}162. \end{aligned}$$

Thus, the Chernobyl nuclear explosion had seismic waves that were 3,162 times more powerful than W_0.

53. **(a)** Since $\log x$ becomes more and more negative as x decreases to 0 from above,

$$\lim_{x \to 0^+} \log x = -\infty.$$

(b) Since $-x$ is positive if x is negative and $-x$ decreases to 0 as x increases to 0 from below,

$$\lim_{x \to 0^-} \ln(-x) = -\infty.$$

Solutions for Section 5.4

Skill Refresher

S1. 1.455×10^6

S5. 3.6×10^{-4}

S9. Since $0.1 < \frac{1}{3} < 1$, $10^{-1} < \frac{1}{3} < 1 = 10^0$.

Exercises

1. Using a linear scale, the wealth of everyone with less than a million dollars would be indistinguishable because all of them are less than one one-thousandth of the wealth of the average billionaire. A log scale is more useful.

5. **(a)**

Table 5.2

n	1	2	3	4	5	6	7	8	9
$\log n$	0	0.3010	0.4771	0.6021	0.6990	0.7782	0.8451	0.9031	0.9542

Table 5.3

n	10	20	30	40	50	60	70	80	90
$\log n$	1	1.3010	1.4771	1.6021	1.6990	1.7782	1.8451	1.9031	1.9542

(b) The first tick mark is at $10^0 = 1$. The dot for the number 2 is placed $\log 2 = 0.3010$ of the distance from 1 to 10. The number 3 is placed at $\log 3 = 0.4771$ units from 1, and so on. The number 30 is placed 1.4771 units from 1, the number 50 is placed 1.6989 units from 1, and so on.

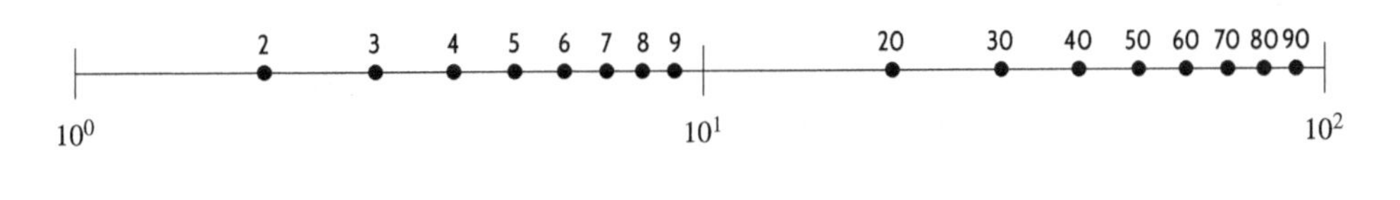

Figure 5.3

9. We have $\log 160 = 2.204$, so this lifespan would be marked at 2.2 inches.

13. **(a)** Run a linear regression on the data. The resulting function is $y = -3582.145 + 236.314x$, with $r \approx 0.7946$. We see from the sketch of the graph of the data that the estimated regression line provides a reasonable but not excellent fit. See Figure 5.4.

(b) If, instead, we compare x and $\ln y$ we get

$$\ln y = 1.568 + 0.200x.$$

We see from the sketch of the graph of the data that the estimated regression line provides an excellent fit with $r \approx 0.9998$. See Figure 5.5. Solving for y, we have

$$\begin{aligned} e^{\ln y} &= e^{1.568+0.200x} \\ y &= e^{1.568}e^{0.200x} \\ y &= 4.797e^{0.200x} \\ \text{or} \quad y &= 4.797(e^{0.200})^x \approx 4.797(1.221)^x. \end{aligned}$$

Figure 5.4 Figure 5.5

(c) The linear equation is a poor fit, and the exponential equation is a better fit.

Problems

17. (a) An appropriate scale is from 0 to 70 at intervals of 10. (Other answers are possible.) See Figure 5.6. The points get more and more spread out as the exponent increases.

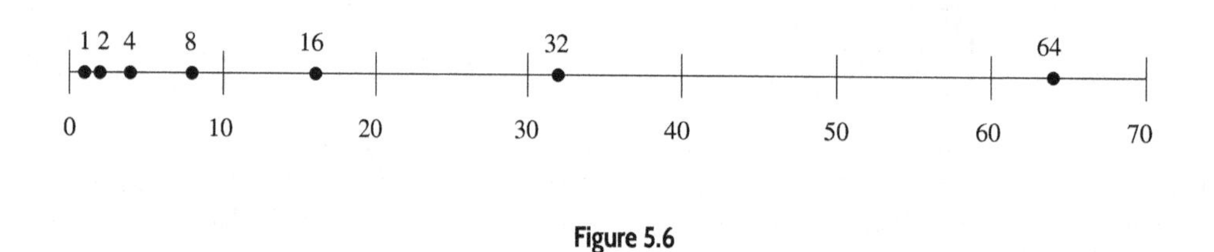

Figure 5.6

(b) If we want to locate 2 on a logarithmic scale, since $2 = 10^{0.3}$, we find $10^{0.3}$. Similarly, $8 = 10^{0.9}$ and $32 = 10^{1.5}$, so 8 is at $10^{0.9}$ and 32 is at $10^{1.5}$. Since the values of the logs go from 0 to 1.8, an appropriate scale is from 0 to 2 at intervals of 0.2. See Figure 5.7. The points are spaced at equal intervals.

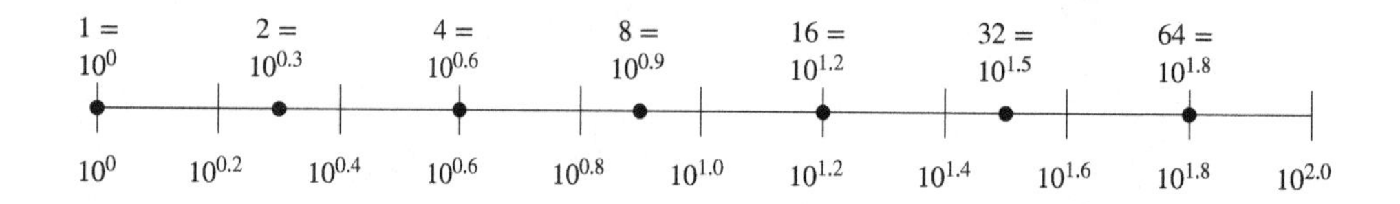

Figure 5.7

21. For the pack of gum, $\log(0.50) = -0.3$, so the pack of gum is plotted at -0.3. For the movie ticket, $\log(8) = 0.9$, so the ticket is plotted at 0.90, and so on. See Figure 5.8.

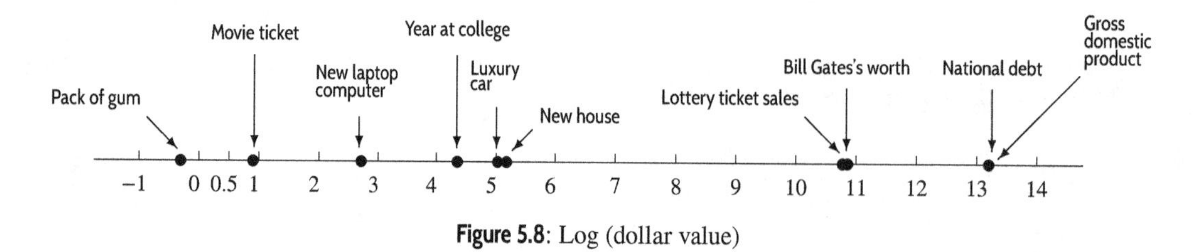

Figure 5.8: Log (dollar value)

25. **(a)**

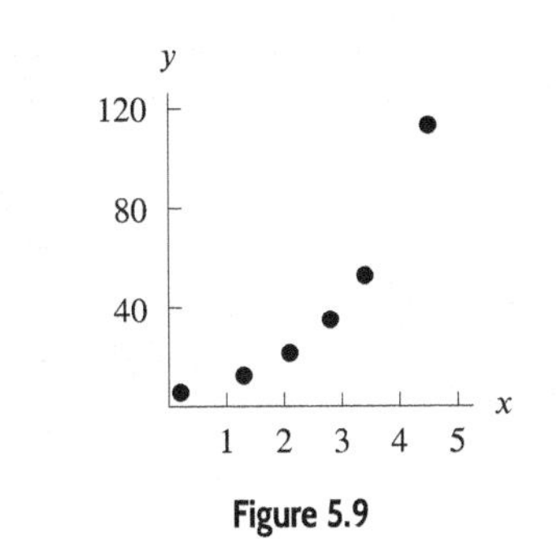

Figure 5.9

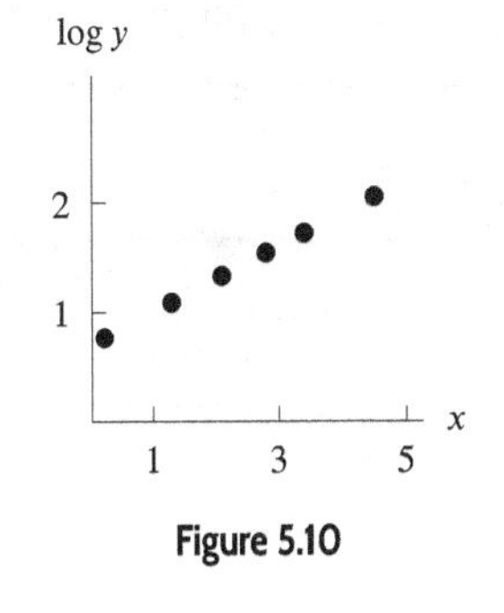

Figure 5.10

(b) The data appear to be exponential.
(c) See Figure 5.10. The data appear to be linear.

Table 5.4

x	0.2	1.3	2.1	2.8	3.4	4.5
$\log y$	.76	1.09	1.33	1.54	1.72	2.05

29. **(a)** Table 5.5 gives values of $L = \ln \ell$ and $W = \ln w$. The data in Table 5.5 have been plotted in Figure 5.11, and a line of best fit has been drawn in. See part (b).
(b) The formula for the line of best fit is $W = 3.06L - 4.54$, as determined using a spreadsheet. However, you could also obtain comparable results by fitting a line by eye.
(c) We have

$$\begin{aligned} W &= 3.06L - 4.54 \\ \ln w &= 3.06 \ln \ell - 4.54 \\ \ln w &= \ln \ell^{3.06} - 4.54 \\ w &= e^{\ln \ell^{3.06} - 4.54} \\ &= \ell^{3.06} e^{-4.54} \approx 0.011 \ell^{3.06}. \end{aligned}$$

(d) Weight tends to be directly proportional to volume, and in many cases volume tends to be proportional to the cube of a linear dimension (e.g., length). Here we see that w is in fact very nearly proportional to the cube of ℓ.

Table 5.5 *$L = \ln \ell$ and $W = \ln w$ for 16 different fish*

Type	1	2	3	4	5	6	7	8
L	2.092	2.208	2.322	2.477	2.501	2.625	2.695	2.754
W	1.841	2.262	2.451	2.918	3.266	3.586	3.691	3.857
Type	9	10	11	12	13	14	15	16
L	2.809	2.874	2.929	2.944	3.025	3.086	3.131	3.157
W	4.184	4.240	4.336	4.413	4.669	4.786	5.131	5.155

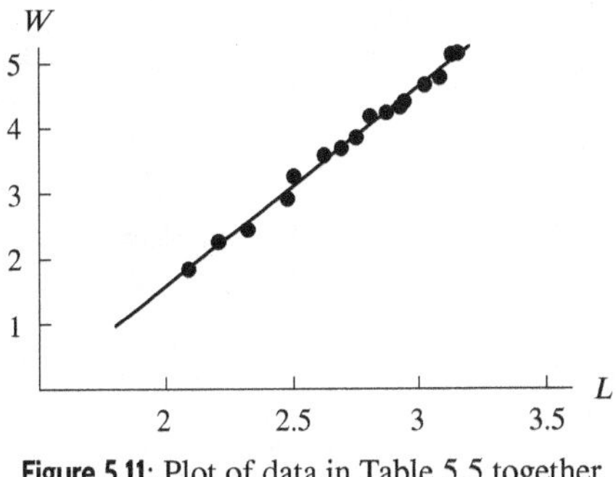

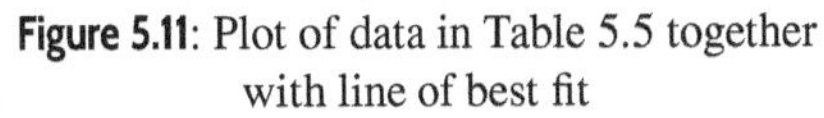

Figure 5.11: Plot of data in Table 5.5 together with line of best fit

Solutions for Chapter 5 Review

Exercises

1. The continuous percent growth rate is the value of k in the equation $Q = ae^{kt}$, which is -10.

To convert to the form $Q = ab^t$, we first say that the right sides of the two equations equal each other (since each equals Q), and then we solve for a and b. Thus, we have $ab^t = 7e^{-10t}$. At $t = 0$, we can solve for a:

$$\begin{aligned} ab^0 &= 7e^{-10\cdot 0} \\ a\cdot 1 &= 7\cdot 1 \\ a &= 7. \end{aligned}$$

Thus, we have $7b^t = 7e^{-10t}$, and we solve for b:

$$\begin{aligned} 7b^t &= e^{-10t} \\ b^t &= e^{-10t} \\ b^t &= \left(e^{-10}\right)^t \\ b &= e^{-10} \approx 0.0000454. \end{aligned}$$

Therefore, the equation is $Q = 7(0.0000454)^t$.

5. To convert to the form $Q = ae^{kt}$, we first say that the right sides of the two equations equal each other (since each equals Q), and then we solve for a and k. Thus, we have $ae^{kt} = 4\cdot 8^{1.3t}$. At $t = 0$, we can solve for a:

$$\begin{aligned} ae^{k\cdot 0} &= 4\cdot 8^0 \\ a\cdot 1 &= 4\cdot 1 \\ a &= 4. \end{aligned}$$

Thus, we have $4e^{kt} = 4\cdot 8^{1.3t}$, and we solve for k:

$$\begin{aligned} 4e^{kt} &= 4\cdot 8^{1.3t} \\ e^{kt} &= \left(8^{1.3}\right)^t \\ \left(e^k\right)^t &= 14.929^t \\ e^k &= 14.929 \\ \ln e^k &= \ln 14.929 \\ k &= 2.703. \end{aligned}$$

Therefore, the equation is $Q = 4e^{2.703t}$.

9. We begin by dividing both sides by 46 to isolate the exponent:

$$\frac{91}{46} = (1.1)^x.$$

We then take the log of both sides and use the rules of logs to solve for x:

$$\begin{aligned} \log\frac{91}{46} &= \log(1.1)^x \\ \log\frac{91}{46} &= x\log(1.1) \\ x &= \frac{\log\frac{91}{46}}{\log(1.1)}. \end{aligned}$$

13.

$$
\begin{aligned}
5(1.031)^x &= 8\\
1.031^x &= \frac{8}{5}\\
\log(1.031)^x &= \log\frac{8}{5}\\
x\log 1.031 &= \log\frac{8}{5} = \log 1.6\\
x &= \frac{\log 1.6}{\log 1.031} = 15.395.
\end{aligned}
$$

Check your answer: $5(1.031)^{15.395} \approx 8$.

17.

$$
\begin{aligned}
3^{(4\log x)} &= 5\\
\log 3^{(4\log x)} &= \log 5\\
(4\log x)\log 3 &= \log 5\\
4\log x &= \frac{\log 5}{\log 3}\\
\log x &= \frac{\log 5}{4\log 3}\\
x &= 10^{\frac{\log 5}{4\log 3}}
\end{aligned}
$$

21. $\log\left(100^{x+1}\right) = \log\left((10^2)^{x+1}\right) = 2(x+1)$.

25.
- The common logarithm is defined only for positive inputs, so the domain of this function is given by

$$
\begin{aligned}
x - 20 &> 0\\
x &> 20.
\end{aligned}
$$

- The graph of $y = \log x$ has an asymptote at $x = 0$, that is, where the input is zero. So the graph of $y = \log(x - 20)$ has a vertical asymptote where its input is zero, at $x = 20$.

29.
- The common logarithm is defined only for positive inputs, so the domain of this function is given by

$$
\begin{aligned}
x + 15 &> 0\\
x &> -15.
\end{aligned}
$$

- The graph of $y = \log x$ has an asymptote at $x = 0$, that is, where the input is zero. So the graph of $y = \log(x + 15)$ has a vertical asymptote where its input is zero, at $x = -15$.

Problems

33. **(a)**

$$
\begin{aligned}
e^{x+3} &= 8\\
\ln e^{x+3} &= \ln 8\\
x + 3 &= \ln 8\\
x &= \ln 8 - 3
\end{aligned}
$$

(b)

$$
\begin{aligned}
4(1.12^x) &= 5\\
1.12^x &= \frac{5}{4} = 1.25
\end{aligned}
$$

$$\begin{aligned} \log 1.12^x &= \log 1.25 \\ x \log 1.12 &= \log 1.25 \\ x &= \frac{\log 1.25}{\log 1.12} \end{aligned}$$

(c)

$$\begin{aligned} e^{-0.13x} &= 4 \\ \ln e^{-0.13x} &= \ln 4 \\ -0.13x &= \ln 4 \\ x &= \frac{\ln 4}{-0.13} \end{aligned}$$

(d)

$$\begin{aligned} \log(x-5) &= 2 \\ x-5 &= 10^2 \\ x &= 10^2+5 = 105 \end{aligned}$$

(e)

$$\begin{aligned} 2\ln(3x)+5 &= 8 \\ 2\ln(3x) &= 3 \\ \ln(3x) &= \frac{3}{2} \\ 3x &= e^{\frac{3}{2}} \\ x &= \frac{e^{\frac{3}{2}}}{3} \end{aligned}$$

(f)

$$\begin{aligned} \ln x - \ln(x-1) &= \frac{1}{2} \\ \ln\left(\frac{x}{x-1}\right) &= \frac{1}{2} \\ \frac{x}{x-1} &= e^{\frac{1}{2}} \\ x &= (x-1)e^{\frac{1}{2}} \\ x &= xe^{\frac{1}{2}} - e^{\frac{1}{2}} \\ e^{\frac{1}{2}} &= xe^{\frac{1}{2}} - x \\ e^{\frac{1}{2}} &= x(e^{\frac{1}{2}}-1) \\ \frac{e^{\frac{1}{2}}}{e^{\frac{1}{2}}-1} &= x \\ x &= \frac{e^{\frac{1}{2}}}{e^{\frac{1}{2}}-1} \end{aligned}$$

Note: (g) (h) and (i) can not be solved analytically, so we use graphs to approximate the solutions.

(g) From Figure 5.12 we can see that $y = e^x$ and $y = 3x+5$ intersect at $(2.534, 12.601)$ and $(-1.599, 0.202)$, so the values of x which satisfy $e^x = 3x+5$ are $x = 2.534$ or $x = -1.599$. We also see that $y_1 \approx 12.601$ and $y_2 \approx 0.202$.

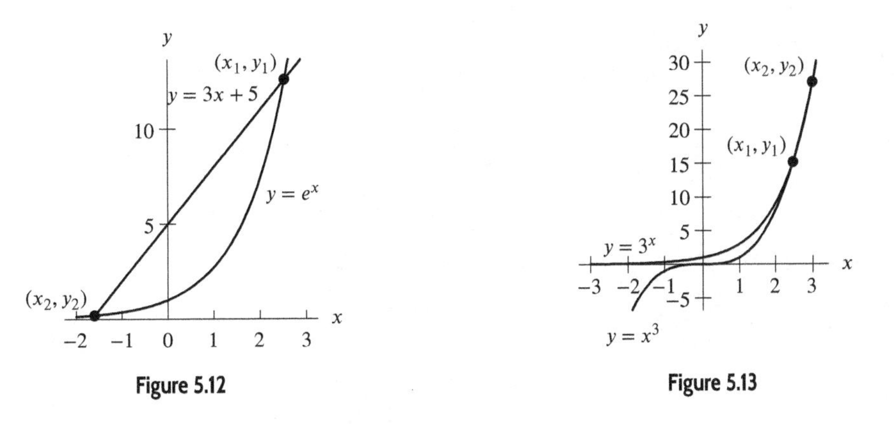

Figure 5.12

Figure 5.13

(h) The graphs of $y = 3^x$ and $y = x^3$ are seen in Figure 5.13. It is very hard to see the points of intersection, though $(3, 27)$ would be an immediately obvious choice (substitute 3 for x in each of the formulas). Using technology, we can find a second point of intersection, $(2.478, 15.216)$. So the solutions for $3^x = x^3$ are $x = 3$ or $x = 2.478$.

Since the points of intersection are very close, it is difficult to see these intersections even by zooming in. So, alternatively, we can find where $y = 3^x - x^3$ crosses the x-axis. See Figure 5.14.

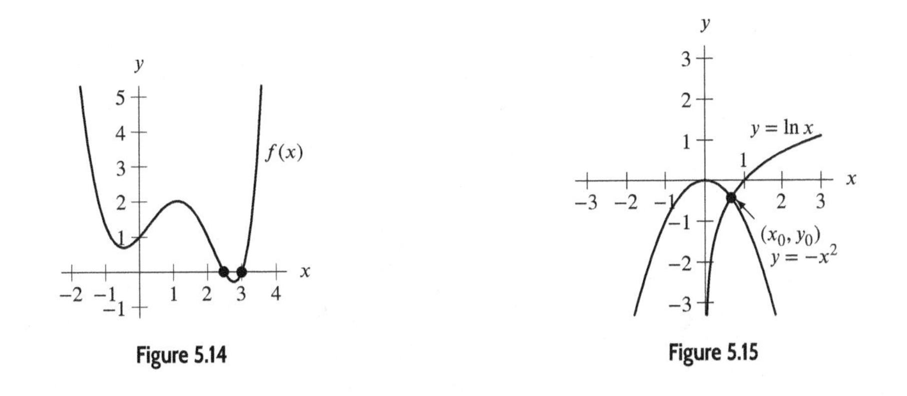

Figure 5.14

Figure 5.15

(i) From the graph in Figure 5.15, we see that $y = \ln x$ and $y = -x^2$ intersect at $(0.6529, -0.4263)$, so $x = 0.6529$ is the solution to $\ln x = -x^2$.

37. We have

$$\begin{aligned} M_2 - M_1 &= \log\left(\frac{W_2}{W_1}\right) \\ 5.6 - 4.4 &= \log\left(\frac{W_2}{W_1}\right) \\ 1.2 &= \log\left(\frac{W_2}{W_1}\right) \\ \frac{W_2}{W_1} &= 10^{1.2} = 15.849 \\ W_2 &= 15.849 \cdot W_1. \end{aligned}$$

The seismic waves of the second earthquake are about 15.85 times larger.

41. (a) The number of bacteria present after 1/2 hour is

$$N = 1000e^{0.69(1/2)} \approx 1412.$$

If you notice that $0.69 \approx \ln 2$, you could also say

$$N = 1000e^{0.69/2} \approx 1000e^{\frac{1}{2}\ln 2} = 1000e^{\ln 2^{1/2}} = 1000e^{\ln\sqrt{2}} = 1000\sqrt{2} \approx 1412.$$

(b) We solve for t in the equation

$$\begin{aligned}1{,}000{,}000 &= 1000e^{0.69t}\\ e^{0.69t} &= 1000\\ 0.69t &= \ln 1000\\ t &= \left(\frac{\ln 1000}{0.69}\right) \approx 10.011 \text{ hours.}\end{aligned}$$

(c) The doubling time is the time t such that $N = 2000$, so

$$\begin{aligned}2000 &= 1000e^{0.69t}\\ e^{0.69t} &= 2\\ 0.69t &= \ln 2\\ t &= \left(\frac{\ln 2}{0.69}\right) \approx 1.005 \text{ hours.}\end{aligned}$$

If you notice that $0.69 \approx \ln 2$, you see why the half-life turns out to be 1 hour:

$$\begin{aligned}e^{0.69t} &= 2\\ e^{t\ln 2} &\approx 2\\ e^{\ln 2^t} &\approx 2\\ 2^t &\approx 2\\ t &\approx 1\end{aligned}$$

45. We have

$$\begin{aligned}v(t) &= 30\\ 20e^{0.2t} &= 30\\ e^{0.2t} &= \frac{30}{20}\\ 0.2t &= \ln\left(\frac{30}{20}\right)\\ t &= \frac{\ln\left(\frac{30}{20}\right)}{0.2}\\ t &= \frac{\ln 1.5}{0.2}\end{aligned}$$

49. (a) For $f(x) = 10^x$,

$$\begin{gathered}\text{Domain of } f(x) \text{ is all } x\\ \text{Range of } f(x) \text{ is all } y > 0.\end{gathered}$$

There is one asymptote, the horizontal line $y = 0$.

(b) Since $g(x) = \log x$ is the inverse function of $f(x) = 10^x$, the domain of $g(x)$ corresponds to range of $f(x)$ and range of $g(x)$ corresponds to domain of $g(x)$.

$$\begin{gathered}\text{Domain of } g(x) \text{ is all } x > 0\\ \text{Range of } g(x) \text{ is all } y.\end{gathered}$$

The asymptote of $f(x)$ becomes the asymptote of $g(x)$ under reflection across the line $y = x$. Thus, $g(x)$ has one asymptote, the line $x = 0$.

53. (a) Since the drug is being metabolized continuously, the formula for describing the amount left in the bloodstream is $Q(t) = Q_0e^{kt}$. We know that we start with 2 mg, so $Q_0 = 2$, and the rate of decay is 4%, so $k = -0.04$. (Why is k negative?) Thus $Q(t) = 2e^{-0.04t}$.

(b) To find the percent decrease in one hour, we need to rewrite our equation in the form $Q = Q_0b^t$, where b gives us the percent left after one hour:

$$Q(t) = 2e^{-0.04t} = 2(e^{-0.04})^t \approx 2(0.96079)^t.$$

We see that $b \approx 0.96079 = 96.079\%$, which is the percent we have left after one hour. Thus, the drug level decreases by about 3.921% each hour.

(c) We want to find out when the drug level reaches 0.25 mg. We therefore ask when $Q(t)$ will equal 0.25.

$$\begin{aligned} 2e^{-0.04t} &= 0.25 \\ e^{-0.04t} &= 0.125 \\ -0.04t &= \ln 0.125 \\ t &= \frac{\ln 0.125}{-0.04} \approx 51.986. \end{aligned}$$

Thus, the second injection is required after about 52 hours.

(d) After the second injection, the drug level is 2.25 mg, which means that Q_0, the initial amount, is now 2.25. The decrease is still 4% per hour, so when will the level reach 0.25 again? We need to solve the equation

$$2.25e^{-0.04t} = 0.25,$$

where t is now the number of hours since the second injection.

$$\begin{aligned} e^{-0.04t} &= \frac{0.25}{2.25} = \frac{1}{9} \\ -0.04t &= \ln(1/9) \\ t &= \frac{\ln(1/9)}{-0.04} \approx 54.931. \end{aligned}$$

Thus the third injection is required about 55 hours after the second injection, or about $52 + 55 = 107$ hours after the first injection.

57. **(a)** We have

$$\begin{aligned} \sqrt{\log(\text{googol})} &= \sqrt{\log\left(10^{100}\right)} \\ &= \sqrt{100} \\ &= 10. \end{aligned}$$

(b) We have

$$\begin{aligned} \log\sqrt{\text{googol}} &= \log\sqrt{10^{100}} \\ &= \log\left(\left(10^{100}\right)^{0.5}\right) \\ &= \log\left(10^{50}\right) \\ &= 50. \end{aligned}$$

(c) We have

$$\begin{aligned} \sqrt{\log(\text{googolplex})} &= \sqrt{\log\left(10^{\text{googol}}\right)} \\ &= \sqrt{\text{googol}} \\ &= \sqrt{10^{100}} \\ &= \left(10^{100}\right)^{0.5} \\ &= 10^{50}. \end{aligned}$$

STRENGTHEN YOUR UNDERSTANDING

1. False. Since the $\log 1000 = \log 10^3 = 3$ we know $\log 2000 > 3$. Or use a calculator to find that $\log 2000$ is about 3.3.

5. True. Comparing the equation, we see $b = e^k$, so $k = \ln b$.

9. True. The log function outputs the power of 10 which in this case is n.

13. False. For example, $\log 10 = 1$, but $\ln 10 \approx 2.3$.

17. True. The two functions are inverses of one another.

21. False. Taking the natural log of both sides, we see $t = \ln 7.32$.

25. True. This is the definition of half-life.

29. True. Solve for t by dividing both sides by Q_0, taking the ln of both sides, and then dividing by k.

33. False. Since $26{,}395{,}630{,}000{,}000 \approx 2.6 \cdot 10^{13}$, we see that it would be between 13 and 14 on a log scale.

37. False. The fit will not be as good as $y = x^3$ but an exponential function can be found.

Solutions to Skills for Chapter 5

1. $\log(\log 10) = \log(1) = 0.$

5. $\dfrac{\log 1}{\log 10^5} = \dfrac{0}{5\log 10} = \dfrac{0}{5} = 0.$

9. The equation $10^{-4} = 0.0001$ is equivalent to $\log 0.0001 = -4$.

13. The equation $\log 0.01 = -2$ is equivalent to $10^{-2} = 0.01$.

17. The expression is not the logarithm of a quotient, so it cannot be rewritten using the properties of logarithms.

21. There is no rule for the logarithm of a sum, so it cannot be rewritten.

25. Rewrite the sum as $\log 12 + \log x = \log 12x$.

29. Rewrite with powers and combine,

$$\begin{aligned}3\left(\log(x+1) + \frac{2}{3}\log(x+4)\right) &= 3\log(x+1) + 2\log(x+4)\\ &= \log(x+1)^3 + \log(x+4)^2\\ &= \log\left((x+1)^3(x+4)^2\right)\end{aligned}$$

33. The logarithm of a sum cannot be simplified.

37. Rewrite as $\log 100^{2z} = 2z\log 100 = 2z(2) = 4z$.

41. We divide both sides by 3 to get

$$5^x = 3.$$

Taking logs of both sides, we get

$$\log(5^x) = \log 3.$$

This gives

$$x\log 5 = \log 3$$

or in other words

$$x = \frac{\log 3}{\log 5} \approx 0.683.$$

45. Taking logs of both sides, we get

$$\log 19^{6x} = \log(77 \cdot 7^{4x}).$$

This gives

$$\begin{aligned}6x\log 19 &= \log 77 + \log 7^{4x}\\ 6x\log 19 &= \log 77 + 4x\log 7\\ 6x\log 19 - 4x\log 7 &= \log 77\\ x(6\log 19 - 4\log 7) &= \log 77\\ x &= \frac{\log 77}{6\log 19 - 4\log 7} \approx 0.440.\end{aligned}$$

49. We first rearrange the equation so that the natural log is alone on one side, and we then convert to exponential form:

$$\begin{aligned}
2\ln(6x-1)+5 &= 7\\
2\ln(6x-1) &= 2\\
\ln(6x-1) &= 1\\
e^{\ln(6x-1)} &= e^1\\
6x-1 &= e\\
6x &= e+1\\
x &= \frac{e+1}{6} \approx 0.620.
\end{aligned}$$

CHAPTER SIX

Solutions for Section 6.1

Skill Refresher

S1. Evaluating $f(x)$ at $x = 3$, we have $f(3) = e^3 = 20.086$.

S5. **(a)** $f(-x) = 2(-x)^2 = 2x^2$
(b) $-f(x) = -\left(2x^2\right) = -2x^2$
Notice since $f(-x) = f(x) = 2x^2$, we see that $f(x)$ is an even function.

S9. **(a)** $f(-x) = 3(-x)^4 - 2(-x) = 3x^4 + 2x$
(b) $-f(x) = -\left(3x^4 - 2x\right) = -3x^4 + 2x$
Notice since $f(-x) \neq -f(x)$ and $f(-x) \neq -f(x)$, we see that $f(x) = 3x^4 - 2x$ is neither an even nor an odd function.

Exercises

1. **(a)** The y-coordinate is unchanged, but the x-coordinate is the same distance to the left of the y-axis, so the point is $(-2, -3)$.
(b) The x-coordinate is unchanged, but the y-coordinate is the same distance above the x-axis, so the point is $(2, 3)$.

5. The negative sign reflects the graph of $Q(t)$ horizontally about the y-axis, so the domain of $y = Q(-t)$ is $t < 0$. A horizontal reflection of the graph of $Q(t)$ about the y-axis will not change the range. The range of $Q(-t)$ is therefore the same as the range of $Q(t)$, $-4 \leq Q(-t) \leq 7$.

9. To reflect about the y-axis, we substitute $-x$ for x in the formula, getting $y = e^{-x}$.

13. We have

$$\begin{aligned} y = m(-n) &= (-n)^2 - 4(-n) + 5 \\ &= n^2 + 4n + 5 \end{aligned}$$

To graph this function, reflect the graph of m across the y-axis. See Figure 6.1.

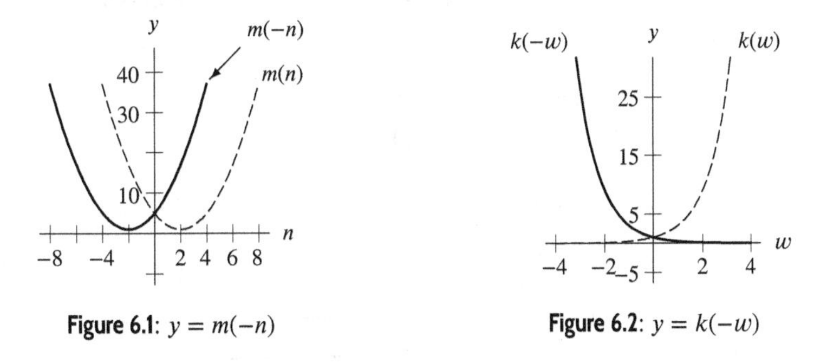

Figure 6.1: $y = m(-n)$

Figure 6.2: $y = k(-w)$

17. We have

$$y = k(-w) = 3^{-w}$$

To graph this function, reflect the graph of k across the y-axis. See Figure 6.2.

21. Since $f(-x) = 7(-x)^2 - 2(-x) + 1 = 7x^2 + 2x + 1$ is equal to neither $f(x)$ or $-f(x)$, the function is neither even nor odd.

Problems

25. **(a)** See Figure 6.3.

(b) See Figure 6.4.

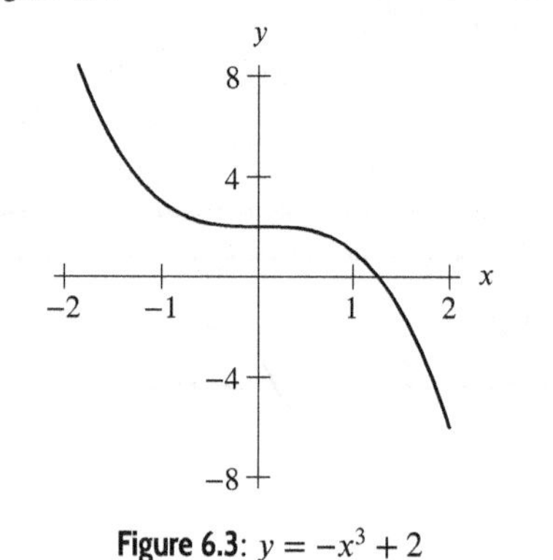

Figure 6.3: $y = -x^3 + 2$

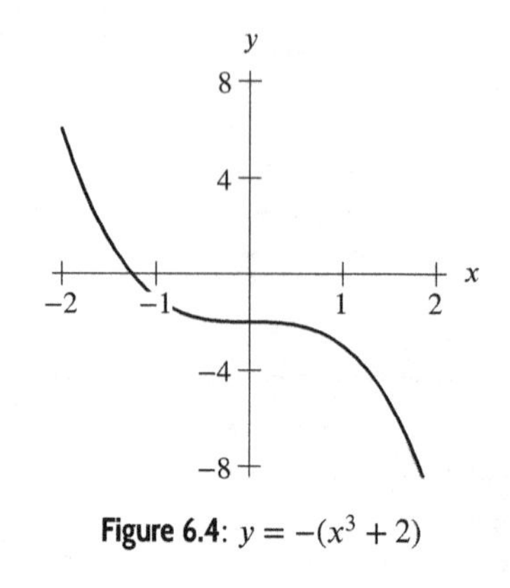

Figure 6.4: $y = -(x^3 + 2)$

(c) The two functions are not the same.

29. The equation of the reflected line is

$$y = b + m(-x) = b - mx.$$

The reflected line has the same y-intercept as the original; that is b. Its slope is $-m$, the negative of the original slope, and its x-intercept is b/m, the negative of the original x-intercept. A possible graph is in Figure 6.5.

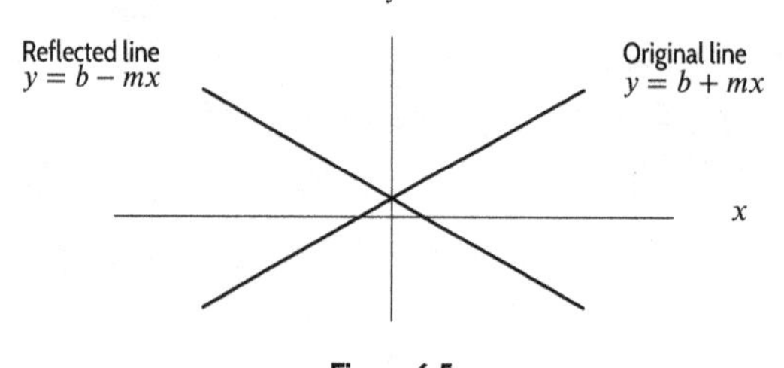

Figure 6.5

33. **(a)** If

$$H(t) = 20 + 54(0.91)^t$$

then $$H(t) + k = (20 + 54(.91)^t) + k = (20 + k) + 54(0.91)^t$$

(b) See Figure 6.6.

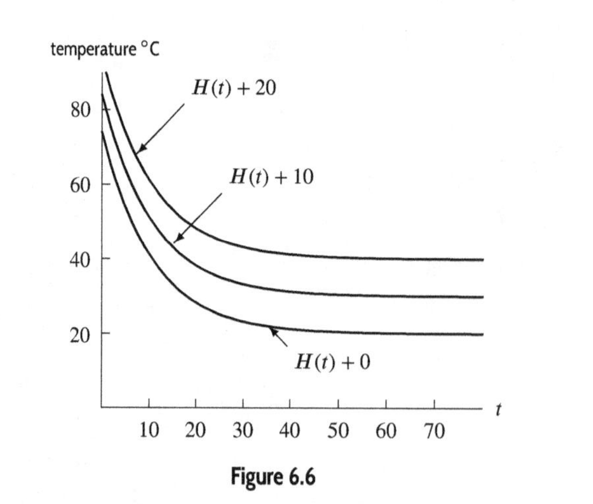

Figure 6.6

(c) As t gets very large, $H(t) + 10$ approaches a final temperature of 30°C. Over time, the soup will gradually cool to the same temperature as its surroundings, so 30°C is the temperature of the room it is in. Similarly, $H(t) + k$ approaches a final temperature of $(20 + k)$°C, which is the temperature of the room the bowl of soup is in.

37. **(a)** Figure 6.7 shows the graph of a function f that is symmetric across the y-axis.
(b) Figure 6.8 shows the graph of function f that is symmetric across the origin.

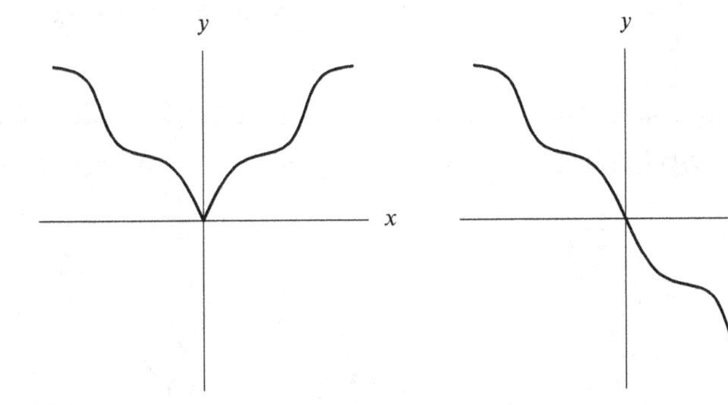

Figure 6.7: The graph of $f(x)$ that is symmetric across the y-axis

Figure 6.8: The graph of $f(x)$ that is symmetric across the origin

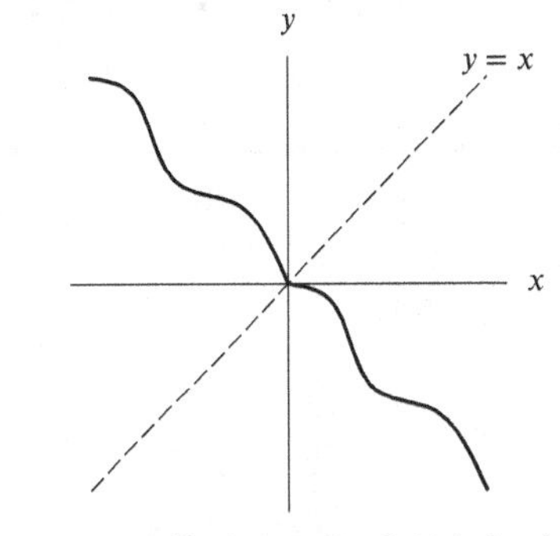

Figure 6.9: The graph of $f(x)$ that is symmetric across the line $y = x$

(c) Figure 6.9 shows the graph of function f that is symmetric across the line $y = x$.

41. See Figure 6.10.

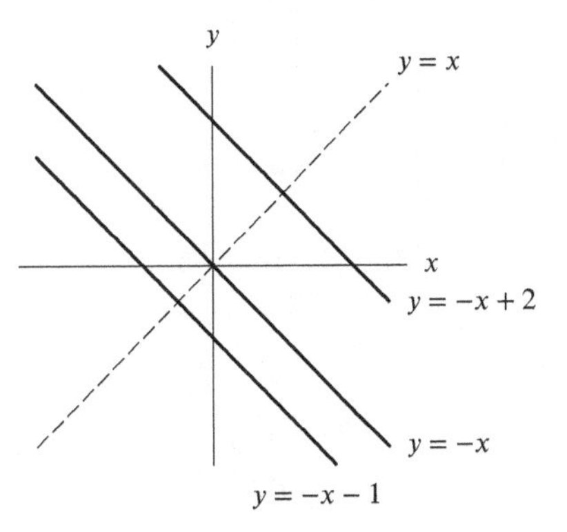

Figure 6.10: The graphs of $y = -x + 2$, $y = -x$, and $y = -x - 1$ are all symmetric across the line $y = x$.

Any straight line perpendicular to $y = x$ is symmetric across $y = x$. Its slope must be -1, so $y = -x + b$, for an arbitrary constant b, is symmetric across $y = x$.

Also, the line $y = x$ is symmetric about itself.

45. No, it is not possible for an odd function to be strictly concave up. If it were concave up in the first or second quadrants, then the fact that it is odd would mean it would have to be symmetric across the origin, and so would be concave down in the third or fourth quadrants.

49. To show that $f(x) = x^{1/3}$ is an odd function, we must show that $f(x) = -f(-x)$:

$$-f(-x) = -(-x)^{1/3} = x^{1/3} = f(x).$$

However, not all power functions are odd. The function $f(x) = x^2$ is an even function because $f(x) = f(-x)$ for all x. Another counterexample is $f(x) = \sqrt{x} = x^{1/2}$. This function is not odd because it is not defined for negative values of x.

Solutions for Section 6.2

Skill Refresher

S1. **(a)** In order to evaluate $2f(6)$, we first evaluate $f(6) = 6^2 = 36$. Then we multiply by 2, and thus we have $2f(6) = 2\cdot 36 = 72$.

(b) Since $f(6) = 36$, we have $-\frac{1}{2}f(6) = -\frac{1}{2} \cdot 36 = -18$.
(c) Since $f(6) = 36$, we have $5f(6) - 3 = 5(36) - 3 = 177$.
(d) In order to evaluate $\frac{1}{4}f(x-1)$ at $x = 6$, we first evaluate $f(6-1) = f(5) = 5^2 = 25$. Next we divide by 4, and thus we have $\frac{1}{4}f(6-1) = \frac{1}{4} \cdot 25 = \frac{25}{4}$.

Exercises

1. To increase by a factor of 10, multiply by 10. The right shift of 2 is made by substituting $x-2$ for x in the function formula. Together they give $y = 10f(x-2)$.

5. See Figure 6.11.

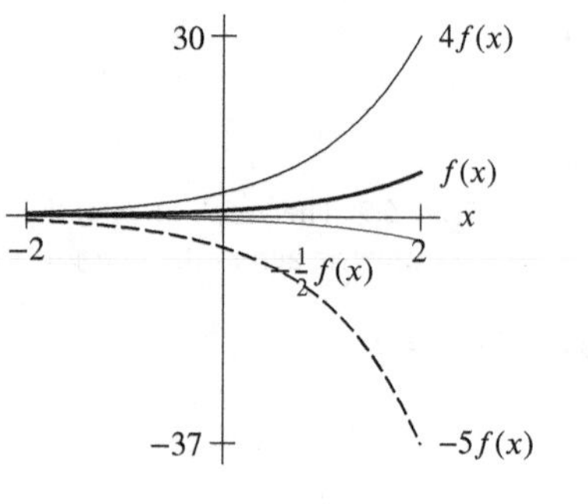

Figure 6.11

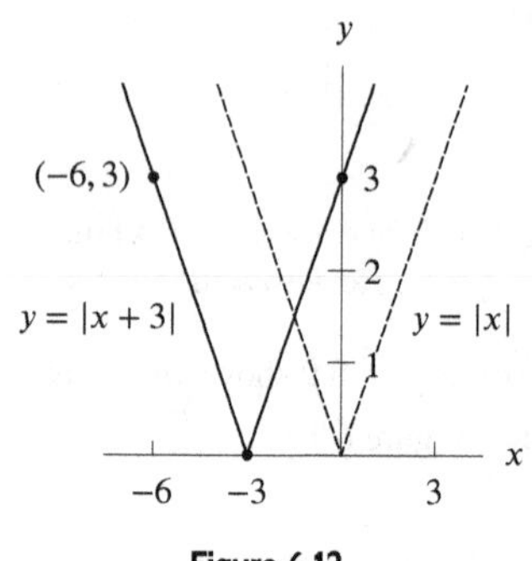

Figure 6.12

9. The function is $y = f(x+3)$. Since $f(x) = |x|$, we want $y = |x+3|$. The transformation shifts the graph of $f(x)$ by 3 units to the left. See Figure 6.12.

13. Since $h(x) = 2^x$, $3h(x) = 3 \cdot 2^x$. The graph of $h(x)$ is stretched vertically by a factor of 3. See Figure 6.13.

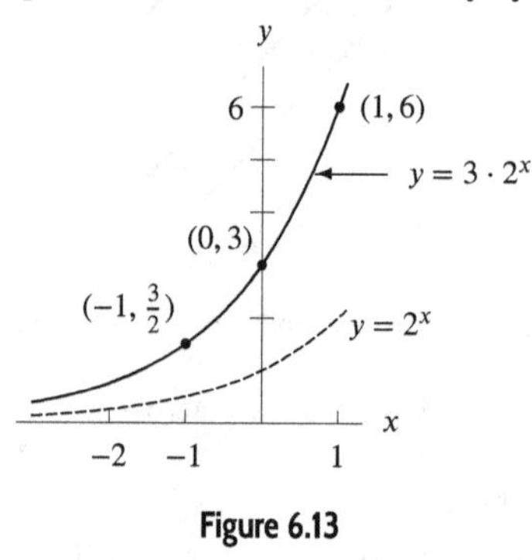

Figure 6.13

Problems

17. The graph of $f(t) = 1/(1+t^2)$ resembles a bell-shaped curve. (It is not, however, a true "bell curve.") See Figure 6.14. The y-axis is the horizontal asymptote.

Figure 6.14

Figure 6.15

21. The graph of $f(t+5) - 5$ is the graph of $f(t)$ shifted to the left by 5 units and then down by 5 units. See Figure 6.15. The horizontal asymptote is at $y = -5$.

25. Since the domain of $R(n)$ is the same as the domain of $P(n)$, no horizontal transformations have been applied. Since the maximum value of $R(n)$ is -5 times the minimum value of $P(n)$, and the minimum value $R(n)$ approaches is -5 times the maximum value $P(n)$ approaches, $P(n)$ has been stretched vertically by a factor of 5 and reflected about the x-axis. Thus, we have

$$R(n) = -5P(n).$$

29. I is (b)
II is (d)
III is (c)
IV is (h)

33. See Figure 6.16. The graph is shifted to the right by 3 units.

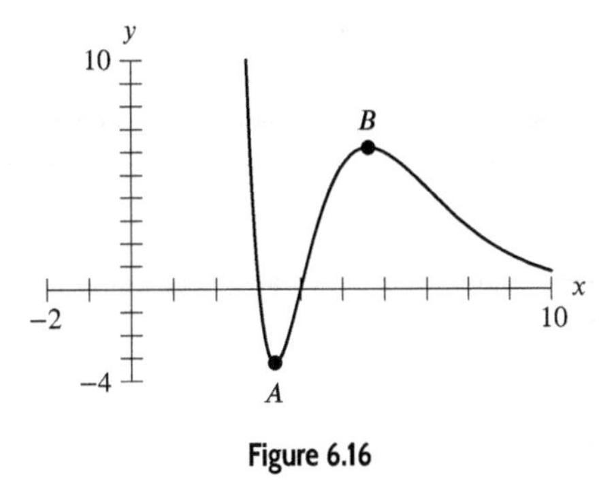

Figure 6.16

37. **(a)** Notice that the value of $h(x)$ at every value of x is one-half the value of $f(x)$ at the same x value. Thus, $f(x)$ has been compressed vertically by a factor of $1/2$, and

$$h(x) = \frac{1}{2}f(x).$$

(b) Observe that $k(-6) = f(6)$, $k(-4) = f(4)$, and so on. Thus, we have

$$k(x) = f(-x).$$

(c) The values of $m(x)$ are 4 less than the values of $f(x)$at the same x value. Thus, we have

$$m(x) = f(x) - 4.$$

41. Figure 6.17 gives a graph of a function $y = f(x)$ together with graphs of $y = \frac{1}{2}f(x)$ and $y = 2f(x)$. All three graphs cross the x-axis at $x = -2$, $x = -1$, and $x = 1$. Likewise, all three functions are increasing and decreasing on the same intervals. Specifically, all three functions are increasing for $x < -1.55$ and for $x > 0.21$ and decreasing for $-1.55 < x < 0.21$.

Even though the stretched and compressed versions of f shown by Figure 6.17 are increasing and decreasing on the same intervals, they are doing so at different rates. You can see this by noticing that, on every interval of x, the graph of $y = \frac{1}{2}f(x)$ is less steep than the graph of $y = f(x)$. Similarly, the graph of $y = 2f(x)$ is steeper than the graph of $y = f(x)$. This indicates that the magnitude of the average rate of change of $y = \frac{1}{2}f(x)$ is less than that of $y = f(x)$, and that the magnitude of the average rate of change of $y = 2f(x)$ is greater than that of $y = f(x)$.

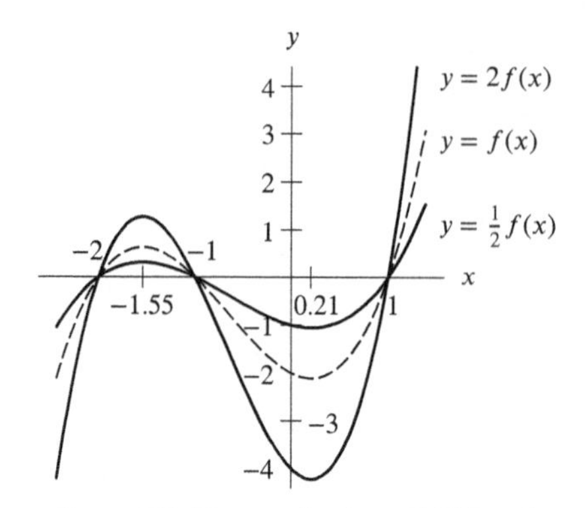

Figure 6.17: The graph of $y = 2f(x)$ and $y = \frac{1}{2}f(x)$ compared to the graph of $f(x)$

Solutions for Section 6.3

Skill Refresher

S1. We have $f(2x) = (2x)^3 - 5 = 8x^3 - 5$.

S5. $Q\left(\frac{1}{3}t\right) = 4e^{6(\frac{1}{3}t)} = 4e^{2t}$.

S9. Factoring 4 out from the left side of the equation, we have

$$4(x + 3) = 4(x - h).$$

Thus, we see $h = -3$.

S13. The function is already in the appropriate form since

$$y = f(-2x) + 9 = 1 \cdot f(-2(x - 0)) + 9.$$

We see from the form above that $A = 1$, $B = -2$, $h = 0$, and $k = 9$.

Exercises

1. **(a)** Since $y = -f(x) + 2$, we first need to reflect the graph of $y = f(x)$ over the x-axis and then shift it upward two units. See Figure 6.18.

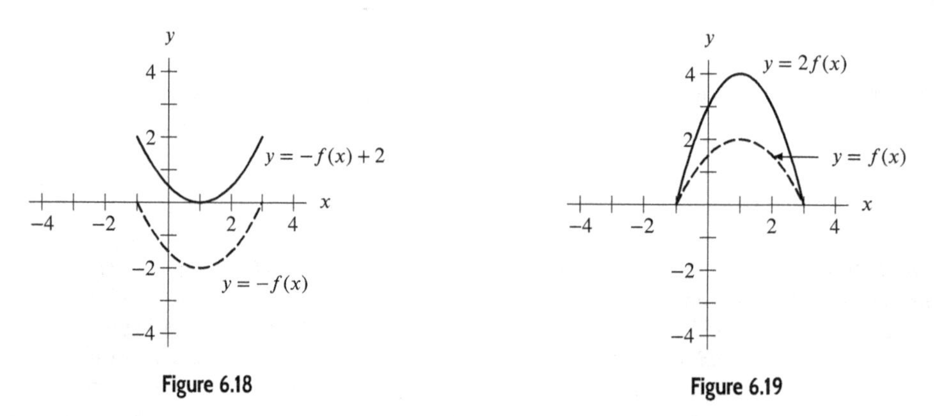

Figure 6.18

Figure 6.19

(b) We need to stretch the graph of $y = f(x)$ vertically by a factor of 2 in order to get the graph of $y = 2f(x)$. See Figure 6.19.

(c) In order to get the graph of $y = f(x-3)$, we will move the graph of $y = f(x)$ to the right by 3 units. See Figure 6.20.

Figure 6.20

Figure 6.21

(d) To get the graph of $y = -\frac{1}{2}f(x+1) - 3$, first vertically compress the graph of $y = f(x)$ by a factor of $1/2$, then reflect about the x-axis, then horizontally shift left 1 unit, and finally vertically shift down 3 units. See Figure 6.21.

5. See Figure 6.22.

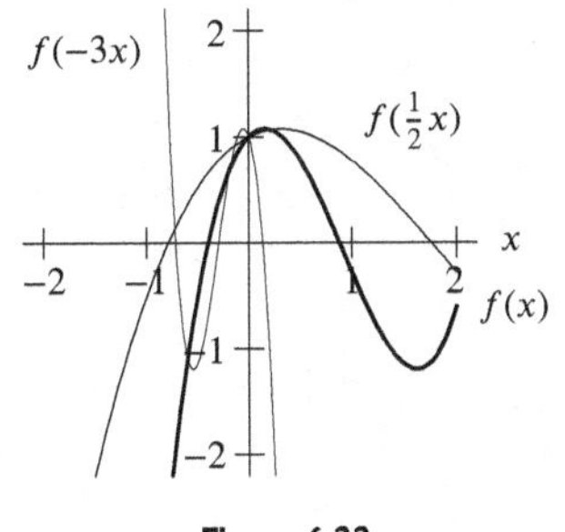

Figure 6.22

9. Since horizontal transformations do not affect features such as horizontal asymptotes or y-intercepts, each of the graphs in (a)-(c) will still have a horizontal asymptote at $y = 2$ and a y-intercept at $(0, -2)$.

(a) The graph of f has been compressed horizontally by a factor of $1/3$. The x-intercepts $(-1, 0)$ and $(3, 0)$ on the graph of f are moved to $(-\frac{1}{3}, 0)$ and $(1, 0)$ respectively on the graph of $y = f(3x)$ in Figure 6.23.

(b) The graph of f has been compressed horizontally by a factor of $1/2$ and reflected about the y-axis. The x-intercepts $(-1, 0)$ and $(3, 0)$ on the graph of f are moved to $(\frac{1}{2}, 0)$ and $(-\frac{3}{2}, 0)$ respectively on the graph of $y = f(-2x)$ in Figure 6.24.

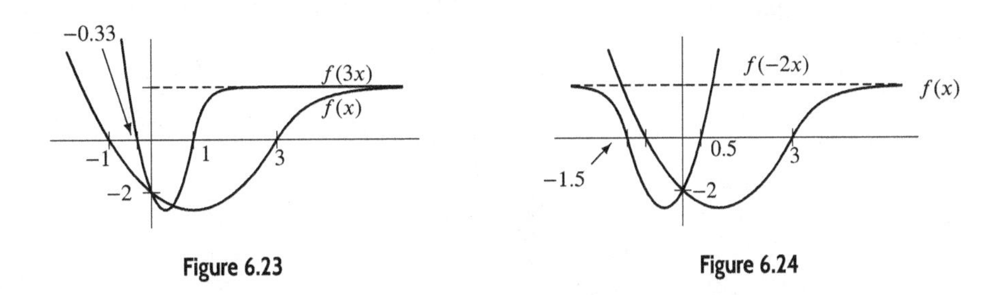

Figure 6.23

Figure 6.24

(c) The graph of f has been stretched horizontally by a factor of 2. The x-intercepts $(-1, 0)$ and $(3, 0)$ on the graph of f are moved to $(-2, 0)$ and $(6, 0)$ respectively on the graph of $y = f(\frac{1}{2}x)$ in Figure 6.25.

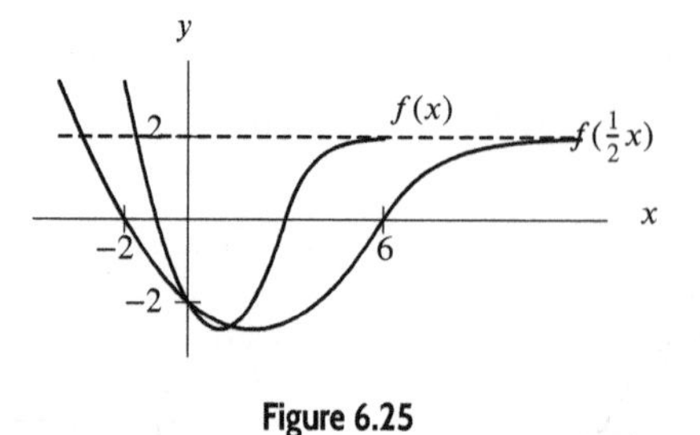

Figure 6.25

13. **(a)** This is a vertical shift of one unit upward. See Figure 6.26.
(b) Writing $h(x) = |x+1|$ as $h(x) = |x-(-1)|$, we see that this is a horizontal shift of one unit to the left. See Figure 6.27.

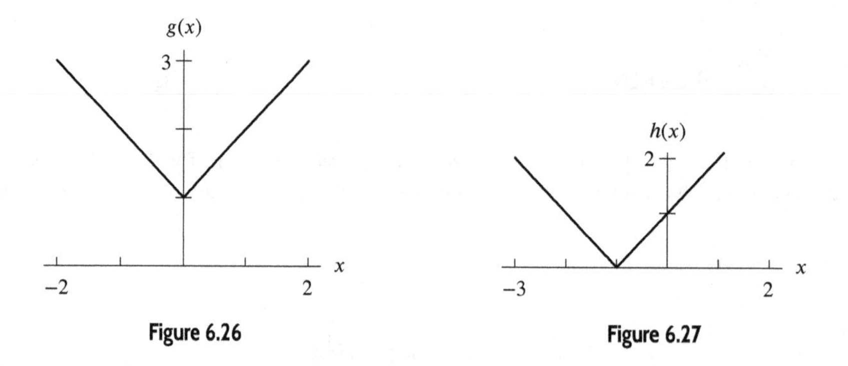

Figure 6.26

Figure 6.27

(c) Writing $j(x) = |2x+1| - 3$ as $j(x) = |2(x+\frac{1}{2})| - 3$, we see that this is a horizontal compression by a factor of $1/2$, then a horizontal shift of $1/2$ unit to the left, and finally a vertical shift down 3 units. See Figure 6.28.

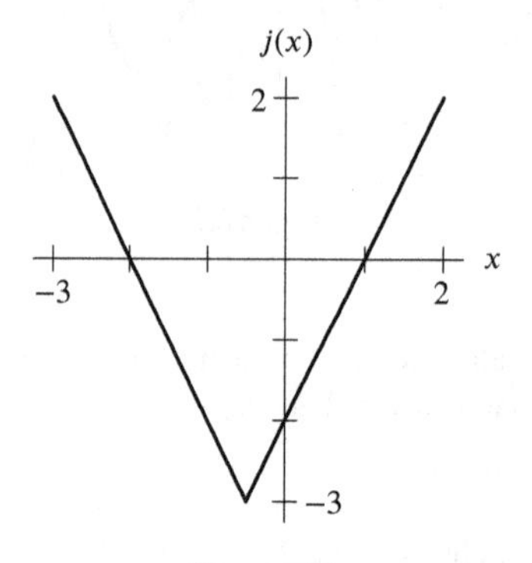

Figure 6.28

(d) Writing $k(x) = \frac{1}{2}|2x-4| + 1$ as $k(x) = \frac{1}{2}|2(x-2)| + 1$, we see that this is a vertical compression by a factor of $1/2$, then a horizontal compression by a factor of $1/2$, then a horizontal shift right 2 units, and finally a vertical shift up 1 unit. See Figure 6.29.

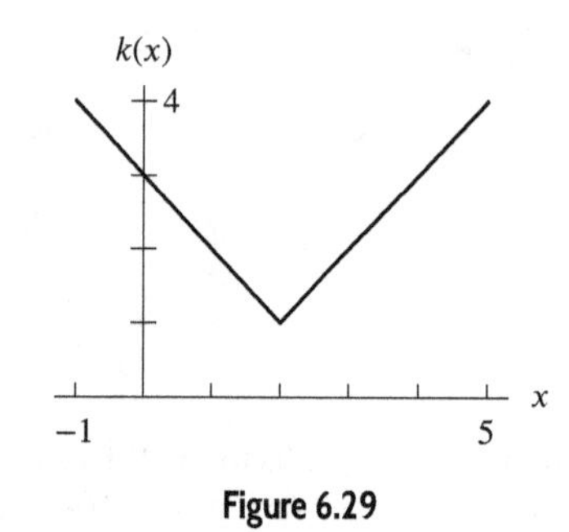

Figure 6.29

(e) Writing $m(x) = -\frac{1}{2}|4x + 12| - 3$ as $m(x) = -\frac{1}{2}|4(x + 3)| - 3$, we see that this is a vertical compression by a factor of $1/2$, then a reflection about the x-axis, then a horizontal compression by a factor of $1/4$, then a horizontal shift left 3 units, and finally a vertical shift down 3 units. See Figure 6.30.

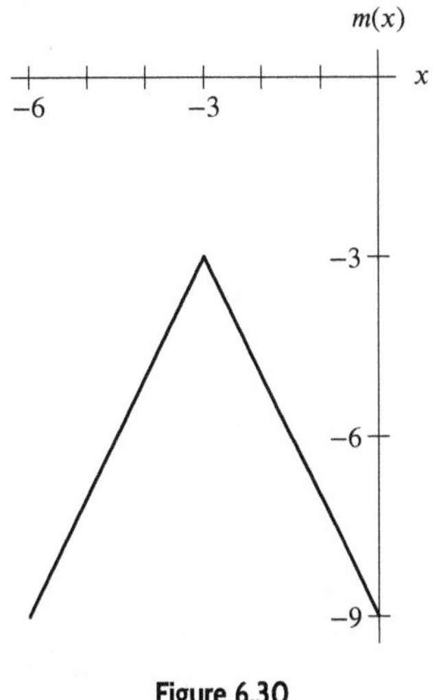

Figure 6.30

17.

Table 6.1

x	−3	−2	−1	0	1	2	3
$f(x)$	−4	−1	2	3	0	−3	−6
$f(\frac{1}{2}x)$	–	2	–	3	–	0	–
$f(2x)$	–	–	−1	3	−3	–	–

Problems

21. The stretch away from the y-axis by a factor of d requires multiplying the x-coordinate by $1/d$, and the upward translation requires adding c to the y-coordinate. This gives $(a/d, b + c)$.

25. **(a)** (ii) The $5 tip is added to the fare $f(x)$, so the total is $f(x) + 5$.
(b) (iv) There were 5 extra miles so the trip was $x + 5$. I paid $f(x + 5)$.
(c) (i) Each trip cost $f(x)$ and I paid for 5 of them, or $5f(x)$.
(d) (iii) The miles were 5 times the usual so $5x$ is the distance, and the cost is $f(5x)$.

29. **(a)** Since I is horizontally stretched compared to one graph and compressed compared to another, it should be $f(x)$.
(b) The most horizontally compressed of the graphs are III and II, so they should be $f(-2x)$ and $f(2x)$. Since III appears to be a compressed version of I reflected across the y-axis, it should be $f(-2x)$.
(c) The most horizontally stretched of the graphs should be $f(-\frac{1}{2}x)$, which is IV.
(d) The most horizontally compressed of the graphs are III and II, so they should be $f(-2x)$ and $f(2x)$. Since II appears to be a compressed version of I, it should be $f(2x)$.

33. We have:

$$\begin{aligned} \text{First step:} \quad & y = -f(x) && \text{reflect} \\ \text{Second step:} \quad & y = -f(x) + 2 && \text{shift up} \end{aligned}$$

Therefore, our final formula is $y = -f(x) + 2$.

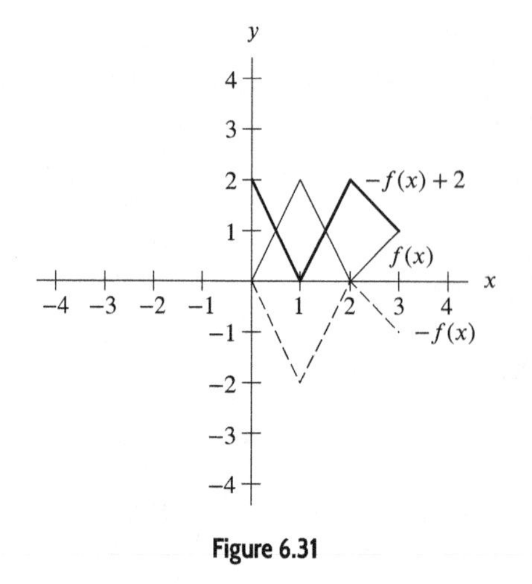

Figure 6.31

37. We have:

First step:	$y = f(x) - 2$	shift down
Second step:	$y = 2(f(x) - 2)$	stretch vertically

Multiplying out, this gives $y = 2f(x) - 4$.

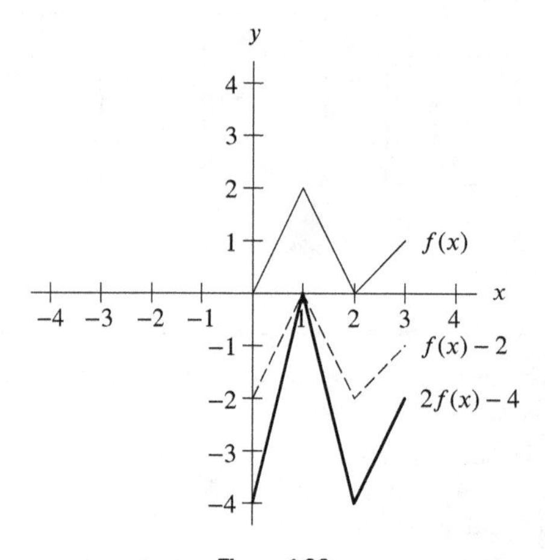

Figure 6.32

41. The function f has been reflected over the x-axis and the y-axis and stretched horizontally by a factor of 2. Thus, $y = -f(-\frac{1}{2}x)$.

45. **(a)** Since $f(x)$ has been shifted left by 2 units, the entire domain of $f(x)$ is shifted to the left by 2. Thus, the domain of $k(x) = f(x + 2)$ is $-8 \leq x \leq 0$.

(b) Since the average rate of change of $f(x)$ over $-6 \leq x \leq 2$ is given as -3, we have

$$\frac{f(2) - f(-6)}{2 - (-6)} = \text{Average rate of change of } f(x)$$
$$\frac{f(2) - f(-6)}{8} = -3$$
$$f(2) - f(-6) = 8(-3) = -24.$$

We now use the fact that $f(2) - f(-6) = -24$ to calculate the average rate of change of $k(x)$ over its domain $-8 \le x \le 0$.

$$\begin{aligned}
\text{Average rate of change of } k(x) &= \frac{\Delta y}{\Delta x} = \frac{k(0) - k(-8)}{0 - (-8)} \\
&= \frac{f(0+2) - f(-8+2)}{8} \quad \text{(since } k(x) = f(x+2)\text{)} \\
&= \frac{f(2) - f(-6)}{8} \\
&= \frac{-24}{8} \quad \text{(since } f(2) - f(-6) = -24\text{)} \\
&= -3.
\end{aligned}$$

49. We have

$$\begin{aligned}
y &= f(x+4) && \text{shift left 4 units} \\
y &= -f(x+4) && \text{then flip vertically} \\
y &= -f(x+4) + 2 && \text{then shift up 2 units} \\
y &= 3\,(-f(x+4) + 2) && \text{then stretch by 3} \\
\text{so} \quad g(x) &= -3f(x+4) + 6.
\end{aligned}$$

53. **(a)** The building is kept at 60° F until 5 am when the heat is turned up. The building heats up at a constant rate until 7 am when it is 68° F. It stays at that temperature until 3 pm when the heat is turned down. The building cools at a constant rate until 5 pm. At that time, the temperature is 60° F and it stays that level through the end of the day.

(b) Since $c(t) = 142 - d(t) = -d(t) + 142$, the graph of $c(t)$ will look like the graph of $d(t)$ that has been first vertically reflected across the t-axis and then vertically shifted up 142 units.

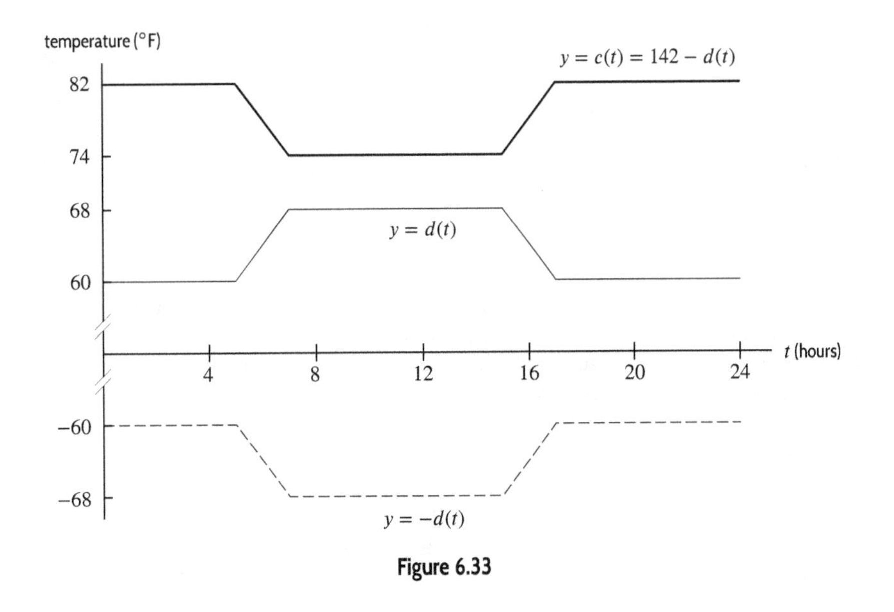

Figure 6.33

(c) This could describe the cooling schedule in the summer months when the temperature is kept at 82° F at night and cooled down to 74° during the day.

Solutions for Chapter 6 Review

Exercises

1. **(a)** The input is $2x = 2 \cdot 2 = 4$.
(b) The input is $\frac{1}{2}x = \frac{1}{2} \cdot 2 = 1$.
(c) The input is $x + 3 = 2 + 3 = 5$.
(d) The input is $-x = -2$.

5. A function is odd if $a(-x) = -a(x)$.

$$\begin{aligned} a(x) &= \frac{1}{x} \\ a(-x) &= \frac{1}{-x} = -\frac{1}{x} \\ -a(x) &= -\frac{1}{x} \end{aligned}$$

Since $a(-x) = -a(x)$, we know that $a(x)$ is an odd function.

9. A function is even if $b(-x) = b(x)$.

$$\begin{aligned} b(x) &= |x| \\ b(-x) &= |-x| = |x| \end{aligned}$$

Since $b(-x) = b(x)$, we know that $b(x)$ is an even function.

Problems

13. The graph is the graph of f shifted to the left by 2 and up by 2. See Figure 6.34.

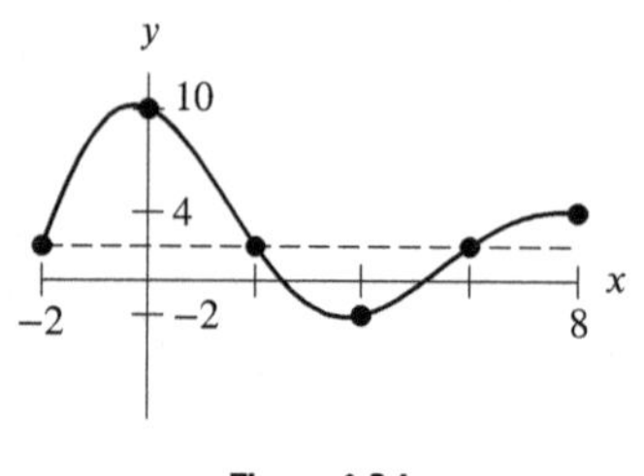

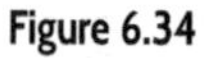

Figure 6.34

17. **(a)** Since the x-coordinate of the point $(-3, 1)$ on the graph of $f(x)$ has been multiplied by -1 in order to obtain the point $(3, 1)$ on the graph of $g(x)$, $g(x)$ must be obtained by reflecting the graph of $f(x)$ horizontally about the y-axis.
(b) Since the y-coordinate remains constant and only the x-coordinate is moved, $f(x)$ must be shifted horizontally. The x-coordinate of the point $(3, 1)$ on the graph of $g(x)$ has been shifted to the right by 6 units from the point $(-3, 1)$ on the graph of $f(x)$.

21. **(a)** Using the formula for $d(t)$, we have

$$\begin{aligned} d(t) - 15 &= (-16t^2 + 38) - 15 \\ &= -16t^2 + 23. \end{aligned}$$

$$\begin{aligned} d(t - 1.5) &= -16(t - 1.5)^2 + 38 \\ &= -16(t^2 - 3t + 2.25) + 38 \\ &= -16t^2 + 48t + 2. \end{aligned}$$

(b) See Figure 6.35.

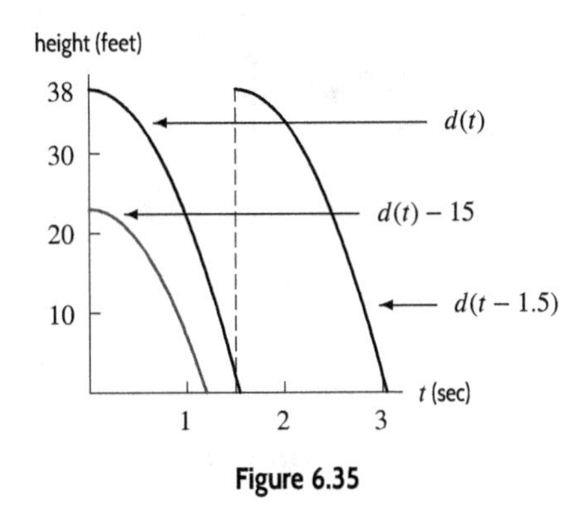

Figure 6.35

(c) $d(t) - 15$ represents the height of a brick which falls from $38 - 15 = 23$ feet above the ground. On the other hand, $d(t - 1.5)$ represents the height of a brick which began to fall from 38 feet above the ground at one and a half seconds after noon.

(d) (i) The brick hits the ground when its height is 0. Thus, if we represent the brick's height above the ground by $d(t)$, we get

$$\begin{aligned} 0 &= d(t) \\ 0 &= -16t^2 + 38 \\ -38 &= -16t^2 \\ t^2 &= \frac{38}{16} \\ t^2 &= 2.375 \\ t &= \pm\sqrt{2.375} \approx \pm 1.541. \end{aligned}$$

We are only interested in positive values of t, so the brick must hit the ground 1.541 seconds after noon.

(ii) If we represent the brick's height above the ground by $d(t) - 15$ we get

$$\begin{aligned} 0 &= d(t) - 15 \\ 0 &= -16t^2 + 23 \\ -23 &= -16t^2 \\ t^2 &= \frac{23}{16} \\ t^2 &= 1.4375 \\ t &= \pm\sqrt{1.4375} \approx \pm 1.199. \end{aligned}$$

Again, we are only interested in positive values of t, so the the brick hits the ground 1.199 seconds after noon.

(e) Since the brick whose height is $d(t - 1.5)$ begins falling 1.5 seconds after the brick whose height is $d(t)$, we expect the brick whose height is $d(t - 1.5)$ to hit the ground 1.5 seconds after the brick whose height is $d(t)$. Thus, the brick should hit the ground $1.5 + 1.541 = 3.041$ seconds after noon.

25. The graph appears to have been shifted to the left 6 units, compressed vertically by a factor of 2, and shifted vertically by 1 unit, so

$$y = \frac{1}{2}h(x + 6) + 1.$$

29.

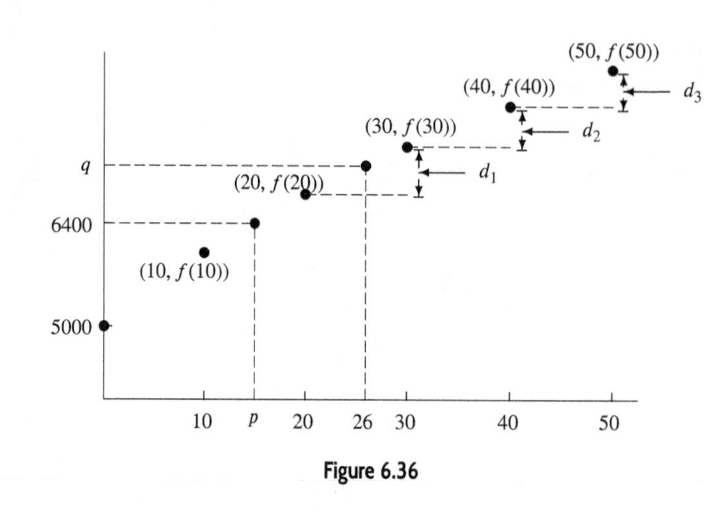

Figure 6.36

33. We have a reflection through the x-axis and a horizontal shift to the right by 1.

$$y = -h(x-1)$$

37. We will reverse Gwendolyn's actions. First, we can shift the parabola back two units to the right by replacing x in $y = (x-1)^2 + 3$ with $(x-2)$. This gives

$$\begin{aligned} y &= ((x-2)-1)^2 + 3 \\ &= (x-3)^2 + 3. \end{aligned}$$

We subtract 3 from this function to move the parabola down three units, so

$$\begin{aligned} y &= (x-3)^2 + 3 - 3 \\ &= (x-3)^2. \end{aligned}$$

Finally, to flip the parabola back across the horizontal axis, we multiply the function by -1. Thus, Gwendolyn's original equation was

$$y = -(x-3)^2.$$

41. Temperatures in this borehole are the same as temperatures 10 meters less deep in the Belleterre borehole. See Table 6.2.

Table 6.2

d	35	60	85	110	135	160	185	210
$m(d)$	5.5	5.2	5.1	5.1	5.3	5.5	5.75	6

STRENGTHEN YOUR UNDERSTANDING

1. True. The graph is shifted down by $|k|$ units.

5. False. The graphs of odd functions are symmetric about the origin.

9. False. Substituting $(x-2)$ in to the formula for g gives $g(x-2) = (x-2)^2 + 4 = x^2 - 4x + 4 + 4 = x^2 - 4x + 8$.

13. True.

17. False. Consider $f(x) = x^2$. Shifting up first and then compressing vertically gives the graph of $g(x) = \frac{1}{2}(x^2+1) = \frac{1}{2}x^2 + \frac{1}{2}$. Compressing first and then shifting gives the graph of $h(x) = \frac{1}{2}x^2 + 1$.

CHAPTER SEVEN

Solutions for Section 7.1

Exercises

1. Graphs (I), (II), and (IV) appear to describe periodic functions.
   (I) This function appears periodic. The rapid variation overlays a slower variation that appears to repeat every 8 units. (It almost appears to repeat every 4 units, but there is subtle difference between consecutive 4-second intervals. Do you see it?)
   (II) This function also appears periodic, again with a period of about 4 units. For instance, the x-intercepts appear to be evenly spaced, at approximately $-11, -7, -3, 1, 5, 9$, and the peaks are also evenly spaced, at $-9, -5, -1, 3, 7, 11$.
   (III) This function does not appear periodic. For instance, the x-intercepts grow increasingly close together (when read from left to right).
   (IV) At first glance this function might appear to vary unpredictably. But on closer inspection we see that the graph repeats the same pattern on the interval $-12 \leq x \leq 0$ and $0 \leq x \leq 12$.
   (V) This function does not appear periodic. The peaks of the graph appear to rise slowly (when read from left to right), and the troughs appear to fall slowly.
   (VI) This function does not appear to be periodic. The peaks and troughs of its graph seem to vary unpredictably, although they are more or less evenly spaced.

5. In the 9 o'clock position, the person is midway between the top and bottom of the wheel. Since the diameter is 150 m, the radius is 75 m, so the person is 75 m below the top, or $165 - 75 = 90$ m above the ground, as in the 3 o'clock position.

9. The graph appears to have a period of b. Every change in the x value of b brings us back to the same y value.

Problems

13. After 24 minutes, the person is three-fourths of the way through one rotation. Since the wheel is turning clockwise, this means she is in the 3 o'clock position, midway between the top and bottom of the wheel. Since the diameter is 150 m, the radius is 75 m, so the person is 75 m below the top, or $165 - 75 = 90$ m above the ground.

17. The wheel will complete two full revolutions after 20 minutes, and the height ranges from $h = 0$ to $h = 600$. So the function is graphed on the interval $0 \leq t \leq 20$. See Figure 7.1.

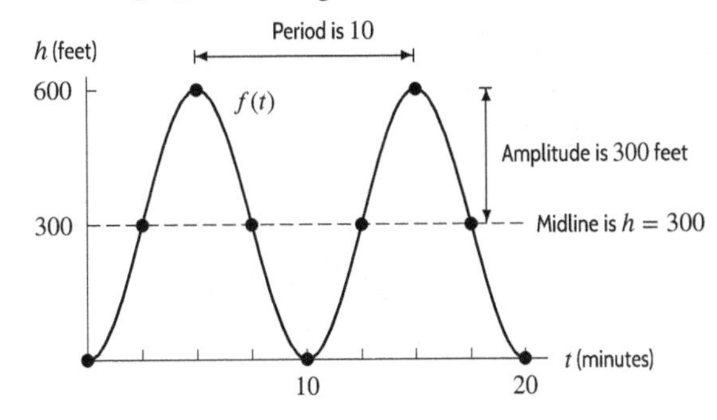

Figure 7.1: Graph of $h = f(t), 0 \leq t \leq 20$

21. The ride is completed in 30 minutes, so the average vertical speed in the last 5 minutes is given by

$$\frac{h(30) - h(25)}{5}.$$

Even though the table does not show this data, we know, due to symmetry, that the passenger is as high 5 minutes before the ride ends as she is 5 minutes into the ride. In other words:

$$h(25) = h(5) = 160 \text{ ft.}$$

Since we are back at the lowest point after 30 minutes, we have $h(30) = h(0) = 30$. Thus:

$$\frac{h(30) - h(25)}{5} = \frac{30 - 160}{5} = -26 \text{ ft/min}.$$

We note the sign is negative since we are moving down.

25. See Figure 7.2.

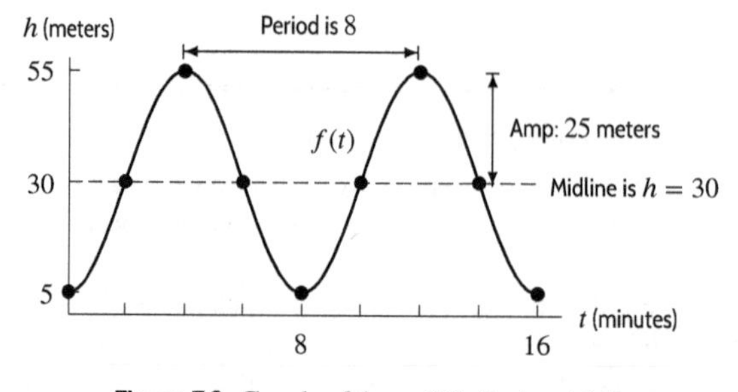

Figure 7.2: Graph of $h = f(t)$, $0 \leq t \leq 16$

29. Your initial position is twelve o'clock, since at $t = 0$, the value of h is at its maximum of 35. The period is 4 because the wheel completes one cycle in 4 minutes. The diameter is 30 meters and the boarding platform is 5 meters above ground. Because you go through 2.5 cycles, the length of time spent on the wheel is 10 minutes.

33. **(a)** Weight B, because the midline is $d = 10$, compared to $d = 20$ for weight A. This means that when the spring is not oscillating, weight B is 10 cm from the ceiling, while weight A is 20 cm from the ceiling.

(b) Weight A, because its amplitude is 10 cm, compared to the amplitude of 5 cm for weight B.

(c) Weight A, because its period is 0.5, compared to the period of 2 for weight B. This means that it takes weight A only half a second to complete one oscillation, whereas weight B completes one oscillation in 2 seconds.

Solutions for Section 7.2

Exercises

1. See Figure 7.3 for the positions of the angles. The coordinates of the points are found using the sine and cosine functions on a calculator.

(a) $(\cos 100°, \sin 100°) = (-0.174, 0.985)$
(b) $(\cos 200°, \sin 200°) = (-0.940, -0.342)$
(c) $(\cos(-200°), \sin(-200°)) = (-0.940, 0.342)$
(d) $(\cos(-45°), \sin(-45°)) = (0.707, -0.707)$
(e) $(\cos 1000°, \sin 1000°) = (0.174, -0.985)$
(f) $(\cos -720°, \sin -720°) = (1, 0)$

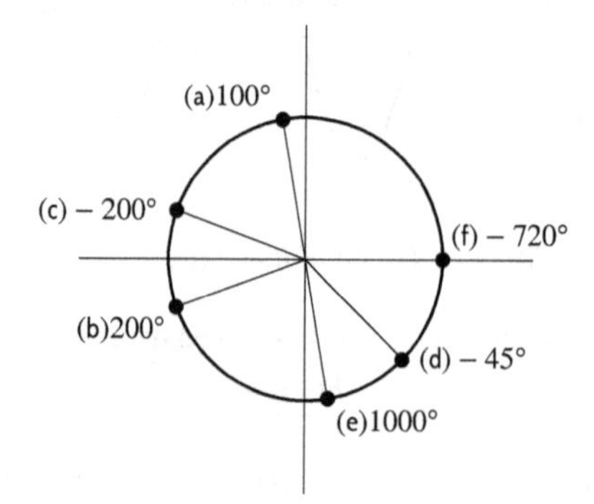

Figure 7.3

5. Since 100° is in the second quadrant, its reference angle is measured from the negative x-axis, corresponding to 180°. We have

$$\text{Reference angle} = 180° - 100° = 80°.$$

9. The car on the Ferris wheel starts at the 3 o'clock position. Let's suppose that you see the wheel rotating counterclockwise. (If not, move to the other side of the wheel.)

The angle $\phi = 420°$ indicates a counterclockwise rotation of the Ferris wheel from the 3 o'clock position all the way around once (360°), and then two-thirds of the way back up to the top (an additional 60°). This leaves you in the 1 o'clock position, or at the angle 60°.

A negative angle represents a rotation in the opposite direction, that is, clockwise. The angle $\theta = -150°$ indicates a rotation from the 3 o'clock position in the clockwise direction, past the 6 o'clock position and two-thirds of the way up to the 9 o'clock position. This leaves you in the 8 o'clock position, or at the angle 210°. (See Figure 7.4.)

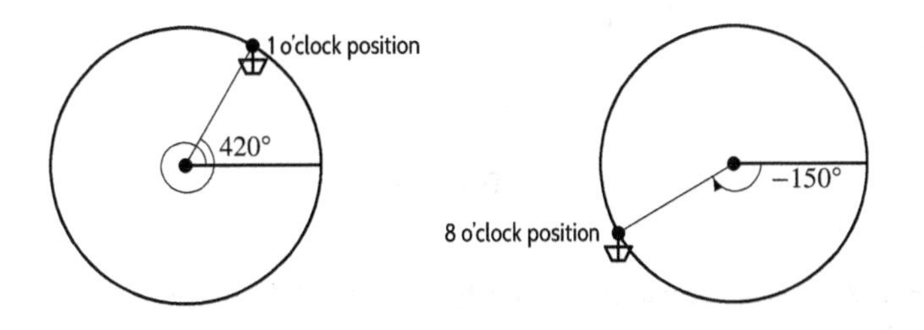

Figure 7.4: The positions and displacements on the Ferris wheel described by 420° and −150°

13. Since the x-coordinate is $r\cos\theta$ and the y-coordinate is $r\sin\theta$ and $r = 3.8$ and $\theta = -270°$, the point is $(3.8\cos(-270°), 3.8\sin(-270°)) = (0, 3.8)$.

17. Since the x-coordinate is $r\cos\theta$ and the y-coordinate is $r\sin\theta$ and $r = 3.8$, the point is $(3.8\cos(-10°), 3.8\sin(-10°)) = (3.742, -0.660)$.

Problems

21. The angles 70°, 180° − 70° = 110°, 180° + 70° = 250°, and 360 − 70° = 290° are in different quadrants and all have the same reference angle 70°. Other answers are possible.

25. **(a)** As we see from Figure 7.5, the angle 135° specifies a point P' on the unit circle directly across the y-axis from the point P. Thus, P' has the same y-coordinate as P, but its x-coordinate is opposite in sign to the x-coordinate of P. Therefore, $\sin 135° = 0.707$, and $\cos 135° = -0.707$.

(b) As we see from Figure 7.6, the angle 285° specifies a point Q' on the unit circle directly across the x-axis from the point Q. Thus, Q' has the same x-coordinate as Q, but its y-coordinate is opposite in sign to the y-coordinate of Q. Therefore, $\sin 285° = -0.966$, and $\cos 285° = 0.259$.

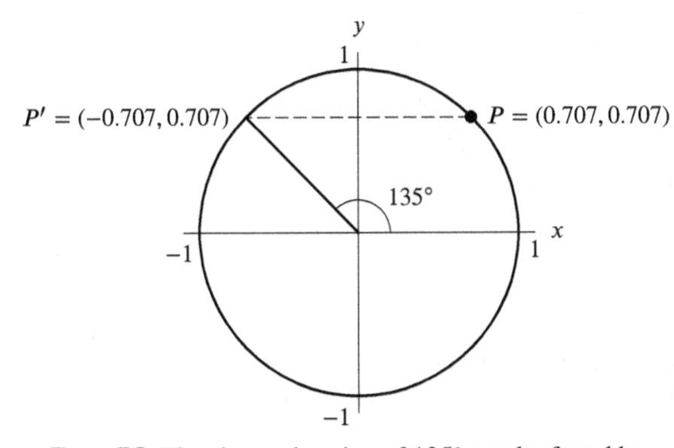

Figure 7.5: The sine and cosine of 135° can be found by referring to the sine and cosine of 45°

y
1
Q = (0.259, 0.966)
285°
x
−1
1
−1
Q′ = (0.259, −0.966)

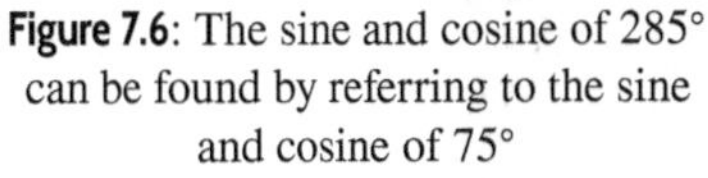

Figure 7.6: The sine and cosine of 285° can be found by referring to the sine and cosine of 75°

29. **(a)** $\sin(\theta + 360^\circ) = \sin\theta = a$, since the sine function is periodic with a period of 360°.

(b) $\sin(\theta + 180^\circ) = -a$. (A point on the unit circle given by the angle $\theta + 180^\circ$ diametrically opposite the point given by the angle θ. So the y-coordinates of these two points are opposite in sign, but equal in magnitude.)

(c) $\cos(90^\circ - \theta) = \sin\theta = a$. This is most easily seen from the right triangles in Figure 7.7.

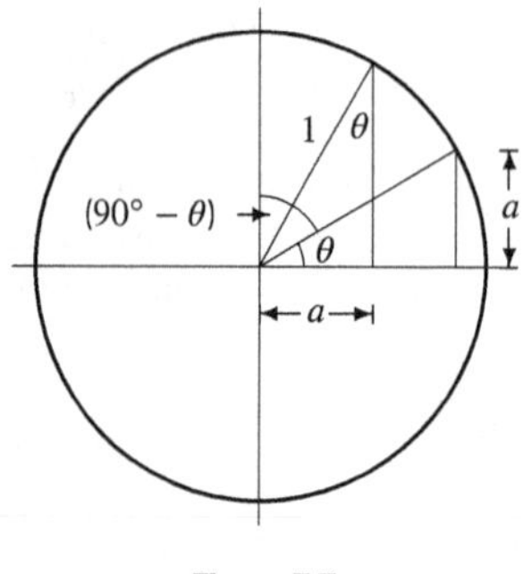

Figure 7.7

(d) $\sin(180^\circ - \theta) = a$. (A point on the unit circle given by the angle $180^\circ - \theta$ has a y-coordinate equal to the y-coordinate of the point on the unit circle given by θ.)

(e) $\sin(360^\circ - \theta) = -a$. (A point on the unit circle given the the angle $360^\circ - \theta$ has a y-coordinate of the same magnitude as the y-coordinate of the point on the unit circle given by θ, but is of opposite sign.)

(f) $\cos(270^\circ - \theta) = -\sin\theta = -a$.

33. **(a)** The five panels split the circle into five equal parts, so the angle between each panel is $360^\circ/5 = 72^\circ$.

(b) Point B is directly across the circle from D, so 180°.

(c) The angle from A to D is the same as the angle from B to C, and the BC angle is the angle between panels, which is 72°. So moving the panel between A and D gives an angle of $(72^\circ)/2 = 36^\circ$. The panel then goes from point D to point B, spanning another 180°. Thus in total the panel traveled $36^\circ + 180^\circ = 216^\circ$.

37. See Figure 7.8. Since the diameter is 120 mm, the radius is 60 mm. The coordinates of the outer edge point, A, on the x-axis is $(60, 0)$. Similarly, the inner edge at point B has coordinates $(7.5, 0)$.

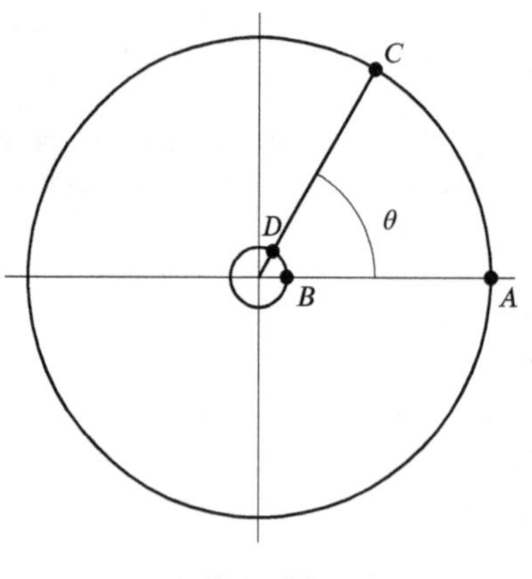

Figure 7.8

Points C and D are at an angle θ from the x-axis and have coordinates of the form $(r\cos\theta, r\sin\theta)$. For the outer edge, $r = 60$ so $C = (60\cos\theta, 60\sin\theta)$. The inner edge has $r = 7.5$, so $D = (7.5\cos\theta, 7.5\sin\theta)$.

Solutions for Section 7.3

Exercises

1. To convert 60° to radians, multiply by $\pi/180°$:

$$60°\left(\frac{\pi}{180°}\right)=\left(\frac{60°}{180°}\right)\pi=\frac{\pi}{3}.$$

We say that the radian measure of a 60° angle is $\pi/3$.

5. In order to change from degrees to radians, we multiply the number of degrees by $\pi/180$, so we have $150\cdot\pi/180$, giving $\frac{5}{6}\pi$ radians.

9. In order to change from radians to degrees, we multiply the number of radians by $180/\pi$, so we have $\frac{7}{2}\pi\cdot 180/\pi$, giving 630 degrees.

13. In order to change from radians to degrees, we multiply the number of radians by $180/\pi$, so we have $45\cdot 180/\pi$, giving $8100/\pi\approx 2578.310$ degrees.

17. If we go around twice, we make two full circles, which is $2\pi\cdot 2=4\pi$ radians. Since we're going around in the negative direction, we have -4π radians.

21. The arc length, s, corresponding to an angle of θ radians in a circle of radius r is $s=r\theta$. In order to change from degrees to radians, we multiply the number of degrees by $\pi/180$, so we have $45\cdot\pi/180$, giving $\frac{\pi}{4}$ radians. Thus, our arc length is $6.2\pi/4\approx 4.869$.

Problems

25. The reference angle between the ray and the negative x-axis is 30°. Since both sine and cosine are negative in the third quadrant, we have

$$x=r\cos\theta=10\cos 210°=10(-\cos 30°)=10(-\sqrt{3}/2)=-5\sqrt{3}$$

and

$$y=r\sin\theta=10\sin 210°=10(-\sin 30°)=10(-1/2)=-5,$$

so the coordinates of W are $(-5\sqrt{3},-5)$.

29. Since the x-coordinate is $r\cos\theta$ and the y-coordinate is $r\sin\theta$ and $r=5$, the point is $(5\cos 135°, 5\sin 135°)=(-5\sqrt{2}/2, 5\sqrt{2}/2)$.

33. The reference angle for $3\pi/4$ is $\pi-3\pi/4=\pi/4$, so $\cos(3\pi/4)=-\cos(\pi/4)=-1/\sqrt{2}$.

37. Using $s=r\theta$, we have $s=8(2)=16$ inches.

41. We have $\theta=22°$ or, in radians,

$$22°\cdot\frac{\pi}{180°}=0.3840.$$

We also know that $r=0.05$, so

$$s=0.05(0.3840)=0.01920.$$

The coordinates of P are $(r\cos\theta, r\sin\theta)=(0.0464, 0.0187)$.

45. **(a)** Yes, in both it seems to be roughly 60°.
(b) Just over 6 arcs fit into the circumference, since the circumference is $2\pi r=2\pi(2)=12.566$.

49. The graphs are found below.

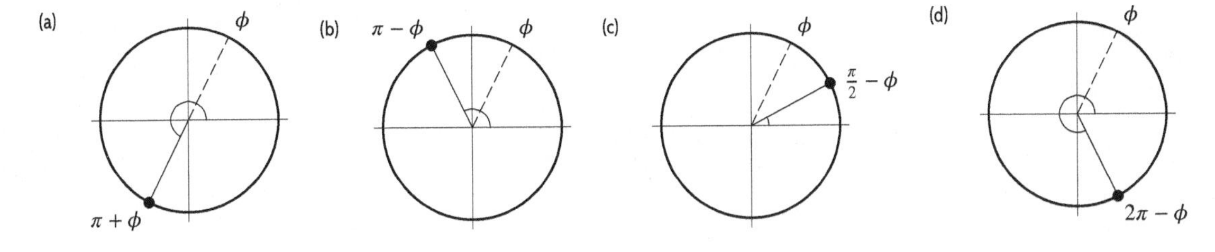

53. The value of t is bigger than the value of $\sin t$ on $0 < t < \pi/2$. On a unit circle, the vertical segment, $\sin t$, is shorter than the arc, t. See Figure 7.9.

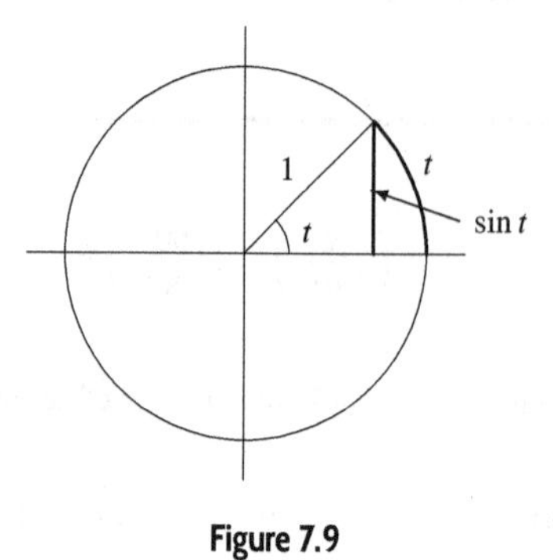

Figure 7.9

Solutions for Section 7.4

Exercises

1. The midline is $y = 3$ and the amplitude is 2.

5. The midline of $f(t)$ is $y = 4$ and the amplitude is $|-2| = 2$. Since the amplitude measures the vertical distance of the minimum and maximum value of $f(t)$ from its midline, the range of $f(t)$ will be $2 \le y \le 6$.

9. Judging from the figure:

- The midline is the dashed horizontal line $y = -1$.
- The vertical distance from the first peak to the midline is 0.5, so the amplitude is 0.5.
- The function starts at its minimum value, so it must be a shift and stretch of $\cos t$.

13. For a function of the form $A \sin t + k$ or $A \cos t + k$, $y = k$ is the midline and $|A|$ is the amplitude. This means possible formulas for this function are $g(t) = 2 \sin t + 3$, $g(t) = -2 \sin t + 3$, $g(t) = 2 \cos t + 3$, or $g(t) = -2 \cos t + 3$. The only one of these to have a maximum at $t = 0$ is $2 \cos t + 3$ and so we have $g(t) = 2 \cos t + 3$.

Problems

17. $f(x) = (\sin x) + 1$
$g(x) = (\sin x) - 1$

21. We can sketch these graphs using a calculator or computer. Figure 7.10 gives a graph of $y = \sin\theta$, together with the graphs of $y = 0.5 \sin\theta$ and $y = -2 \sin\theta$, where θ is in degrees and $0 \le \theta \le 360°$.

These graphs are similar but not the same. The amplitude of $y = 0.5 \sin\theta$ is 0.5 and the amplitude of $y = -2 \sin\theta$ is 2. The graph of $y = -2 \sin\theta$ is vertically reflected relative to the other two graphs. These observations are consistent with the fact that the constant A in the equation

$$y = A \sin\theta$$

may result in a vertical stretching or shrinking and/or a reflection over the x-axis. Note that all three graphs have a period of 360°.

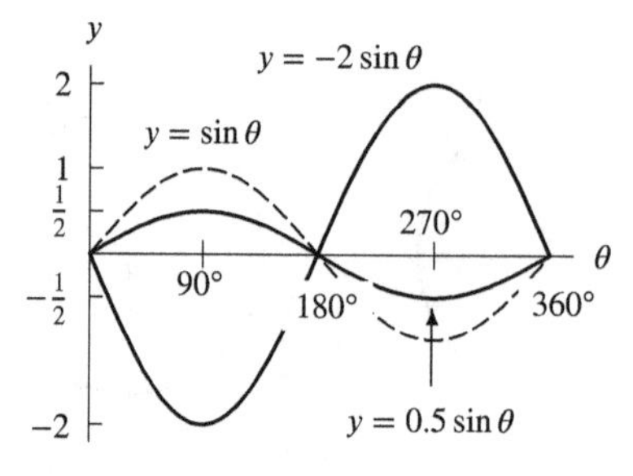

Figure 7.10

25. The data indicates that $g(x)$ is increasing as it passes through $x = 0$. Since $B\cos x + k$ has either a maximum or minimum at $x = 0$, the formula must be of the form

$$g(x) = A\sin x + k.$$

Since $A\sin x + k$ takes on its midline value of k at $x = 0$, we have $k = -3.3$.

We know $|A|$ is equal to the amplitude which is the distance from the midline to the peak of the graph. In this case, $A\sin x + k$ takes its maximum value of -1.5 at $x = \pi/2$, so we have

$$|A| = -1.5 - (-3.3) = 1.8,$$

Since the function is increasing as it passes through $x = 0$, we have $A > 0$ so

$$g(x) = 1.8\sin x - 3.3.$$

29. This function resembles a cosine graph with an amplitude of 4 and a midline of $y = -4$. Thus $y = 4g(x) - 4$.

33.
- At $\theta = 0°$ the tack is at the height of the hub, 370 mm above the ground.
- At $\theta = 90°$ the tack is at the top of the wheel, or 740 mm above the ground.
- At $\theta = 180°$ the tack is halfway back down to the ground, at a height of 370 mm.
- At $\theta = 270°$ the tack is touching the ground, at a height of 0 mm.
- At $\theta = 360°$ the tack is again at the height of the hub, 370 mm above the ground.
- This pattern repeats for $\theta = 450°, 540°, 630°, 720°$. See Table 7.1.

Plotting the points from Table 7.1, we see from the figure that $f(\theta)$ is a sine function with amplitude $A = 370$ and midline $k = 370$, so $s = 370 + 370\sin\theta$. A graph of this function is shown in Figure 7.11.

Table 7.1

θ (°)	0	90	180	270	360	450	540	630	720
s (mm)	370	740	370	0	370	740	370	0	370

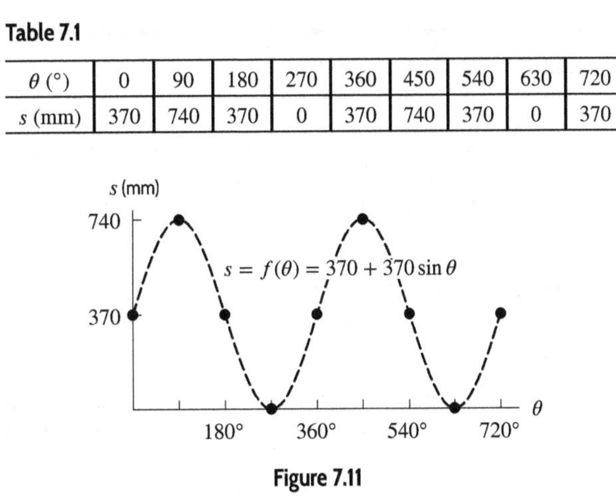

Figure 7.11

37. Make a table, such as Table 7.2, using your calculator to see that $\cos t$ is decreasing and the values of t are increasing.

Table 7.2

t	0	0.1	0.2	0.3	0.4	0.5	0.6	0.7	0.8	0.9
$\cos t$	1	0.995	0.980	0.953	0.921	0.878	0.825	0.765	0.697	0.622

Use a more refined table to see $t \approx 0.74$. (See Table 7.3.) Further refinements lead to $t \approx 0.739$.

Table 7.3

t	0.70	0.71	0.72	0.73	0.74	0.75	0.76
$\cos t$	0.765	0.758	0.752	0.745	0.738	0.732	0.725

Alternatively, consider the graphs of $y = t$ and $y = \cos t$ in Figure 7.12. They intersect at a point in the first quadrant, so for the t-coordinate of this point, $t = \cos t$. Trace with a calculator to find $t \approx 0.739$.

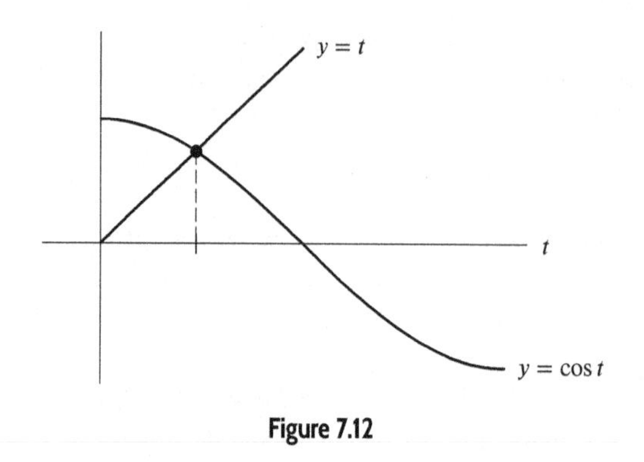

Figure 7.12

Solutions for Section 7.5

Exercises

1. The midline is $y = 0$. The amplitude is 6. The period is 2π.

5. Judging from the figure:

- The curve looks the same from $t = 0$ to $t = 8$ as from $t = 8$ to $t = 16$, so it repeats with a period of 8.
- The midline is the dashed horizontal line $y = 30$.
- The vertical distance from the first peak to the midline is 20, so the amplitude is 20.

9. Judging from the figure:

- The curve looks the same from $t = 0$ to $t = 0.5$ as from $t = 0.5$ to $t = 1$, so it repeats with a period of 0.5.
- The midline is the dashed horizontal line $y = 0.5$.
- The vertical distance from the first peak to the midline is 0.5, so the amplitude is 0.5.

13. This function resembles a sine curve in that it passes through the origin and then proceeds to grow from there. We know that the smallest value it attains is -4, and the largest it attains is 4; thus its amplitude is 4, with a midline of 0. It has a period of 1. Thus in the equation

$$g(t) = A\sin(Bt)$$

we know that $A = 4$ and

$$1 = \text{period} = \frac{2\pi}{B}.$$

So $B = 2\pi$, and then

$$h(t) = 4\sin(2\pi t).$$

17. The midline is $y = 4000$. The amplitude is $8000 - 4000 = 4000$. The period is 60, so B is $2\pi/60$. The graph at $x = 0$ rises from its midline, so we use the sine. Thus,

$$y = 4000 + 4000\sin\left(\frac{2\pi}{60}x\right).$$

Problems

21. See Figure 7.13.

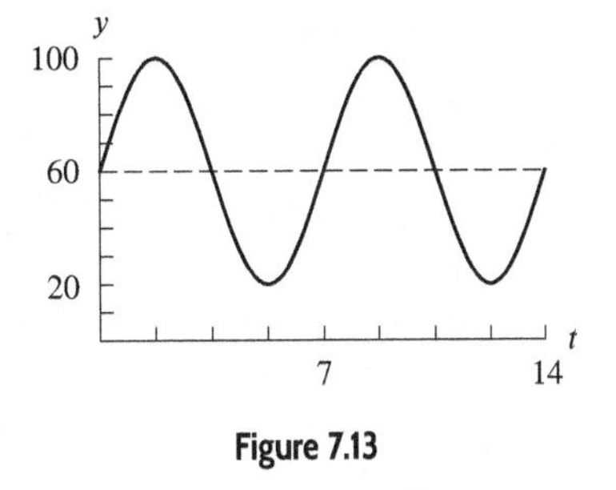

Figure 7.13

25. Because the period of $\sin x$ is 2π, and the period of $\sin 2x$ is π, so from the figure in the problem we see that

$$f(x) = \sin x.$$

The points on the graph are $a = \pi/2$, $b = \pi$, $c = 3\pi/2$, $d = 2\pi$, and $e = 1$.

29. **(a)** The population has initial value 1500 and grows at a constant rate of 200 animals per year.
(b) The population has initial value 2700 and decreases at a constant rate of 80 animals per year.
(c) The population has initial value 1800 and increases at the constant percent rate of 3% per year.
(d) The population has initial value 800 and decreases at the *continuous* percent rate of 4% per year.
(e) The population has initial value 3800, climbs to $3800 + 230 = 4030$, drops to $3800 - 230 = 3570$, and climbs back to 3800 over a 7-year period. This pattern keeps repeating itself.

33. This function has an amplitude of 3 and a period 1, and resembles a sine graph. Thus $y = 3f(x)$.

37.

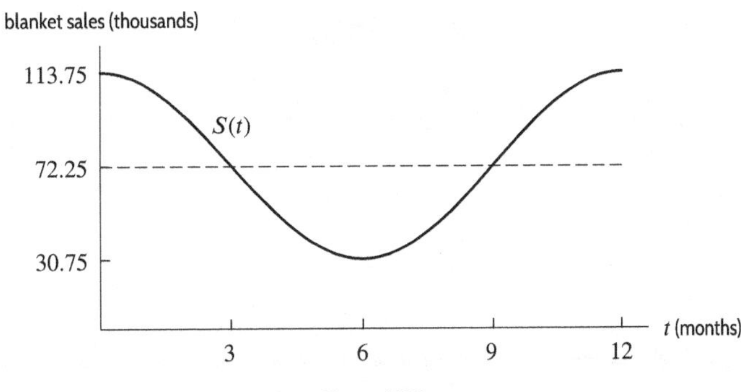

Figure 7.14

The amplitude of this graph is 41.5. The period is $P = 2\pi/B = (2\pi)/(\pi/6) = 12$ months. The amplitude of 41.5 tells us that during winter months sales of electric blankets are 41,500 above the average. Similarly, sales reach a minimum of 41,500 below average in the summer months. The period of one year indicates that this seasonal sales pattern repeats annually.

41. **(a)** See Figure 7.15, where January is represented by $t = 0$.

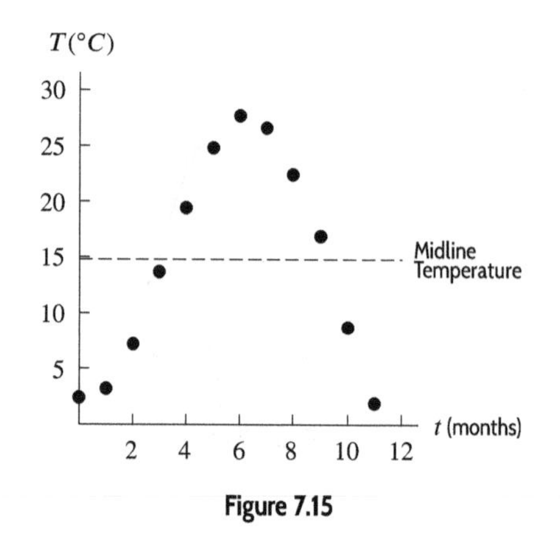

Figure 7.15

(b) The midline temperature is approximately $(81.8+35.4)/2 = 58.6$ degrees. The amplitude of the temperature function is then $81.8 - 58.6 = 23.2$ degrees. The period equals 12 months.

(c) We choose the approximating function $T = f(t) = -A\cos(Bt) + D$. Since the graph resembles an inverted cosine curve, we know $A = 23.2$ and $D = 58.6$. Since the period is 12, $B = 2\pi/12 = \pi/6$. Thus

$$T = f(t) = -23.2\cos\left(\frac{\pi}{6}t\right) + 58.6$$

is a good approximation, though it does not exactly agree with all the data.

(d) In October, $T = f(9) = -23.2\cos((\pi/6)9) + 58.6 \approx 58.6$ degrees, while the table shows an October value of 62.5.

45. We see that the phase shift is 13; the shift is to the left. To find the horizontal shift, we factor out a 7 within the cosine function, giving

$$y = -4\cos\left(7\left(t + \frac{13}{7}\right)\right) - 5.$$

Thus, the horizontal shift is $-13/7$.

49. The midline is 20, the amplitude is 15, and the period is 4. The graph is a sine curve shifted a quarter period to the left, so the phase shift is $\frac{1}{4}(2\pi) = \pi/2$. Thus

$$h = 20 + 15\sin\left(\frac{\pi}{2}t + \frac{\pi}{2}\right).$$

Solutions for Section 7.6

Exercises

1. $\sin 0° = 0$, $\cos 0° = 1$, $\tan 0° = \sin 0°/\cos 0° = 0/1 = 0$.

5. We see that $\tan 540° = \tan 180° = \dfrac{\sin 180°}{\cos 180°} = \dfrac{0}{-1} = 0$.

9. Using the exact values of sine and cosine for special angles, we have

$$\tan\frac{\pi}{3} = \frac{\sin(\pi/3)}{\cos(\pi/3)} = \frac{\sqrt{3}/2}{1/2} = \sqrt{3}$$

13. The graph of $\tan\theta$ has asymptotes at $\theta = \pm\pi/2, \pm 3\pi/2, \pm 5\pi/2, \ldots$. Since the graph of $h(\theta)$ is the graph of $\tan\theta$ vertically shifted up by 2 units, it has asymptotes at the same places. Of these, the only ones in the interval $0 \le \theta \le 2\pi$ are $\theta = \pi/2$ and $\theta = 3\pi/2$.

17. See Figure 7.16. The graph has been shifted up by 1.

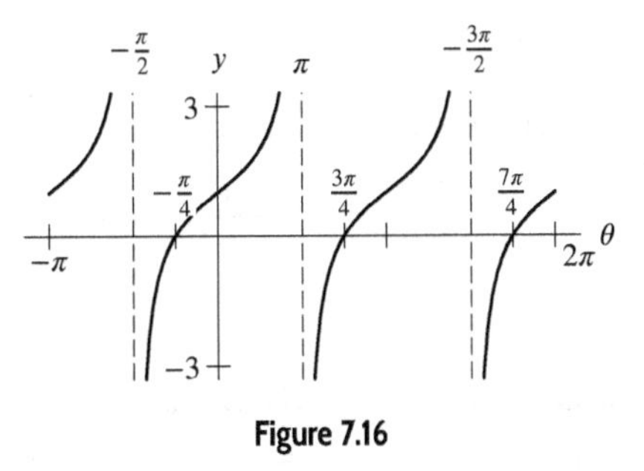

Figure 7.16

Problems

21. We have:

$$\text{Slope} = \frac{\Delta y}{\Delta x} = \frac{y-0}{x-0} = \frac{\sin(6\pi/7)}{\cos(6\pi/7)} = \tan(6\pi/7) = -0.482.$$

So, the slope of this line is -0.482. This makes sense since $6\pi/7$ is more than 90° but less than 180° and the line's slope is then negative.

25. This looks like a tangent graph. At $\pi/4$, we have $\tan\theta = 1$. On this graph $y = 1/2$ if $\theta = \pi/4$, and since it appears to have the same period as $\tan\theta$ without a horizontal or vertical shift, a possible formula is $y = \frac{1}{2}\tan\theta$.

29. **(a)** For a possible graph see Figure 7.17.

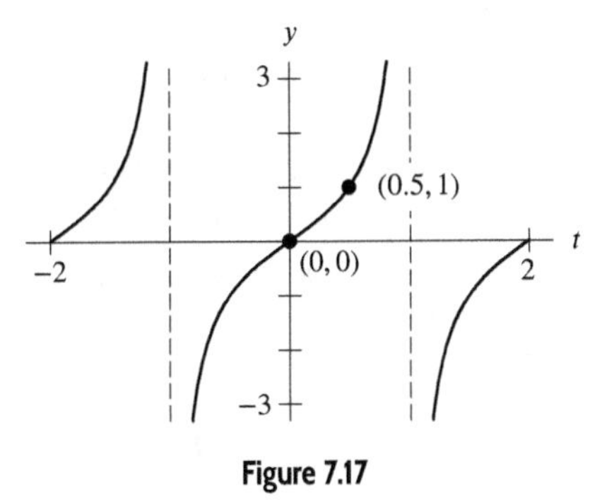

Figure 7.17

(b) We can see from the graph that this function has period 2. Since the graph passes through (0, 0), which lies directly in the middle of the period, there is no vertical shift. Therefore a formula is of the form $f(t) = A\tan\left(\frac{\pi}{2}t\right)$ for some A. Since $f(t)$ passes through (0.5, 1), we have

$$f(0.5) = A\tan\left(\frac{\pi}{4}\right) = A = 1.$$

Therefore, a possible formula for $f(t)$ is $f(t) = \tan\left(\frac{\pi}{2}t\right)$.

33. Your friend is incorrect. The function $f(t) = \tan t$ is an example of a function that is increasing everywhere it is defined, yet it is periodic.

Solutions for Section 7.7

Exercises

1. $-\sqrt{3}$

5. Since $\cot(5\pi/3) = 1/(\tan(5\pi/3) = 1/(\sin(5\pi/3)/\cos(5\pi/3)) = \cos(5\pi/3)/\sin(5\pi/3)$, we know that $\cot(5\pi/3) = (1/2)/(-\sqrt{3}/2) = -1/\sqrt{3}$.

9. Let $x = 2\theta$. Then $(\cos(2\theta))^2 + (\sin(2\theta))^2 = (\cos x)^2 + (\sin x)^2 = 1$, using the Pythagorean identity.

13. Writing $1 - \sin^2 t = \cos^2 t$ from the Pythagorean identity, we have

$$\frac{1-\sin^2 t}{\cos t} = \frac{\cos^2 t}{\cos t} = \cos t.$$

Problems

17. **(a)** Angle π is a half revolution so the point Q at angle $t + \pi$ is the point on the unit circle at the opposite end from P of the diameter that passes through P. We have $Q = (-x, -y)$.

(b) By definition of cosine, $\cos t$ is the first coordinate of the point P so $\cos t = x$, and $\cos(t + \pi)$ is the first coordinate of Q so $\cos(t + \pi) = -x$. Thus $\cos(t + \pi) = -\cos t$.

(c) By definition of sine, $\sin t$ is the second coordinate of the point P so $\sin t = y$, and $\sin(t + \pi)$ is the second coordinate of Q so $\sin(t + \pi) = -y$. Thus $\sin(t + \pi) = -\sin t$.

(d) By definition of tangent, $\tan t$ is the ratio of the second coordinate to the first coordinate of the point P so $\tan t = y/x$, and $\sin(t + \pi)$ is the ratio of the coordinates of Q so $\tan(t + \pi) = -y/(-x) = y/x$. Thus $\tan(t + \pi) = \tan t$.

21. Since $\sec\theta = 1/\cos\theta$, we begin with the Pythagorean identity, $\cos^2\theta + \sin^2\theta = 1$. We have

$$\begin{aligned}\cos^2\theta + \left(\frac{1}{3}\right)^2 &= 1\\ \cos^2\theta &= 1 - \frac{1}{9}\\ \cos\theta &= \pm\sqrt{\frac{8}{9}} = \pm\frac{\sqrt{8}}{3}.\end{aligned}$$

Since $0 \le \theta \le \pi/2$, we know that $\cos\theta \ge 0$, so $\cos\theta = \sqrt{8}/3$. Therefore, $\sec\theta = 1/(\sqrt{8}/3) = 3/\sqrt{8}$.

Since $\tan\theta = \sin\theta/\cos\theta$, we have $\tan\theta = (1/3)/(\sqrt{8}/3) = 1/\sqrt{8}$.

25. **(a)** $\sin^2\phi = 1 - \cos^2\phi = 1 - (0.4626)^2$ and $\sin\phi$ is negative, so $\sin\phi = -\sqrt{1-(0.4626)^2} = -0.8866$. Thus $\tan\phi = (\sin\phi)/(\cos\phi) = (-0.8866)/(0.4626) = -1.9166$.

(b) $\cos^2\theta = 1 - \sin^2\theta = 1 - (-0.5917)^2$ and $\cos\theta$ is negative, so $\cos\theta = -\sqrt{1-(-0.5917)^2} = -0.8062$. Thus $\tan\theta = (\sin\theta)/(\cos\theta) = (-0.5917)/(-0.8062) = 0.7339$.

29. First notice that $\tan\theta = \frac{x}{9}$ so $\tan\theta = \sin\theta/\cos\theta = x/9$, so $\sin\theta = x/9 \cdot \cos\theta$. Now to find $\cos\theta$ by using $1 = \sin^2\theta + \cos^2\theta = (x^2/81)\cos^2\theta + \cos^2\theta = \cos^2\theta(x^2/81 + 1)$, so $\cos^2\theta = 81/(x^2+81)$ and $\cos\theta = 9/\sqrt{x^2+81}$. Thus, $\sin\theta = (x/9)\cdot(9/\sqrt{x^2+81}) = x/\sqrt{x^2+81}$.

33. Divide both sides of $\cos^2\theta + \sin^2\theta = 1$ by $\cos^2\theta$. For $\cos\theta \neq 0$,

$$\begin{aligned}\frac{\cos^2\theta}{\cos^2\theta} + \frac{\sin^2\theta}{\cos^2\theta} &= \frac{1}{\cos^2\theta}\\ 1 + \left(\frac{\sin\theta}{\cos\theta}\right)^2 &= \left(\frac{1}{\cos\theta}\right)^2\\ 1 + \tan^2\theta &= \sec^2\theta.\end{aligned}$$

Solutions for Section 7.8

Exercises

1. **(a)** We are looking for the value of the sine of $\left(\frac{1}{2}\right)^\circ$. Using a calculator or computer, we have $\sin\left(\frac{1}{2}\right)^\circ = 0.009$.

(b) We are looking for the angle between $-\pi/2$ and $\pi/2$ whose sine is $\frac{1}{2}$. Therefore, we have $\sin^{-1}\left(\frac{1}{2}\right) = \pi/6 = 30°$.

(c) We are looking for the reciprocal of the sine of $\left(\frac{1}{2}\right)^{\circ}$. We have

$$\begin{aligned}(\sin x)^{-1} &= \left(\sin\frac{1}{2}\right)^{-1}\\ &= \frac{1}{\sin\frac{1}{2}}\\ &= 114.593.\end{aligned}$$

5. We have

$$\begin{aligned}6\cos\theta - 2 &= 3\\ 6\cos\theta &= 5\\ \cos\theta &= \left(\frac{5}{6}\right)\\ \theta &= \cos^{-1}\left(\frac{5}{6}\right)\\ \theta &= 0.58569 = 33.557°.\end{aligned}$$

9. We use the inverse sine function on a calculator to get $\theta = 0.608$.

13. Since the angles have a negative cosine they must be in the second or third quadrant. An angle in the second quadrant with a reference angle of $\pi/3$ must measure

$$\pi - \frac{\pi}{3} = \frac{2\pi}{3}.$$

An angle in the third quadrant with this same reference angle must measure

$$\pi + \frac{\pi}{3} = \frac{4\pi}{3}.$$

So $\theta = 2\pi/3, 4\pi/3$ are two of many possible answers.

Problems

17. Since $\cos t = -1$, we have $t = \pi$.

21. Since $\tan t = \sqrt{3}$, we have $t = \pi/3$ and $t = 4\pi/3$.

25. Since $\cos^{-1} a = z$ means $\cos z = a$, the angle is z and the value is a.

29. Since $\tan^{-1}\left(\frac{1}{y}\right)$ means $\tan z = \frac{1}{y}$, the angle is z and the value is $\frac{1}{y}$.

33. **(a)** The domain of f is $-1 \le x \le 1$, so $\sin^{-1}(x)$ is defined only if x is between -1 and 1 inclusive. The range is between $-\pi/2$ and $\pi/2$ inclusive.

(b) The domain of g is $-1 \le x \le 1$, so $\cos^{-1}(x)$ is defined only if x is between -1 and 1 inclusive. The range is between 0 and π inclusive.

(c) The domain of h is all real numbers, so $\tan^{-1}(x)$ is defined for all values of x. The range is between but not including $-\pi/2$ and $\pi/2$.

Solutions for Chapter 7 Review

Exercises

1. This function appears to be periodic because it repeats regularly.

5. This function does not appear to be periodic. Though it does rise and fall, it does not do so regularly.

9.

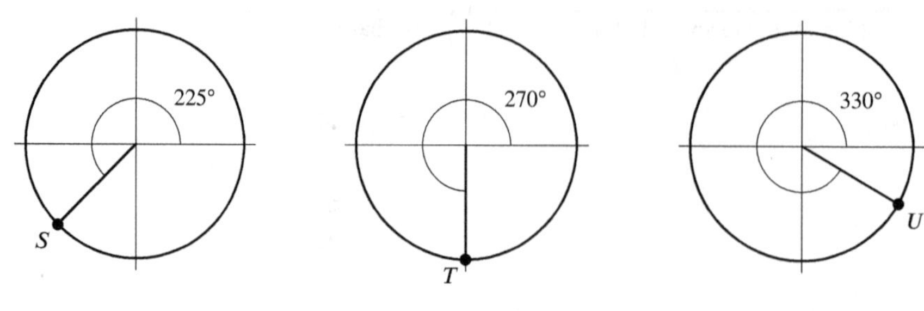

Figure 7.18

$S = (-0.707, -0.707)$, $T = (0, -1)$, $U = (0.866, -0.5)$

13.

$$x = r\cos\theta = 16\cos(-72°) \approx 4.944$$

and

$$y = r\sin\theta = 16\sin(-72°) \approx -15.217,$$

so the approximate coordinates of Z are $(4.944, -15.217)$.

17. Since $\tan^{-1}\left(c^{-1}\right) = d$ means $\tan d = \frac{1}{c}$, the angle is d and the value is $\frac{1}{c}$.

21. $\sin 270° = -1$

25. In order to change from degrees to radians, we multiply the number of degrees by $\pi/180$, so we have $330 \cdot \pi/180$, giving $\frac{11}{6}\pi$ radians.

29. In order to change from radians to degrees, we multiply the number of radians by $180/\pi$, so we have $\frac{3}{2}\pi \cdot 180/\pi$, giving 270 degrees.

33. If we go around six times, we make six full circles, which is $2\pi \cdot 6 = 12\pi$ radians. Since we're going in the negative direction, we have -12π radians.

37. The arc length, s, corresponding to an angle of θ radians in a circle of radius r is $s = r\theta$. In order to change from degrees to radians, we multiply the number of degrees by $\pi/180$, so we have $-585 \cdot \pi/180$, giving $-\frac{13}{4}\pi$ radians. Since length cannot be negative, we take the absolute value, giving us $13\pi/4$ radians. Thus, our arc length is $6.2 \cdot 13\pi/4 \approx 63.303$.

41. The midline is $y = 7$. The amplitude is 1. The period is 2π.

Problems

45. We need a b to the right of -2 that also satisfies

$$\frac{\Delta y}{\Delta t} = \frac{g(b) - g(-2)}{b+2} > 0.$$

Since $b > -2$, the difference quotient will be positive as long as $g(b) > g(-2)$. One possibility is $b = 1$, since the graph suggests

$$g(b) = g(1) \approx 3 > 1 = g(-2).$$

Thus, a possible answer is $-2 \leq t \leq 1$. There are many others.

49. We need a value of b to the right of -2 that also satisfies

$$\frac{\Delta y}{\Delta t} = \frac{g(b) - g(-2)}{b+2} = 2.$$

When $b = 0$

$$g(b) - g(-2) = 5 - 1 = 4 \quad \text{and} \quad b + 2 = 2,$$

and so,

$$\frac{\Delta y}{\Delta t} = \frac{4}{2} = 2.$$

Thus, a possible answer is $-2 \leq t \leq 0$.

53. By plotting the data in Figure 7.19, we can see that the midline is at $h = 2$ (approximately). Since the maximum value is 3 and the minimum value is 1, we have

$$\text{Amplitude} = 2 - 1 = 1.$$

Finally, we can see from the graph that one cycle has been completed from time $t = 0$ to time $t = 1$, so the period is 1 second.

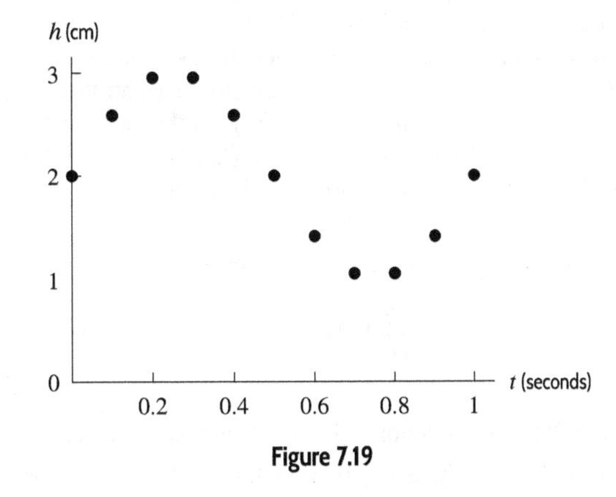

Figure 7.19

57. The period of the Ferris wheel is 30 minutes, so in 17.5 minutes you will travel $17.5/30$, or $7/12$, of a complete revolution, which is $7/12(360°) = 210°$ from your starting position (the 6 o'clock position). This location is $120°$ from the horizontal line through the center of the wheel. Thus, your height is $200 + 200 \sin 120° = 373.205$ feet above the platform.

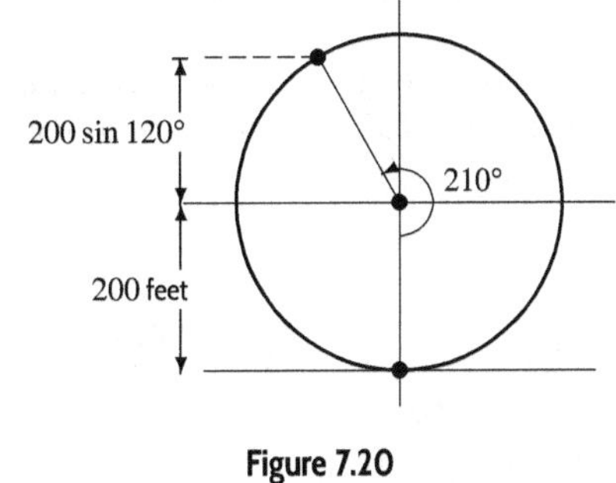

Figure 7.20

61. Amplitude is 40; midline is $y = 50$; period is 16.

65. See Figure 7.21.

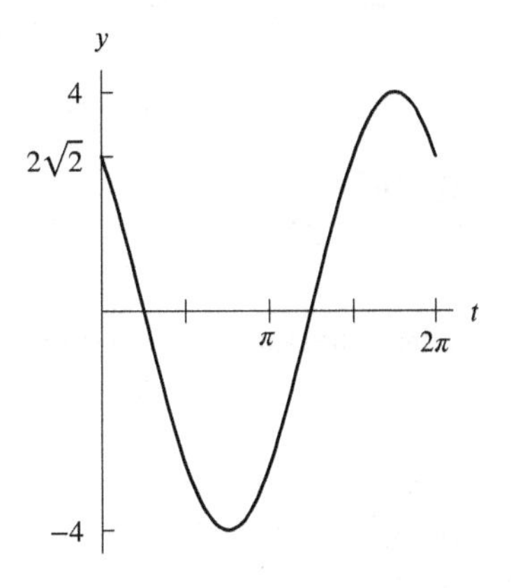

Figure 7.21: $y = 4\cos(t + \pi/4)$

69. $\cos 540° = \cos(360° + 180°) = \cos 180° = -1.$

73. We use the inverse tangent function on a calculator to get $\theta - 1 = 0.168$, which gives us $\theta = 1.168$.

77. First, by looking at the graph of f, we note that its amplitude is 3, and its midline is given by $y = -3$. Therefore, we have $A = \pm 3$ and $k = -3$. Also, since the graph of f completes one full cycle between $x = -3$ and $x = 5$, we see that the period of f is 8, so we have $B = 2\pi/8 = \pi/4$. Combining these observations, we see that we can take the four formulas to be horizontal translations of the functions $y_1 = 3\cos((\pi/4)x) - 3$, $y_2 = -3\cos((\pi/4)x) - 3$, $y_3 = 3\sin((\pi/4)x) - 3$, and $y_4 = -3\sin((\pi/4)x) - 3$, all of which have the same amplitude, period, and midline as f. After sketching y_1, we see that we can obtain the graph of f by shifting the graph of y_1 3 units to the right; therefore, $f_1(x) = 3\cos((\pi/4)(x-3)) - 3$. Similarly, we can obtain the graph of f by shifting the graph of y_2 to the left 1 unit, so $f_2(x) = -3\cos((\pi/4)(x+1)) - 3$, or by shifting the graph of y_3 to the right 1 unit, so $f_3(x) = 3\sin((\pi/4)(x-1)) - 3$. Finally, a sketch of y_4 reveals that we can obtain the graph of f by shifting the graph of y_4 to the right 5 units, so $f_4(x) = -3\sin((\pi/4)(x-5)) - 3$. Answers may vary.

81. The arc length is equal to the radius times the radian measure, so

$$d = (2)\left(\frac{87}{60}\right)(2\pi) = 5.8\pi \approx 18.221 \text{ inches.}$$

85. The function has a maximum of 3000, a minimum of 1200, which means the upward shift is $\frac{3000+1200}{2} = 2100$. A period of eight years gives $B = \frac{2\pi}{8} = \frac{\pi}{4}$. The amplitude is $|A| = 3000 - 2100 = 900$. Thus a function for the population would be an inverted cosine and $f(t) = -900\cos((\pi/4)t) + 2100$.

89. The amplitude is 3, the period is $\dfrac{2\pi}{4\pi} = \dfrac{1}{2}$, the phase shift is 6π, and

$$\text{Horizontal shift} = -\frac{6\pi}{4\pi} = -\frac{3}{2}.$$

Since the horizontal shift is negative, the graph of $y = 3\sin(4\pi t)$ is shifted $3/2$ units to the left to give the graph in Figure 7.22. Note that a shift of $3/2$ units produces the same graph as the unshifted graph. We expect this since $y = 3\sin(4\pi t + 6\pi) = 3\sin(4\pi t)$. See Figure 7.22.

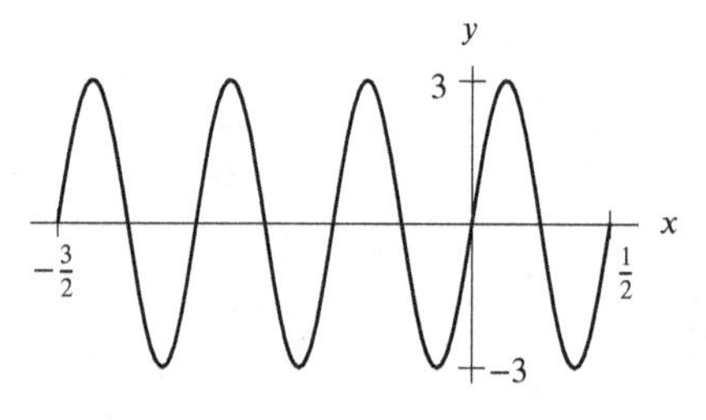

Figure 7.22: $y = 3\sin(4\pi t + 6\pi)$

STRENGTHEN YOUR UNDERSTANDING

1. True, because $\sin x$ is an odd function.

5. True. Because the angles θ and $\theta + \pi$ determine the same line through the origin and hence have the same slope, which is the tangent.

9. False. A unit circle must have a radius of 1.

13. True. This is the definition of radian measure.

17. True. Because P and R have the same y-coordinates.

21. False. Use $s = r\theta$ to get, $s = 3 \cdot (\pi/3) = \pi$.

25. False. The amplitude is half the difference between its maximum and minimum values.

29. True. The graph of a periodic function is obtained by starting with the graph on one period and horizontally shifting it by all multiples of the period. The entire graph is determined by the graph of a single period. Since the graphs of the periodic functions $f(t)$ and $g(t)$ are the same on the period $0 \le t < A$, their graphs are the same everywhere.

33. False. The amplitude is (maximum – minimum)/2 = $(6-2)/2 = 2$.

37. False. The amplitude is 10.

41. False. The midline equation is $y = 25$

45. False. The cosine function has been vertically stretched and then shifted downward by 4. It has not been reflected.

49. True. The conditions for A, B and C in $y = A\cos(Bx) + C$ are perfectly met.

53. False. It is undefined at $\pi/2$ and $3\pi/2$.

57. True. We have $\sec\pi = 1/\cos\pi = 1/(-1) = -1$.

61. True. It takes 60° to reach the x-axis from 300°.

65. True. If $\cos t = 1$, then $\sin t = 0$ so $\tan t = 0/1 = 0$.

69. True; because $\tan A = \tan B$ means $A = B + k\pi$, for some integer k. Thus $\frac{A-B}{\pi} = k$ is an integer.

73. False. $\sin\frac{\pi}{6} = \frac{1}{2}$.

77. True. $\sin(\pi/3) = \sqrt{3}/2$.

Solutions to Skills for Chapter 7

Exercises

1. We have $\sin 30^\circ = 1/2$.

5. Since we know that the x-coordinate on the unit circle at -60° is the same as the x-coordinate at 60°, we know that $\cos(-60^\circ) = \cos 60^\circ = 1/2$.

9. Since 135° is in the second quadrant,
$$\sin 135^\circ = \sin 45^\circ = \frac{1}{\sqrt{2}}.$$

13. Since 405° is in the first quadrant,
$$\sin 405^\circ = \sin 45^\circ = \frac{1}{\sqrt{2}}.$$

17. The reference angle for 300° is $360^\circ - 300^\circ = 60^\circ$, so $\sin 300^\circ = -\sin 60^\circ = -\sqrt{3}/2$.

Problems

21. Since $\sin 30^\circ = x/10$, we have $x = 10(1/2) = 5$.

25. In a 45°-45°-90° triangle, the two legs are equal and the hypotenuse is $\sqrt{2}$ times the length of a leg. So the sides are 5 and 5 and $5\sqrt{2}$.

29. A right triangle with two equal sides is a 45°-45°-90° triangle. In such triangles the length of the hypotenuse side is $\sqrt{2}$ times the length of a leg, so the third side is $4\sqrt{2}$.

CHAPTER EIGHT

Solutions for Section 8.1

Exercises

1. By the Pythagorean theorem, the hypotenuse has length $\sqrt{1^2+2^2}=\sqrt{5}$.

(a) $\tan\theta = \dfrac{\text{opposite}}{\text{adjacent}} = \dfrac{2}{1} = 2.$

(b) $\sin\theta = \dfrac{\text{opposite}}{\text{hypotenuse}} = \dfrac{2}{\sqrt{5}}.$

(c) $\cos\theta = \dfrac{\text{adjacent}}{\text{hypotenuse}} = \dfrac{1}{\sqrt{5}}.$

5. We use the Pythagorean theorem to find the length of the hypotenuse:

$$\begin{aligned}\text{Hypotenuse}^2 &= (0.1)^2 + (0.2)^2 = 0.01 + 0.04 = 0.05\\ \text{Hypotenuse} &= \sqrt{0.05}.\end{aligned}$$

(a) We have

$$\sin\theta = \frac{\text{Side opposite}}{\text{Hypotenuse}} = \frac{0.1}{\sqrt{0.05}} = 0.447.$$

(b) We have

$$\cos\theta = \frac{\text{Side adjacent}}{\text{Hypotenuse}} = \frac{0.2}{\sqrt{0.05}} = 0.894.$$

9. Since $\cos 37^\circ = 6/r$, we have $r = 6/\cos 37^\circ$. Similarly, since $\tan 37^\circ = y/6$, we have $y = 6\tan 37^\circ$.

13. We have $\sin\theta = 0.876$, so $\theta = \sin^{-1} 0.876 = 61.164^\circ$.

17. We have $\sin\theta = 0.999$, so $\theta = \sin^{-1} 0.999 = 89.190^\circ$.

21. We have

$$\begin{aligned}B &= 90^\circ - A = 62^\circ\\ a &= c\cdot\sin A = 20\sin 28^\circ = 9.389\\ b &= c\cdot\sin B = 20\sin 62^\circ = 17.659.\end{aligned}$$

Problems

25. We have $c = \sqrt{a^2+a^2} = \sqrt{2}\,a$. Hence

(a) $\sin 45^\circ = \dfrac{\text{opposite}}{\text{hypotenuse}} = \dfrac{a}{\sqrt{2}a} = \dfrac{1}{\sqrt{2}} = \dfrac{\sqrt{2}}{2}.$

(b) $\cos 45^\circ = \dfrac{\text{adjacent}}{\text{hypotenuse}} = \dfrac{a}{\sqrt{2}a} = \dfrac{1}{\sqrt{2}} = \dfrac{\sqrt{2}}{2}.$

(c) $\tan 45^\circ = \dfrac{\text{opposite}}{\text{adjacent}} = \dfrac{a}{a} = 1.$

29. Let the origin be at the rendezvous point and the positive x-axis point east. Then, as she reaches the river, the volunteer's coordinates are $(a, 1.3)$, where a is her distance east of the rendezvous point, in miles.

Since the slope of the line traced by the volunteer's path is

$$\text{Slope} = \frac{\Delta y}{\Delta x} = \frac{1.3 - 0}{a - 0} = \frac{1.3}{a},$$

and the line forms a 75° angle with the positive x-axis, we have:

$$\tan 75^\circ = \frac{1.3}{a} \quad \text{or} \quad a = \frac{1.3}{\tan 75^\circ} = 0.348 \text{ miles}.$$

Thus, the volunteer is located about 0.35 miles east of the rendezvous point as she reaches the river.

33. Since $(0, 0)$ and $(p, -1)$ are points on the line,

$$\text{Slope of line} = \frac{\Delta y}{\Delta x} = \frac{-1 - 0}{p - 0} = -\frac{1}{p}.$$

Since the line forms an angle of $5\pi/6$, we also have

$$\text{Slope of line} = \tan\frac{5\pi}{6}.$$

The two expressions together give

$$-\frac{1}{p} = \tan\frac{5\pi}{6} \quad \text{which means} \quad p = -\frac{1}{\tan(5\pi/6)} = 1.732.$$

This answer makes sense because a point left of the y-axis on a line through the origin with a negative slope should be above the x-axis.

37. (a) Since the grade of the ramp is $7\% = 7/100$, this means that a 7-foot height difference occurs over a horizontal distance of 100 feet. So we have $\tan\theta = \frac{7}{100}$. Using the $\tan^{-1}$ button on the calculator, we get

$$\theta = \tan^{-1}\left(\frac{7}{100}\right) = 4.00417^\circ.$$

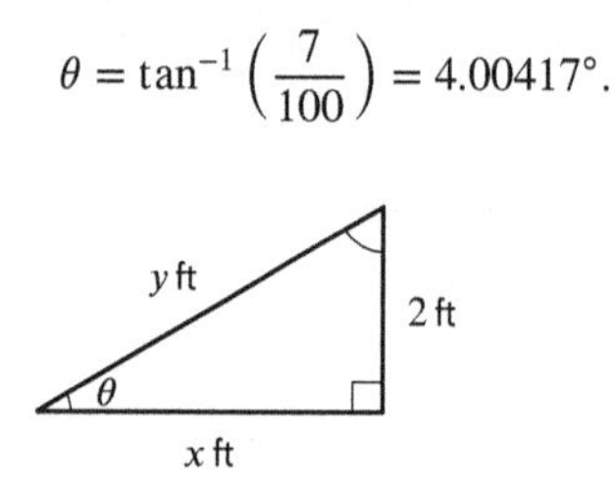

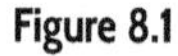
Figure 8.1

(b) From the right triangle representing the ramp we see that one leg represents the height difference between the driveway and the front door, which is 2 ft. The other leg represents the driveway, of which we would like to find the length x. Then $2/x = \tan 4.00417^\circ$. Solving this equation for x, we have

$$x = \frac{2}{\tan 4.00417^\circ} = 28.57.$$

So the driveway has to be 28.57 feet long. (We can also use similar triangles: $2/7 = x/100$.)

(c) The ramp is represented by the hypotenuse y of the right triangle. Using the Pythagorean theorem we have $y = \sqrt{28.57^2 + 2^2} = 28.64$ feet. (We can also use $\sin 4.00417^\circ = 2/y$.)

41. To solve for the distance x, we use $\tan 53^\circ = \frac{954}{x}$ and solve for x:

$$x = 954/\tan 53^\circ = 718.891 \text{ ft}.$$

To solve for the height of the Seafirst Tower, we can use $\tan 37^\circ = y/x$ and solve for y:

$$y = 718.891 \tan 37^\circ = 541.723 \text{ ft}.$$

(The actual height of the Seafirst Tower is 543 ft.)

45. In the figure, we see that $\tan\alpha = 2/10 = 0.2$ and $\tan\beta = (2+1)/10 = 0.3$. Thus,

$$\tan\alpha = 0.2, \quad \text{so} \quad \alpha = \tan^{-1}(0.2) \approx 11.310^\circ$$
$$\tan\beta = 0.3, \quad \text{so} \quad \beta = \tan^{-1}(0.3) \approx 16.699^\circ$$

Solutions for Section 8.2

Exercises

1. By the Law of Sines, we have

$$\frac{x}{\sin 100^\circ} = \frac{6}{\sin 18^\circ}$$
$$x = 6\left(\frac{\sin 100^\circ}{\sin 18^\circ}\right) \approx 19.121.$$

5. In Figure 8.2, we have

$$\theta = 180^\circ - 90^\circ - 10^\circ$$
$$\theta = 80^\circ.$$

$$a = 12\cos 10^\circ$$
$$a \approx 12(0.985)$$
$$a \approx 11.818.$$

$$b = 12\sin 10^\circ$$
$$b \approx 12(0.174)$$
$$b \approx 2.084.$$

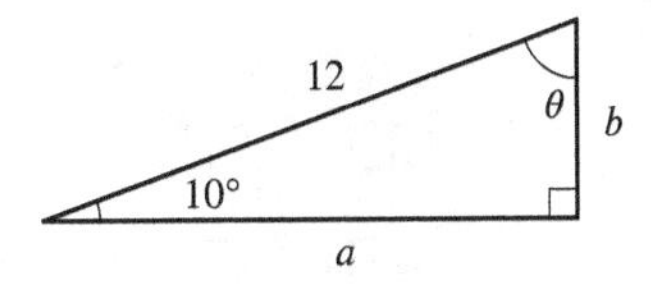

Figure 8.2

9. Using the Law of Cosines, we have

$$41^2 = 20^2 + 28^2 - 2\cdot 20\cdot 28\cos C$$
$$497 = -1120\cos C$$
$$C = \cos^{-1}\left(-\frac{497}{1120}\right)$$
$$C = 116.343^\circ.$$

Using the Law of Cosines again (though we could use the Law of Sines), we have

$$20^2 = 41^2 + 28^2 - 2\cdot 41\cdot 28\cos A$$
$$-2065 = -2296\cos A$$
$$A = \cos^{-1}\left(\frac{2065}{2296}\right)$$
$$A = 25.922^\circ.$$

Thus, $B = 180 - A - C = 37.735^\circ$.

13. We begin by using the Law of Cosines to find side c:

$$\begin{aligned} c^2 &= 9^2 + 8^2 - 2 \cdot 9 \cdot 8 \cos 80^\circ \\ c^2 &= 119.995 \\ c &= 10.954. \end{aligned}$$

We can now use the Law of Sines to find the other two angles.

$$\begin{aligned} \frac{\sin B}{8} &= \frac{\sin 80^\circ}{10.954} \\ \sin B &= 8\frac{\sin 80^\circ}{10.954} \\ B &= \sin^{-1} 0.719 \\ B &= 45.990^\circ. \end{aligned}$$

Therefore, $A = 180^\circ - 80^\circ - 45.990 = 54.010^\circ$.

17. From the Law of Sines, we have

$$\begin{aligned} \frac{\sin A}{a} &= \frac{\sin C}{c} \\ \frac{\sin A}{8} &= \frac{\sin 98^\circ}{17} \\ A &= \sin^{-1}\left(\frac{8 \sin 98^\circ}{17}\right) \\ &= 27.7755^\circ. \end{aligned}$$

Since C is obtuse, A and B are acute, so this is the correct value of A. (Had A been obtuse, we would have had to subtract this value from 180° to compensate.) Now we have $B = 180^\circ - A - C = 54.2245^\circ$. Once again, we can use the Law of Sines to solve for b:

$$\begin{aligned} \frac{b}{\sin B} &= \frac{c}{\sin C} \\ \frac{b}{\sin 54.2245} &= \frac{17}{\sin 98^\circ} \\ b &= \frac{17 \sin 54.2245}{\sin 98^\circ} \\ &= 13.9279. \end{aligned}$$

21. First, we recognize that it is possible that there are two triangles, since we may have the ambiguous case. However, since 95° is greater than 90°, there are no other obtuse angles possible, so there is but one possible triangle.

We begin by finding the angle C using the Law of Sines:

$$\begin{aligned} \frac{\sin C}{10} &= \frac{\sin 95^\circ}{5} \\ \sin C &= 10 \cdot \frac{\sin 95^\circ}{5} \\ C &= \sin^{-1} 1.992. \end{aligned}$$

Since there is no arcsine of 1.992, we notice that there is a problem. There are no solutions. We could have seen this before because the longest side is always across from the largest angle, and since C cannot be greater than 95°, the side across from it (10) cannot be longer than the side across from 95°. Since it is bigger, no triangle fulfills the conditions given.

25. We begin by finding the angle C, which is

$$\pi - \frac{3\pi}{5} - \frac{\pi}{20} = \frac{20\pi - 12\pi - \pi}{20} = \frac{7\pi}{20} \text{ radians.}$$

We can now use the Law of Sines to find the other two sides.

$$\frac{b}{\sin(\pi/20)} = \frac{15}{\sin(7\pi/20)}$$
$$b = \sin(\pi/20) \cdot \frac{15}{\sin(7\pi/20)}$$
$$b = 2.634.$$

Similarly,

$$\frac{a}{\sin(3\pi/5)} = \frac{15}{\sin(7\pi/20)}$$
$$a = \sin(3\pi/5) \cdot \frac{15}{\sin(7\pi/20)}$$
$$a = 16.011.$$

29.

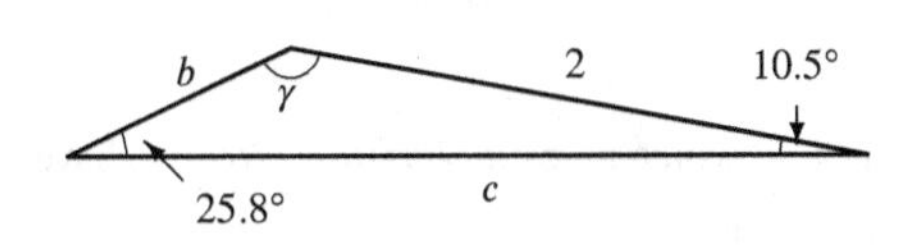

Figure 8.3

The Law of Sines tells us that $\frac{b}{\sin 10.5°} = \frac{2}{\sin 25.8°}$, so $b = 0.837$ m. We have $\gamma = 180° - 10.5° - 25.8° = 143.7°$. We use this value to find side length c. We have $c^2 = 2^2 + (.837)^2 - 2(2)(.837)\cos 143.7° \approx 7.401$, or $c = 2.720$ m.

Problems

33. (a) By the Law of Sines, we have

$$\frac{\sin\theta}{3} = \frac{\sin 110°}{10}$$
$$\sin\theta = \left(\frac{3}{10}\right)\sin 110° \approx 0.282.$$

(b) If $\sin\theta = 0.282$, then $\theta \approx 16.374°$ (as found on a calculator) or $\theta \approx 180° - 16.374° \approx 163.626°$. Since the triangle already has a $110°$ angle, $\theta \approx 16.374°$. (The $163.626°$ angle would be too large.)

(c) The height of the triangle is $10\sin\theta = 10 \cdot 0.282 = 2.819$ cm. Since the sum of the angles of a triangle is $180°$, and we know two of the angles, $\theta = 16.374°$ and $110°$, so the third angle is $180° - 16.374° - 110° = 53.626°$. By the Law of Sines, we have

$$\frac{\sin 110°}{10} = \frac{\sin 53.626°}{\text{Base}}$$
$$\text{Base} = \frac{10\sin 53.626°}{\sin 110°} \approx 8.568 \text{ cm}.$$

Thus, the triangle has

$$\text{Area} = \frac{1}{2}\text{Base} \cdot \text{Height} = \frac{1}{2} \cdot 8.568 \cdot 2.819 = 12.077 \text{ cm}^2.$$

37. Figure 8.4 shows the triangle; we want to find x. The other two angles are $(180 - 39)°/2 = 70.5°$. Using the Law of Sines:

$$\frac{425}{\sin 39°} = \frac{x}{\sin 70.5°}$$
$$x = \frac{425\sin 70.5°}{\sin 39°} = 636.596 \text{ feet}.$$

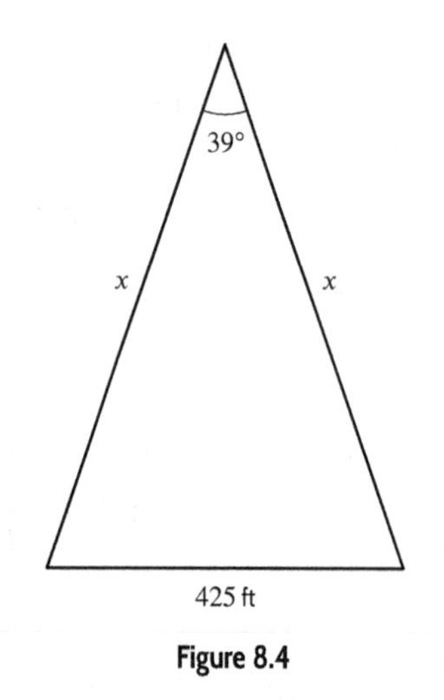

Figure 8.4

41. From the figure, we see that $\sin A = \dfrac{h}{b}$, which gives $h = b \sin A$. We also have $\sin B = \dfrac{h}{a}$, which gives $h = a \sin B$. Thus, $b \sin A = a \sin B$, which gives the Law of Sines:

$$\frac{\sin A}{a} = \frac{\sin B}{b}.$$

45. One way to organize this situation is to use the abbreviations from high school geometry. The six possibilities are {SSS,SAS,SSA,ASA,AAS,AAA}.

SSS Knowing all three sides allows us to find the angles by using the Law of Cosines.

SAS Knowing two sides and the included angle allows us to find the third side length by using the Law of Cosines. We can then use the SSS procedure.

SSA Knowing two sides but not the included angle is called the ambiguous case, because there could be two different solutions. Use the Law of Sines to find one of the missing angles, which, because we use the arcsine, may give two values. Or, use the Law of Cosines, which produces a quadratic equation that may also give two values. Treating these cases separately, we can continue to find all sides and angles using the SAS procedure.

ASA Knowing two angles allows us to easily find the third angle. Use the Law of Sines to find each side.

AAS Find the third angle and then use the Law of Sines to find each side.

AAA This has an infinite number of solutions because of similarity of triangles. Once one side is known, then the ASA or AAS procedure can be followed.

49.

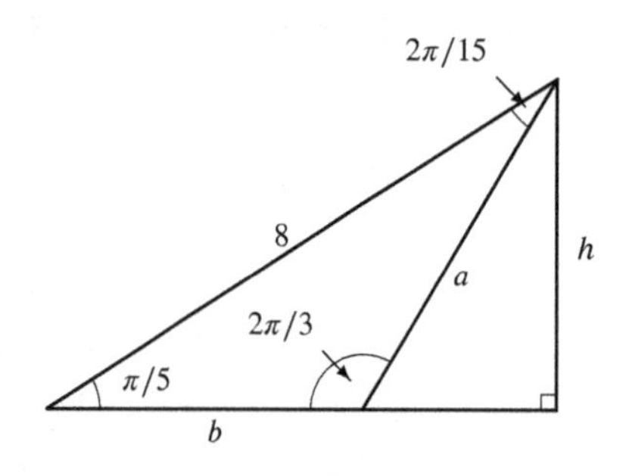

Figure 8.5

(a) $\dfrac{\sin(2\pi/3)}{8} = \dfrac{\sin(\pi/5)}{a}$, so $a = \dfrac{8\sin(\pi/5)}{\sin(2\pi/3)} = 5.420$. Similarly, $b = \dfrac{8\sin(2\pi/15)}{\sin(2\pi/3)} = 3.757$.

(b) Construct an altitude h as in Figure 8.5. We have $\sin(\pi/5) = \dfrac{h}{8}$, so $h = 8\sin(\pi/5) = 4.702$. Then area of the triangle is $\frac{1}{2}(3.757)(4.702) = 8.833$.

Solutions for Section 8.3

Exercises

1. Quadrant IV.

5. Since $3.2\pi = 2\pi + 1.2\pi$, such a point is in Quadrant III.

9. Since -7 is an angle of -7 radians, corresponding to a rotation of just over 2π, or one full revolution, in the clockwise direction, such a point is in Quadrant IV.

13. We have $180^\circ < \theta < 270^\circ$. See Figure 8.6.

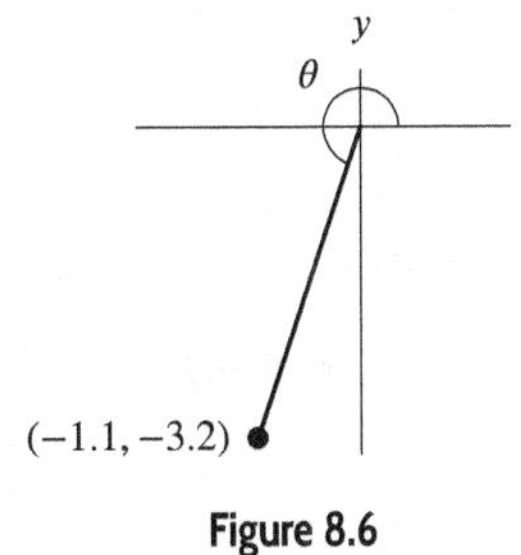

Figure 8.6

17. With $x = -\sqrt{3}$ and $y = 1$, find $r = \sqrt{(-\sqrt{3})^2 + 1^2} = \sqrt{4} = 2$. Find θ from $\tan\theta = y/x = 1/(-\sqrt{3})$. Thus, $\theta = \tan^{-1}(-1/\sqrt{3}) = -\pi/6$. Since $(-\sqrt{3}, 1)$ is in the second quadrant, $\theta = -\pi/6 + \pi = 5\pi/6$. The polar coordinates are $(2, 5\pi/6)$.

21. With $r = 2\sqrt{3}$ and $\theta = -\pi/6$, we find $x = r\cos\theta = 2\sqrt{3}\cos(-\pi/6) = 2\sqrt{3}\cdot\sqrt{3}/2 = 3$ and $y = r\sin\theta = 2\sqrt{3}\sin(-\pi/6) = 2\sqrt{3}(-1/2) = -\sqrt{3}$.

The rectangular coordinates are $(3, -\sqrt{3})$.

Problems

25. Since $\theta = \tan^{-1}(y/x)$, write $\tan^{-1}(y/x) = \frac{\pi}{4}$. Taking the tangent of both sides, we get $y/x = \tan(\pi/4) = 1$. In rectangular coordinates, the equation is $y = x$.

29. By substituting $x = r\cos\theta$ and $y = r\sin\theta$, the equation becomes $r\sin\theta = (r\cos\theta)^2$. This could also be written as $r = \dfrac{\sin\theta}{\cos^2\theta}$.

33. Figure 8.7 shows that at 9 am, we have:
In Cartesian coordinates, $H = (-3, 0)$. In polar coordinates, $H = (3, \pi)$; that is $r = 3, \theta = \pi$. In Cartesian coordinates, $M = (0, 4)$. In polar coordinates, $M = (4, \pi/2)$, that is $r = 4, \theta = \pi/2$.

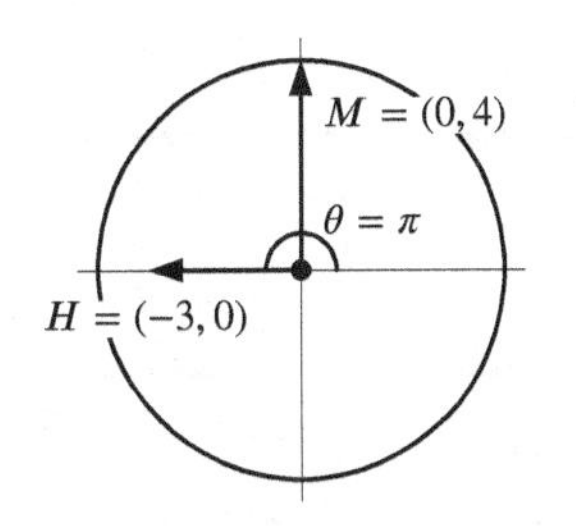

Figure 8.7

37. Figure 8.8 shows that at 3:30 pm, the polar coordinates of the point H (halfway between 3 and 4 on the clock face) are $r = 3$ and $\theta = 75° + 270° = 23\pi/12$. Thus, the Cartesian coordinates of H are given by

$$x = 3\cos\left(\frac{23\pi}{12}\right) \approx 2.898, \quad y = 3\sin\left(\frac{23\pi}{12}\right) \approx -0.776.$$

We have:

In Cartesian coordinates, $H \approx (2.898, -0.776)$; $M = (0, -4)$.
In polar coordinates, $H = (3, 23\pi/12)$; $M = (4, 3\pi/2)$.

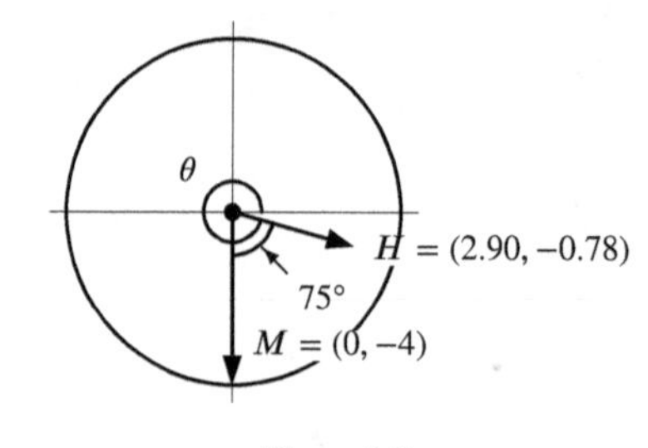

Figure 8.8

41. The circular arc has equation $r = 1$, for $0 \le \theta \le \pi/2$. The vertical line $x = 2$ has polar equation $r\cos\theta = 2$, or $r = 2/\cos\theta$. So the region is described by $0 \le \theta \le \pi/2$ and $1 \le r \le 2/\cos\theta$.

45. The curve will be a smaller loop inside a larger loop with an intersection point at the origin. Larger n values increase the size of the loops. See Figures 8.9-8.11.

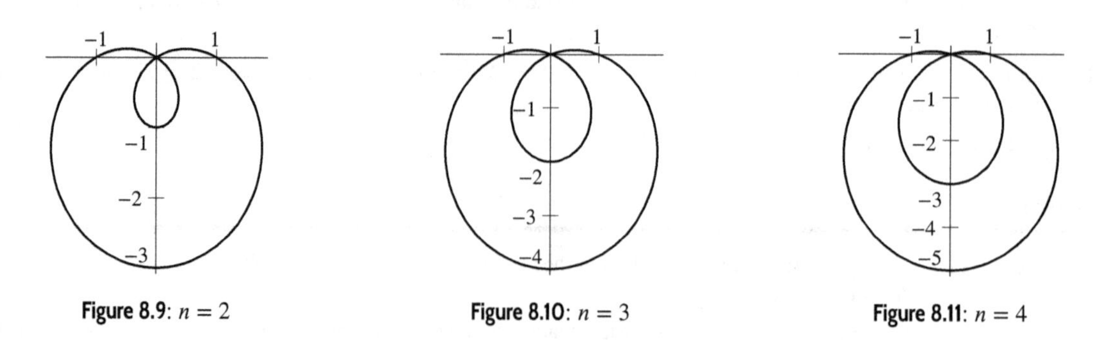

Figure 8.9: $n = 2$ Figure 8.10: $n = 3$ Figure 8.11: $n = 4$

49. **(a)** Let $0 \le \theta \le \pi/4$ and $0 \le r \le 1$.
(b) Break the region into two pieces: one with $0 \le x \le \sqrt{2}/2$ and $0 \le y \le x$, the other with $\sqrt{2}/2 \le x \le 1$ and $0 \le y \le \sqrt{1-x^2}$.

Solutions for Chapter 8 Review

Exercises

1. By the Pythagorean theorem, we know that the third side must be $\sqrt{7^2 - 2^2} = \sqrt{45}$.

(a) Since $\sin\theta$ is opposite side over hypotenuse, we have $\sin\theta = \sqrt{45}/7$.
(b) Since $\cos\theta$ is adjacent side over hypotenuse, we have $\cos\theta = 2/7$.
(c) Since $\tan\theta$ is opposite side over adjacent side, we have $\tan\theta = \sqrt{45}/2$.

5. By the Pythagorean theorem, we know that the third side must be $\sqrt{11^2 - 2^2} = \sqrt{117}$.

(a) Since $\sin\theta$ is opposite side over hypotenuse, we have $\sin\theta = \sqrt{117}/11$.
(b) Since $\cos\theta$ is adjacent side over hypotenuse, we have $\cos\theta = 2/11$.
(c) Since $\tan\theta$ is opposite side over adjacent side, we have $\tan\theta = \sqrt{117}/2$.

9. Since we are looking for the angle θ, we have $\tan\theta = \frac{5}{3}$, so $\theta = \tan^{-1}\frac{5}{3} = 59.036°$.

13. We know that $\cos 45° = \dfrac{\sqrt{2}}{2}$. Therefore, $\theta = 45°$.

17. Since $-7.7\pi = -3 \cdot 2\pi - 1.7\pi$, such a point is in Quadrant I.

21. With $r = 0$, the point specified is the origin, no matter what the angle measure. So $x = r\cos\theta = 0$ and $y = r\sin\theta = 0$. The rectangular coordinates are $(0, 0)$.

Problems

25. See Figure 8.12. By the Pythagorean theorem, $x = \sqrt{13^2 - 12^2} = 5$.

$$\begin{aligned}\sin\theta &= \frac{12}{13}\\ \theta &= \sin^{-1}\left(\frac{12}{13}\right)\\ \theta &\approx 67.380°.\end{aligned}$$

Thus $\varphi = 90° - \theta \approx 22.620°$.

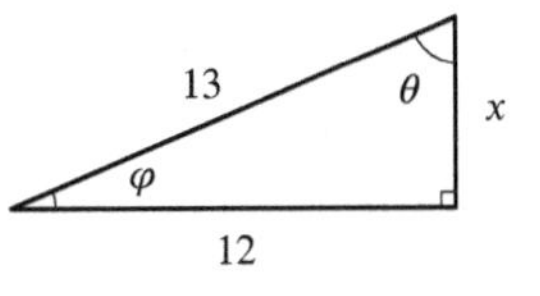

Figure 8.12

29. We have

$$\tan 28° = \frac{\overbrace{\text{Leg opposite}}^{x}}{\underbrace{\text{Leg adjacent}}_{4}} = \frac{x}{4}$$

so

$$\begin{aligned}x &= 4\tan 28°\\ &= 4(0.5317) \qquad \text{using a calculator}\\ &= 2.1268 \text{ miles.}\end{aligned}$$

33. Let d be the distance from Hampton to the point where the beam strikes the shore. Then, $\tan\phi = d/3$, so $d = 3\tan\phi$ miles.

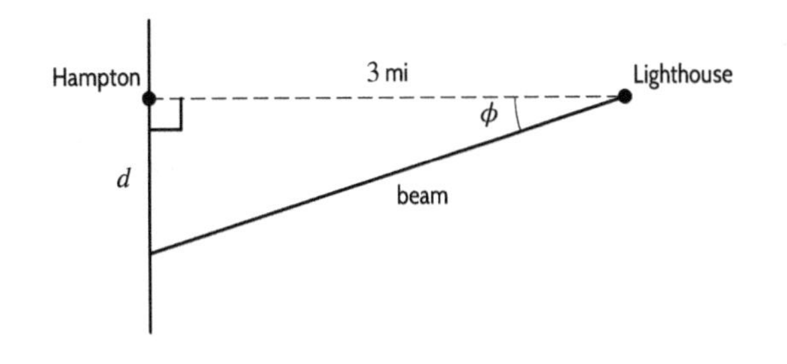

Figure 8.13

37. **(a)** Clearly $(x-1)^2 + y^2 = 1$ is a circle with center $(1,0)$. To convert this to polar, use $x = r\cos\theta$ and $y = r\sin\theta$. Then $(r\cos\theta - 1)^2 + (r\sin\theta)^2 = 1$ or $r^2\cos^2\theta - 2r\cos\theta + 1 + r^2\sin^2\theta = 1$. This means $r^2(\cos^2\theta + \sin^2\theta) = 2r\cos\theta$, or $r = 2\cos\theta$.

(b) 12 o'clock $\rightarrow$ $(x,y) = (1,1)$ and $(r,\theta) = (\sqrt{2}, \pi/4)$,
3 o'clock $\rightarrow$ $(x,y) = (2,0)$ and $(r,\theta) = (2,0)$,
6 o'clock $\rightarrow$ $(x,y) = (1,-1)$ and $(r,\theta) = (\sqrt{2}, -\pi/4)$,
9 o'clock $\rightarrow$ $(x,y) = (0,0)$ and $(r,\theta) = (0$, any angle $)$.

STRENGTHEN YOUR UNDERSTANDING

1. False. Both acute angles are 45 degrees, and $\sin 45° = \sqrt{2}/2$.

5. True. By the Law of Cosines, we have $p^2 = n^2 + r^2 - 2nr\cos P$, so $\cos P = (n^2 + r^2 - p^2)/(2nr)$.

9. True. Identify the opposite angles as B and L and use the Law of Sines to obtain $\dfrac{LA}{\sin B} = \dfrac{BA}{\sin L}$. Thus $\dfrac{LA}{BA} = \dfrac{\sin B}{\sin L}$.

13. False. The graph of $r = 1$ is the unit circle.

17. True. The point is on the y axis three units down from the origin. Thus $r = 3$ and $\theta = 3\pi/2$. In polar coordinates this is $(3, 3\pi/2)$.

CHAPTER NINE

Solutions for Section 9.1

Exercises

1. We draw a graph of $y = \cos t$ for $-\pi \leq t \leq 3\pi$ and trace along it on a calculator to find points at which $y = 0.4$. We read off the t-values at the points t_0, t_1, t_2, t_3 in Figure 9.1. If t is in radians, we find $t_0 = -1.159$, $t_1 = 1.159$, $t_2 = 5.124$, $t_3 = 7.442$. We can check these values by evaluating:

$$\cos(-1.159) = 0.40, \quad \cos(1.159) = 0.40, \quad \cos(5.124) = 0.40, \quad \cos(7.442) = 0.40.$$

Notice that because the cosine function is periodic, the equation $\cos t = 0.4$ has infinitely many solutions. The symmetry of the graph suggests that the solutions are related.

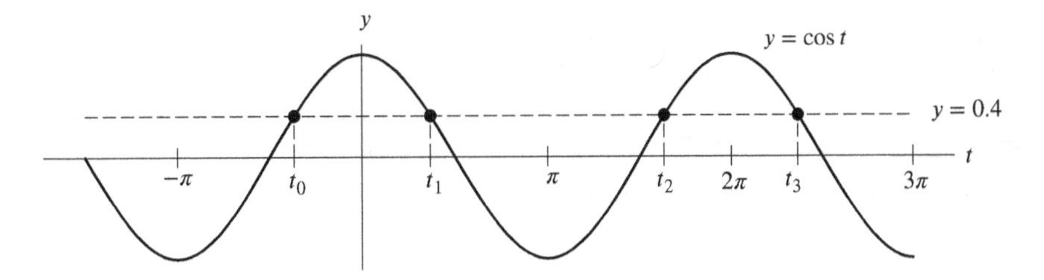

Figure 9.1: The points t_0, t_1, t_2, t_3 are solutions to the equation $\cos t = 0.4$

5. Graph $y = \cos t$ on $0 \leq t \leq 2\pi$ and locate the two points with y-coordinate -0.24. The t-coordinates of these points are approximately $t = 1.813$ and $t = 4.473$. See Figure 9.2.

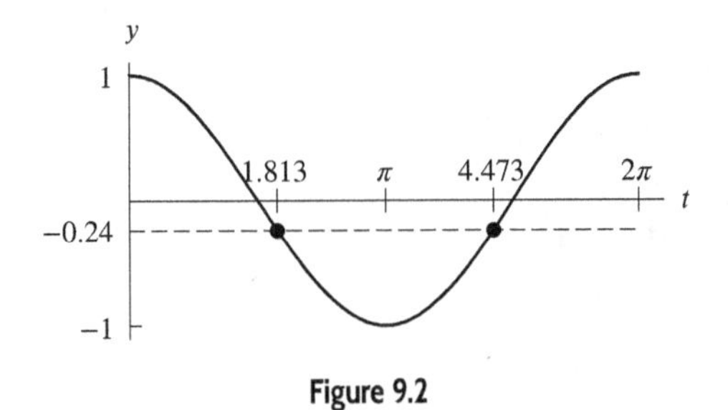

Figure 9.2

9. We have

$$\begin{aligned} 3\sin\theta + 3 &= 5\sin\theta + 2 \\ 1 &= 2\sin\theta \\ \frac{1}{2} &= \sin\theta \\ \sin^{-1}\left(\frac{1}{2}\right) &= \theta \\ 0.5236 = 30^\circ &= \theta. \end{aligned}$$

13. We use the inverse tangent function on a calculator to get $5\theta + 7 = -0.236$. Solving for θ, we get $\theta = -1.447$.

Problems

17. We have $x = \cos^{-1}(0.6) = 0.927$. A graph of $\cos x$ shows that the second solution is $x = 2\pi - 0.927 = 5.356$.

21. Since $\sin(x - 1) = 0.25$, we know

$$x - 1 = \sin^{-1}(0.25) = 0.253.$$
$$x = 1.253.$$

Another solution for $x - 1$ is given by

$$x - 1 = \pi - 0.253 = 2.889$$
$$x = 3.889.$$

25. By sketching a graph, we see that there are four solutions (see Figure 9.3). The first solution is given by $x = \cos^{-1}(0.6) = 0.927$, which is equivalent to the length labeled "b" in Figure 9.3. Next, note that, by the symmetry of the graph of the cosine function, we can obtain a second solution by subtracting the length b from 2π. Therefore, a second solution to the equation is given by $x = 2\pi - 0.927 = 5.356$. Similarly, our final two solutions are given by $x = 2\pi + 0.927 = 7.210$ and $x = 4\pi - 0.927 = 11.639$.

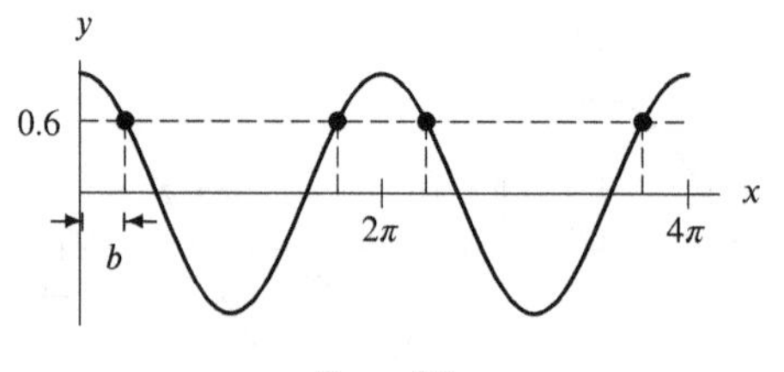

Figure 9.3

29. One solution is $\theta = \sin^{-1}(-\sqrt{2}/2) = -\pi/4$, and a second solution is $5\pi/4$, since $\sin(5\pi/4) = -\sqrt{2}/2$. All other solutions are found by adding integer multiples of 2π to these two solutions. See Figure 9.4.

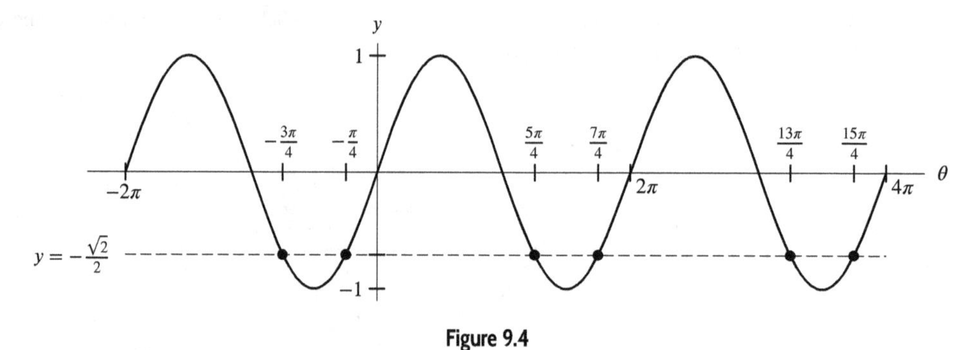

Figure 9.4

33. From Figure 9.5 we can see that the solutions lie on the intervals $\frac{\pi}{8} < t < \frac{\pi}{4}$, $\frac{3\pi}{4} < t < \frac{7\pi}{8}$, $\frac{9\pi}{8} < t < \frac{5\pi}{4}$ and $\frac{7\pi}{4} < t < \frac{15\pi}{8}$. Using the trace mode on a calculator, we can find approximate solutions $t = 0.52$, $t = 2.62$, $t = 3.67$ and $t = 5.76$.

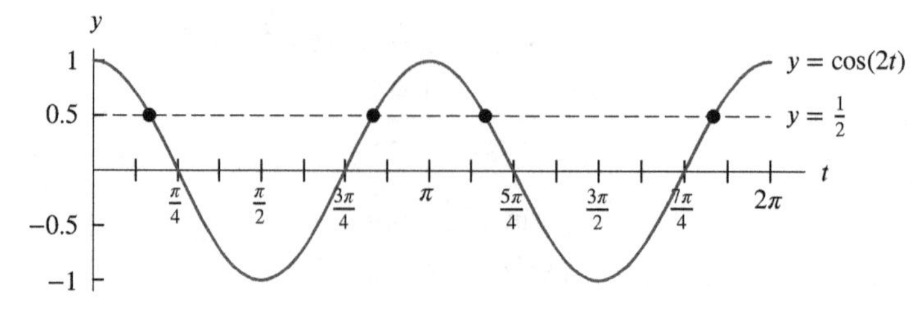

Figure 9.5

For a more precise answer we solve $\cos(2t) = \frac{1}{2}$ algebraically, giving $2t = \arccos(1/2)$. One solution is $2t = \pi/3$. But $2t = 5\pi/3, 7\pi/3$, and $11\pi/3$ are also angles that have a cosine of $1/2$. Thus $t = \pi/6, 5\pi/6, 7\pi/6$, and $11\pi/6$ are the solutions between 0 and 2π.

37. Graph $y = 12 - 4\cos(3t)$ on $0 \leq t \leq 2\pi/3$ and locate the two points with y-coordinate 14. (See Figure 9.6.) These points have t-coordinates of approximately $t = 0.698$ and $t = 1.396$. There are six solutions in three cycles of the graph between 0 and 2π.

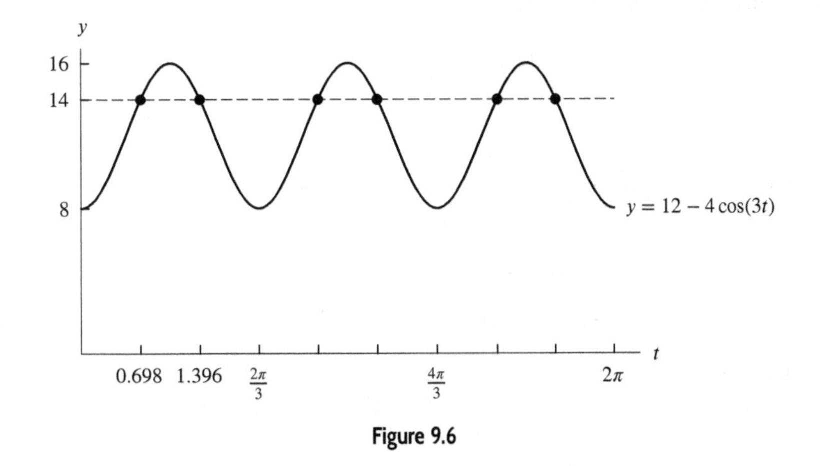

Figure 9.6

41. The curve is a sine curve with an amplitude of 5, a period of 8 and a vertical shift of −3. Thus the equation for the curve is $y = 5\sin\left(\frac{\pi}{4}x\right) - 3$. Solving for $y = 0$, we have

$$\begin{aligned} 5\sin\left(\frac{\pi}{4}x\right) &= 3 \\ \sin\left(\frac{\pi}{4}x\right) &= \frac{3}{5} \\ \frac{\pi}{4}x &= \sin^{-1}\left(\frac{3}{5}\right) \\ x &= \frac{4}{\pi}\sin^{-1}\left(\frac{3}{5}\right) \approx 0.819. \end{aligned}$$

This is the x-coordinate of P. The x-coordinate of Q is to the left of 4 by the same distance P is to the right of O, by the symmetry of the sine curve. Therefore,

$$x \approx 4 - 0.819 = 3.181$$

is the x-coordinate of Q.

Solutions for Section 9.2

Exercises

1. No, these expressions are not equal everywhere. They have different amplitudes (1 and 3) and different periods ($2\pi/3$ and 2π).

The value of the two functions are different at $t = \pi/2$, since $\sin(3\pi/2) = -1$ and $3\sin(\pi/2) = 3$.

5. We use the relationship $\sin^2\theta + \cos^2\theta = 1$ to find $\cos\theta$. Substitute $\sin\theta = 1/4$:

$$\left(\frac{1}{4}\right)^2 + \cos^2\theta = 1$$

$$\frac{1}{16} + \cos^2\theta = 1$$
$$\cos^2\theta = 1 - \frac{1}{16} = \frac{15}{16}$$
$$\cos\theta = \pm\sqrt{\frac{15}{16}} = \pm\frac{\sqrt{15}}{4}.$$

Because θ is in the first quadrant, $\cos\theta$ is positive, so $\cos\theta = \sqrt{15}/4$. To find $\tan\theta$, use the relationship

$$\tan\theta = \frac{\sin\theta}{\cos\theta} = \frac{1/4}{\sqrt{15}/4} = \frac{1}{\sqrt{15}}.$$

9. We have:

$$\tan t\cos t - \frac{\sin t}{\tan t} = \frac{\sin t}{\cos t}\cdot\cos t - \frac{\sin t}{\left(\frac{\sin t}{\cos t}\right)} \quad \text{because } \tan t = \frac{\sin t}{\cos t}$$
$$= \sin t - \sin t\cdot\frac{\cos t}{\sin t}$$
$$= \sin t - \cos t.$$

13. Writing $\sin 2\alpha = 2\sin\alpha\cos\alpha$, we have

$$\frac{\sin 2\alpha}{\cos\alpha} = \frac{2\sin\alpha\cos\alpha}{\cos\alpha} = 2\sin\alpha.$$

17. Combining terms and using $\cos^2\phi + \sin^2\phi = 1$, we have

$$\frac{\cos\phi - 1}{\sin\phi} + \frac{\sin\phi}{\cos\phi + 1} = \frac{(\cos\phi - 1)(\cos\phi + 1) + \sin^2\phi}{\sin\phi(\cos\phi + 1)} = \frac{\cos^2\phi - 1 + \sin^2\phi}{\sin\phi(\cos\phi + 1)} = \frac{0}{\sin\phi(\cos\phi + 1)} = 0$$

21. We have:

$$\frac{3\sin(\phi + 1)}{4\cos(\phi + 1)} = \frac{3}{4}\cdot\frac{\sin(\phi + 1)}{\cos(\phi + 1)}$$
$$= \frac{3}{4}\tan(\phi + 1).$$

25. We solve the Pythagorean identity for $\sin\theta$.

$$\sin^2\theta + \cos^2\theta = 1$$
$$\sin^2\theta = 1 - \cos^2\theta$$
$$(\sin\theta)^2 = 1 - \cos^2\theta.$$

If $\sin\theta \geq 0$,

$$\sin\theta = \sqrt{1 - \cos^2\theta}.$$

If $\sin\theta < 0$,

$$\sin\theta = -\sqrt{1 - \cos^2\theta}.$$

Problems

29. We have

$$\tan 2\theta = \frac{\sin 2\theta}{\cos 2\theta} = \frac{2\sin\theta\cos\theta}{\cos^2\theta - \sin^2\theta}.$$

Dividing both top and bottom by $\cos^2\theta$ gives

$$\tan 2\theta = \frac{\frac{2\sin\theta\cos\theta}{\cos^2\theta}}{\frac{\cos^2\theta - \sin^2\theta}{\cos^2\theta}} = \frac{2\tan\theta}{1 - \tan^2\theta}.$$

33. In order to get tan to appear, divide by $\cos x \cos y$:

$$\frac{\sin x \cos y + \cos x \sin y}{\cos x \cos y - \sin x \sin y} = \frac{\dfrac{\sin x \cos y}{\cos x \cos y} + \dfrac{\cos x \sin y}{\cos x \cos y}}{\dfrac{\cos x \cos y}{\cos x \cos y} - \dfrac{\sin x \sin y}{\cos x \cos y}} = \frac{\tan x + \tan y}{1 - \tan x \tan y}$$

37. Using the trigonometric identity $\tan(2\theta) = 2\tan\theta/(1 - \tan^2\theta)$, we have

$$\begin{aligned}
\tan(2\theta) + \tan\theta &= 0\\
\frac{2\tan\theta}{1-\tan^2\theta} &= -\tan\theta\\
2\tan\theta &= -\tan\theta + \tan^3\theta\\
\tan^3\theta - 3\tan\theta &= 0\\
\tan\theta(\tan^2\theta - 3) &= 0\\
\tan\theta = 0 \quad &\text{or} \quad \tan\theta = \pm\sqrt{3}.
\end{aligned}$$

If $\tan\theta = 0$, then we have three solutions: $\theta = 0$ and $\theta = \pi$ and $\theta = 2\pi$. On the other hand, if $\tan\theta = \sqrt{3}$, we first calculate the associated reference angle, which is $\tan^{-1}(\sqrt{3}) = \pi/3$. Using a graph of the tangent function on the interval $0 \le \theta \le 2\pi$, we see that the two solutions to $\tan\theta = \sqrt{3}$ are given by $\theta = \pi/3$ and $\theta = \pi + \pi/3 = 4\pi/3$. Finally, if $\tan\theta = -\sqrt{3}$, we again have a reference angle of $\pi/3$, and the two solutions to $\tan\theta = -\sqrt{3}$ are given by $\theta = \pi - \pi/3 = 2\pi/3$ and $\theta = 2\pi - \pi/3 = 5\pi/3$. Combining the above observations, we see that there are seven solutions to the original equation: $0, \pi, \pi/3, 4\pi/3, 2\pi/3, 5\pi/3$, and 2π.

41. Not an identity. False for $x = \pi/2$.

45. If we let $A = 1$, then we have

$$\frac{\sin(2A)}{\cos(2A)} = \frac{\sin 2}{\cos 2} = -2.185 \neq 3.115 = 2\tan 1 = 2\tan A.$$

Therefore, since the equation is not true for $A = 1$, it is not an identity.

49. Working on the left side, we have

$$\begin{aligned}
\tan t + \frac{1}{\tan t} &= \frac{\sin t}{\cos t} + \frac{1}{\sin t/\cos t}\\
&= \frac{\sin t}{\cos t} + \frac{\cos t}{\sin t}\\
&= \frac{\sin^2 t + \cos^2 t}{\cos t \sin t}\\
&= \frac{1}{\sin t \cos t}.
\end{aligned}$$

Therefore, the left side equals the right side and the equation is an identity.

53. Identity. $\dfrac{1-\tan^2 x}{1+\tan^2 x}\cdot\dfrac{\cos^2 x}{\cos^2 x} = \dfrac{\cos^2 x - \sin^2 x}{\cos^2 x + \sin^2 x} = \dfrac{\cos 2x}{1} = \cos 2x.$

57. We have $\cos\theta = x/3$, so $\sin\theta = \sqrt{1-(x/3)^2} = \frac{\sqrt{9-x^2}}{3}$. Therefore,

$$\sin 2\theta = 2\sin\theta\cos\theta = 2\left(\frac{\sqrt{9-x^2}}{3}\right)\left(\frac{x}{3}\right) = \frac{2x}{9}\sqrt{9-x^2}.$$

61. First use $\sin(2x) = 2\sin x\cos x$, where $x = 2\theta$. Then

$$\sin(4\theta) = \sin(2x) = 2\sin(2\theta)\cos(2\theta).$$

Since $\sin(2\theta) = 2\sin\theta\cos\theta$ and $\cos(2\theta) = 2\cos^2\theta - 1$, we have

$$\sin 4\theta = 2(2\sin\theta\cos\theta)(2\cos^2\theta - 1).$$

65. **(a)** Since $-\pi \leq t < 0$ we have $0 < -t \leq \pi$ so the double-angle formula for cosine can be used for the angle $\theta = -t$. Therefore $\cos 2\theta = 1 - 2\sin^2\theta$ tells us that

$$\cos(-2t) = 1 - 2\sin^2(-t).$$

(b) Since cosine is even we have $\cos(-2t) = \cos 2t$. Since sine is odd we have

$$-2\sin^2(-t) = 1 - 2(-\sin t)^2 = 1 - 2\sin^2 t.$$

Substitution of these results into the results of part (a) gives

$$\cos 2t = 1 - 2\sin^2 t.$$

Solutions for Section 9.3

Exercises

1. Applying the sum-of-angles formula for sine, we have

$$\begin{aligned}\sin(A+B) &= \sin A\cos B + \cos A\sin B\\ &= (0.84)(0.39) + (0.54)(0.92) = 0.8244.\end{aligned}$$

5. Applying the sum-of-angles formula for sine, we have

$$\begin{aligned}\sin(S+T) &= \sin S\cos T + \cos S\sin T\\ &= \left(\frac{7}{25}\right)\left(\frac{8}{17}\right) + \left(\frac{24}{25}\right)\left(\frac{15}{17}\right) = \frac{416}{425}.\end{aligned}$$

9. Write $\sin 15^\circ = \sin(45^\circ - 30^\circ)$, and then apply the appropriate trigonometric identity.

$$\begin{aligned}\sin 15^\circ &= \sin(45^\circ - 30^\circ)\\ &= \sin 45^\circ\cos 30^\circ - \sin 30^\circ\cos 45^\circ\\ &= \frac{\sqrt{6}}{4} - \frac{\sqrt{2}}{4}\end{aligned}$$

Similarly, $\sin 75^\circ = \sin(45^\circ + 30^\circ)$.

$$\begin{aligned}\sin 75^\circ &= \sin(45^\circ + 30^\circ)\\ &= \sin 45^\circ\cos 30^\circ + \sin 30^\circ\cos 45^\circ\\ &= \frac{\sqrt{6}}{4} + \frac{\sqrt{2}}{4}\end{aligned}$$

Also, note that $\cos 75^\circ = \sin(90^\circ - 75^\circ) = \sin 15^\circ$, and $\cos 15^\circ = \sin(90^\circ - 15^\circ) = \sin 75^\circ$.

13. **(a)** We have $\sin(15^\circ + 42^\circ) = \sin 15^\circ\cos 42^\circ + \sin 42^\circ\cos 15^\circ = 0.839$.

(b) See Figure 9.7.

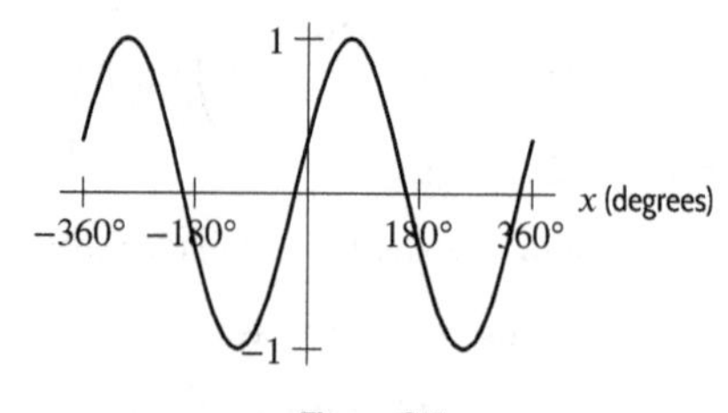

Figure 9.7

Problems

17. First, we note that the unlabeled side of the triangle has length $\sqrt{4-y^2}$.

(a) Using an angle difference formula, we have

$$\begin{aligned}\cos(\theta-\phi) &= \cos\theta\cos\phi+\sin\theta\sin\phi \\ &= \frac{\sqrt{4-y^2}}{2}\cdot\frac{y}{2}+\frac{y}{2}\cdot\frac{\sqrt{4-y^2}}{2} \\ &= \frac{y\sqrt{4-y^2}}{2}.\end{aligned}$$

(b) Using an angle difference formula, we have

$$\begin{aligned}\sin(\theta-\phi) &= \sin\theta\cos\phi-\cos\theta\sin\phi \\ &= \frac{y}{2}\cdot\frac{y}{2}-\frac{\sqrt{4-y^2}}{2}\cdot\frac{\sqrt{4-y^2}}{2} \\ &= \frac{y^2-(4-y^2)}{4} \\ &= \frac{y^2-2}{2}.\end{aligned}$$

(c) We have

$$\begin{aligned}\cos\theta-\cos\phi &= \frac{\sqrt{4-y^2}}{2}-\frac{y}{2} \\ &= \frac{\sqrt{4-y^2}-y}{2}.\end{aligned}$$

(d) Since $\theta+\phi=\pi/2$, we have

$$\sin(\theta+\phi)=\sin\left(\frac{\pi}{2}\right)=1.$$

21. We have

$$\begin{aligned}\cos 3t &= \cos(2t+t) \\ &= \cos 2t\cos t-\sin 2t\sin t \\ &= (2\cos^2 t-1)\cos t-(2\sin t\cos t)\sin t \\ &= \cos t((2\cos^2 t-1)-2\sin^2 t) \\ &= \cos t(2\cos^2 t-1-2(1-\cos^2 t)) \\ &= \cos t(2\cos^2 t-1-2+2\cos^2 t) \\ &= \cos t(4\cos^2 t-3) \\ &= 4\cos^3 t-3\cos t,\end{aligned}$$

as required.

25. We manipulate the equation for the average rate of change as follows:

$$\begin{aligned}\frac{\tan(x+h)-\tan x}{h} &= \frac{\dfrac{\tan x+\tan h}{1-\tan x\tan h}-\tan x}{h} \\ &= \frac{(\tan x+\tan h-\tan x+\tan^2 x\tan h)/(1-\tan x\tan h)}{h} \\ &= \frac{\tan h+\tan^2 x\tan h}{(1-\tan x\tan h)\cdot h} \\ &= \frac{\dfrac{\sin h}{\cos h}+\tan^2 x\cdot\dfrac{\sin h}{\cos h}}{\left(1-\tan x\cdot\dfrac{\sin h}{\cos h}\right)\cdot h}\end{aligned}$$

$$= \frac{(1+\tan^2 x)\dfrac{\sin h}{\cos h}}{\left(1-\tan x\dfrac{\sin h}{\cos h}\right)\cdot h}$$

$$= \frac{\left(\dfrac{1}{\cos^2 x}\right)\cdot\dfrac{\sin h}{\cos h}}{\left(1-\tan x\cdot\dfrac{\sin h}{\cos h}\right)\cdot h}$$

$$= \frac{\dfrac{1}{\cos^2 x}\cdot\sin h}{(\cos h-\tan x\sin h)\cdot h}$$

$$= \frac{\dfrac{1}{\cos^2 x}\cdot\sin h}{\cos h-\sin h\tan x}\cdot\left(\frac{1}{h}\right)$$

$$= \frac{1}{\cos^2 x}\,\frac{\sin h}{h}\cdot\frac{1}{\cos h-\sin h\tan x}.$$

29. **(a)** The coordinates of P_1 are $(\cos\theta,\sin\theta)$; for P_2 they are $(\cos(-\phi),\sin(-\phi))=(\cos\phi,-\sin\phi)$; for P_3 they are $(\cos(\theta+\phi),\sin(\theta+\phi))$; and for P_4 they are $(1,0)$.

(b) The triangles P_1OP_2 and P_3OP_4 are congruent by the side-angle-side property because $\angle P_1OP_2=\theta+\phi=\angle P_3OP_4$. Therefore their corresponding sides P_1P_2 and P_3P_4 are equal.

(c) We have

$$\begin{aligned}(P_1P_2)^2 &= (\cos\theta-\cos\phi)^2+(\sin\theta+\sin\phi)^2\\ &= \cos^2\theta-2\cos\theta\cos\phi+\cos^2\phi+\sin^2\theta+2\sin\theta\sin\phi+\sin^2\phi\\ &= \cos^2\theta+\sin^2\theta+\cos^2\phi+\sin^2\phi-2\cos\theta\cos\phi+2\sin\theta\sin\phi\\ &= 2-2(\cos\theta\cos\phi-\sin\theta\sin\phi)\end{aligned}$$

We also have

$$\begin{aligned}(P_3P_4)^2 &= (\cos(\theta+\phi)-1)^2+(\sin(\theta+\phi)-0)^2\\ &= \cos^2(\theta+\phi)-2\cos(\theta+\phi)+1+\sin^2(\theta+\phi)\\ &= 2-2\cos(\theta+\phi)\end{aligned}$$

The distances P_1P_2 and P_3P_4 are the square roots of these expressions (but we will use the squares of the distances).

(d) $(P_3P_4)^2=(P_1P_2)^2$ by part (b), so

$$\begin{aligned}2-2\cos(\theta+\phi) &= 2-2(\cos\theta\cos\phi-\sin\theta\sin\phi)\\ \cos(\theta+\phi) &= \cos\theta\cos\phi-\sin\theta\sin\phi.\end{aligned}$$

Solutions for Section 9.4

Exercises

1. We have $A=\sqrt{8^2+(-6)^2}=\sqrt{100}=10$. Since $\cos\phi=8/10=0.8$ and $\sin\phi=-6/10=-0.6$, we know that ϕ is in the fourth quadrant. Thus,

$$\tan\phi=-\frac{6}{8}=-0.75 \qquad \text{and} \qquad \phi=\tan^{-1}(-0.75)=-0.644,$$

so $8\sin t-6\cos t=10\sin(t-0.644)$.

5.

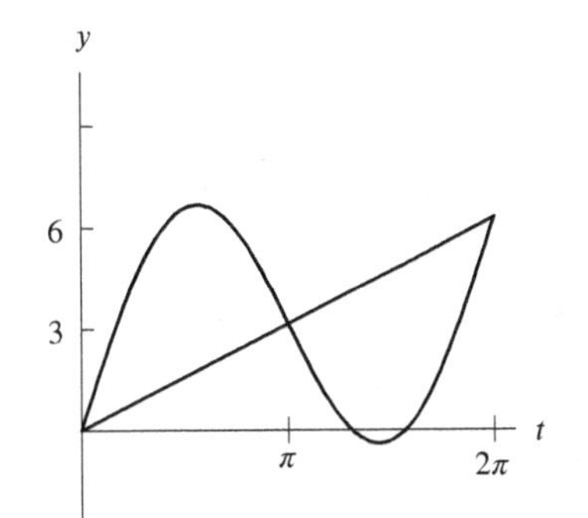

For the graphs to intersect, $t + 5\sin t = t$. So $\sin t = 0$, or $t =$ any integer multiple of π.

9. The function $y = (10 - t)\cos 2\pi t$ matches graph (II). We see that the oscillation is damped equally each period, so the amplitude (the function multiplying the cosine) must be decreasing and linear. Only $10 - t$ has that characteristic.

Problems

13. **(a)** This is linear growth: $P = 5000 + 300t$.

(b) This is exponential growth: $P = 3200(1.04)^t$.

(c) The population is oscillating, so we can use a trigonometric model. Since the population starts at its lowest value, we use a reflected cosine function of the form

$$P(t) = -A\cos(Bt) + k.$$

Low $= 1200$; High $= 3000$.
Midline $k = \frac{1}{2}(3000 + 1200) = 2100$.
Amplitude $A = 3000 - 2100 = 900$.
The period is 5, so $5 = 2\pi/B$, and we have $B = 2\pi/5$. Thus,

$$P(t) = -900\cos\left(\frac{2\pi}{5}t\right) + 2100.$$

17. **(a)** As $x \to \infty$, $\frac{1}{x} \to 0$ and we know $\sin 0 = 0$. Thus, $y = 0$ is the equation of the asymptote.

(b) As $x \to 0$ and $x > 0$, we have $\frac{1}{x} \to \infty$. This means that for small changes of x the change in $\frac{1}{x}$ is large. Since $\frac{1}{x}$ is a large number of radians, the function will oscillate more and more frequently as x becomes smaller.

(c) No, because the interval on which $f(x)$ completes a full cycle is not constant as x increases.

(d) $\sin\left(\frac{1}{x}\right) = 0$ means that $\frac{1}{x} = \sin^{-1}(0) + k\pi$ for k equal to some integer. Therefore, $x = \frac{1}{k\pi}$, and the greatest zero of $f(x) = \sin\frac{1}{x}$ corresponds to the smallest k, that is, $k = 1$. Thus, $z_1 = \frac{1}{\pi}$.

(e) There are an infinite number of zeros because $z = \frac{1}{k\pi}$ for all $k > 0$ are zeros.

(f) If $a = \frac{1}{k\pi}$ then the largest zero of $f(x)$ less then a would be $b = \frac{1}{(k+1)\pi}$.

21. **(a)** We have:

$$\begin{array}{c}\text{Average change in } C(t)\\ \text{over Dec'05–Jun'06}\end{array} = \frac{C(6) - C(0)}{6 - 0} = \frac{382 - 381}{6} = \frac{1}{6} \text{ ppm/month.}$$

Also,

$$\begin{array}{c}\text{Average change in } C(t)\\ \text{over Dec'10–Jun'11}\end{array} = \frac{C(66) - C(60)}{66 - 60} = \frac{392 - 391}{6} = \frac{1}{6} \text{ ppm/month.}$$

And lastly,

$$\begin{array}{c}\text{Average change in } C(t)\\ \text{over Mar–Sep'12}\end{array} = \frac{C(81) - C(75)}{81 - 75} = \frac{391 - 397}{6} = -1 \text{ ppm/month.}$$

So, on average, the concentration of carbon dioxide increased by 0.167 parts per million per month during the first half of 2006 and 2011. It also decreased, on average, by 1 ppm per month during the middle months of 2012.

(b) $S(t)$ is a periodic function which measures the variation in the concentration of carbon dioxide due to the seasons. It does not take into account the long-term changes in the concentration of carbon dioxide.

(c) We have:

$$\begin{array}{c}\text{Average change in } S(t)\\ \text{over Dec'05–Jun'06}\end{array} = \frac{S(6)-S(0)}{6-0} = \frac{0-0}{6} = 0 \text{ ppm/month.}$$

Also,

$$\begin{array}{c}\text{Average change in } S(t)\\ \text{over Dec'10–Jun'11}\end{array} = \frac{S(66)-S(60)}{66-60} = \frac{0-0}{6} = 0 \text{ ppm/month.}$$

And lastly,

$$\begin{array}{c}\text{Average change in } S(t)\\ \text{over Mar–Sep'12}\end{array} = \frac{S(81)-S(75)}{81-75} = \frac{-3.5-3.5}{6} = \frac{-7}{6} \text{ ppm/month.}$$

So, on average, there was no change in the concentration of carbon dioxide due to seasonal variations during the first half of 2006 and 2011. This makes sense, because the carbon dioxide concentration grows and decays by 3.5 ppm during these periods. (Note that 3.5 ppm is the amplitude of $S(t)$.)

Also, seasonal changes caused an average decrease in carbon dioxide of about -1.167 ppm per month during the middle months of 2012.

(d) We have:

$$\begin{array}{c}\text{Average change in } C(t)\\ \text{over Dec'05–Jun'06}\end{array} - \begin{array}{c}\text{Average change in } S(t)\\ \text{over Dec'05–Jun'06}\end{array} = \frac{1}{6} - 0 = \frac{1}{6} \text{ ppm/month.}$$

Also,

$$\begin{array}{c}\text{Average change in } C(t)\\ \text{over Dec'10–Jun'11}\end{array} - \begin{array}{c}\text{Average change in } S(t)\\ \text{over Dec'10–Jun'11}\end{array} = \frac{1}{6} - 0 = \frac{1}{6} \text{ ppm/month.}$$

And lastly,

$$\begin{array}{c}\text{Average change in } C(t)\\ \text{over Mar–Sep'12}\end{array} - \begin{array}{c}\text{Average change in } S(t)\\ \text{over Mar–Sep'12}\end{array} = -1 + \frac{7}{6} = \frac{1}{6} \text{ ppm/month.}$$

So, the difference between the average concentration and average the seasonal variation is $1/6$ ppm/month over each of these common periods.

(e) The common difference between the two averages calculated in part (d), $1/6$ ppm/month, gives the average increase in the concentration of carbon dioxide per month. This value is the slope of the rising midline,

$$M(t) = 381 + t/6,$$

the portion of the concentration model $y = C(t) = S(t) + M(t)$ that captures the variation in the concentration of carbon dioxide not due to the seasons.

The slope of the rising midline tells us that, independently of seasonal changes, the concentration of carbon dioxide has been increasing by about $1/6$ ppm per month, on average, during the last few years.

Solutions for Section 9.5

Exercises

1. Though we could use the sum-of-angle and difference-of-angle formulas for cosine on each of the two parts, we can also use the formula for the sum of cosines:

$$\cos 165^\circ - \cos 75^\circ = -2\sin\frac{165+75}{2}\sin\frac{165-75}{2} = -2\sin 120\sin 45 = -2\frac{\sqrt{3}}{2}\frac{\sqrt{2}}{2} = -\frac{\sqrt{6}}{2}.$$

5. These sine functions have the same amplitude and we use the relationship

$$\sin u + \sin v = 2\sin\frac{u+v}{2}\cos\frac{u-v}{2}.$$

Letting $u = 7t$ and $v = 3t$, we have

$$\sin(7t) + \sin(3t) = 2\sin\frac{7t+3t}{2}\cos\frac{7t-3t}{2} = 2\sin(5t)\cos(2t).$$

9. $\sin 35^\circ + \sin 40^\circ = 2\sin\left(\frac{35^\circ + 40^\circ}{2}\right)\cos\left(\frac{35^\circ - 40^\circ}{2}\right) = 1.216$. See Figure 9.8.

Figure 9.8

Problems

13. We start with

$$\sin u + \sin v = 2\sin\left(\frac{u+v}{2}\right)\cos\left(\frac{u-v}{2}\right).$$

Since $-\sin v = \sin(-v)$, we can write

$$\begin{aligned}\sin u - \sin v &= \sin u + \sin(-v)\\ &= 2\sin\left(\frac{u+(-v)}{2}\right)\cos\left(\frac{u-(-v)}{2}\right)\\ &= 2\sin\left(\frac{u-v}{2}\right)\cos\left(\frac{u+v}{2}\right)\\ &= 2\cos\left(\frac{u+v}{2}\right)\sin\left(\frac{u-v}{2}\right).\end{aligned}$$

Solutions for Section 9.6

Exercises

1. $5e^{i\pi}$

5. We have $(-3)^2 + (-4)^2 = 25$, and $\arctan(4/3) \approx 4.069$. So the number is $5e^{i4.069}$.

9. $-3 - 4i$

13. We have $\sqrt{e^{i\pi/3}} = e^{(i\pi/3)/2} = e^{i\pi/6}$, thus $\cos\frac{\pi}{6} + i\sin\frac{\pi}{6} = \frac{\sqrt{3}}{2} + \frac{i}{2}$.

Problems

17. One value of $\sqrt[3]{i}$ is $\sqrt[3]{e^{i\frac{\pi}{2}}} = (e^{i\frac{\pi}{2}})^{\frac{1}{3}} = e^{i\frac{\pi}{6}} = \cos\frac{\pi}{6} + i\sin\frac{\pi}{6} = \frac{\sqrt{3}}{2} + \frac{i}{2}$

21. One value of $(\sqrt{3}+i)^{1/2}$ is
$(2e^{i\frac{\pi}{6}})^{1/2} = \sqrt{2}e^{i\frac{\pi}{12}} = \sqrt{2}\cos\frac{\pi}{12} + i\sqrt{2}\sin\frac{\pi}{12} \approx 1.366 + 0.366i$

25. Substituting $A_2 = i - A_1$ into the second equation gives

$$iA_1 - (i - A_1) = 3,$$

so

$$\begin{aligned}iA_1 + A_1 &= 3 + i\\ A_1 &= \frac{3+i}{1+i} = \frac{3+i}{1+i}\cdot\frac{1-i}{1-i} = \frac{3-3i+i-i^2}{2}\\ &= 2 - i\end{aligned}$$

Therefore $A_2 = i - (2 - i) = -2 + 2i$.

29. Using Euler's formula, we have:

$$e^{i(2\theta)} = \cos 2\theta + i\sin 2\theta$$

On the other hand,

$$e^{i(2\theta)} = \left(e^{i\theta}\right)^2 = (\cos\theta + i\sin\theta)^2 = (\cos^2\theta - \sin^2\theta) + i(2\cos\theta\sin\theta)$$

Equating real parts, we find

$$\cos 2\theta = \cos^2\theta - \sin^2\theta.$$

33. One polar form for $z = -8$ is $z = 8e^{i\pi}$ with $(r,\theta) = (8,\pi)$. Two more sets of polar coordinates for z are $(8, 3\pi)$, and $(8, 5\pi)$. Three cube roots of z are given by

$$\begin{aligned}
\left(8e^{\pi i}\right)^{1/3} &= 8^{1/3}e^{1/3\cdot\pi i} = 2e^{\pi i/3} = 2\cos(\pi/3) + i2\sin(\pi/3)\\
&= 1 + 1.732i.\\
\left(8e^{3\pi i}\right)^{1/3} &= 8^{1/3}e^{1/3\cdot 3\pi i} = 2e^{\pi i} = 2\cos\pi + i2\sin\pi\\
&= -2.\\
\left(8e^{5\pi i}\right)^{1/3} &= 8^{1/3}e^{1/3\cdot 5\pi i} = 2e^{5\pi i/3} = 2\cos(5\pi/3) + i2\sin(5\pi/3)\\
&= 1 - 1.732i.
\end{aligned}$$

37. By de Moivre's formula we have

$$(\cos 2\pi/3 + i\sin 2\pi/3)^3 = \cos(3\cdot 2\pi/3) + i\sin(3\cdot 2\pi/3) = 1 + i0 = 1.$$

41. Using the exponent rules, we see from Euler's formula that

$$\begin{aligned}
e^{i(\theta+\phi)} &= e^{i\theta}\cdot e^{i\phi}\\
&= (\cos\theta + i\sin\theta)(\cos\phi + i\sin\phi)\\
&= \cos\theta\cos\phi + \underbrace{i\cos\theta\sin\phi + i\sin\theta\cos\phi}_{i(\cos\theta\sin\phi+\sin\theta\cos\phi)} + \underbrace{i^2\sin\theta\sin\phi}_{-\sin\theta\sin\phi}\\
&= \underbrace{\cos\theta\cos\phi - \sin\theta\sin\phi}_{\text{Real part}} + i\underbrace{(\sin\theta\cos\phi + \cos\theta\sin\phi)}_{\text{Imaginary part}}.
\end{aligned}$$

But Euler's formula also gives

$$e^{i(\theta+\phi)} = \underbrace{\cos(\theta+\phi)}_{\text{Real part}} + i\underbrace{\sin(\theta+\phi)}_{\text{Imaginary part}}.$$

Two complex numbers are equal only if their real and imaginary parts are equal. Setting real parts equal gives

$$\cos(\theta+\phi) = \cos\theta\cos\phi - \sin\theta\sin\phi.$$

Setting imaginary parts equal gives

$$\sin(\theta+\phi) = \sin\theta\cos\phi + \sin\phi\cos\theta.$$

Solutions for Chapter 9 Review

Exercises

1. We have:

$$\begin{aligned}
(1-\sin t)(1-\cos t) - \cos t\sin t &= \underbrace{1 - \cos t - \sin t + \sin t\cos t}_{(1-\sin t)(1-\cos t)} - \cos t\sin t \quad \text{multiply out}\\
&= 1 - \cos t - \sin t.
\end{aligned}$$

5. We have:

$$\begin{aligned}\frac{\sec t}{\csc t} &= \frac{\left(\frac{1}{\cos t}\right)}{\left(\frac{1}{\sin t}\right)} \\ &= \frac{1}{\cos t}\cdot \sin t \\ &= \tan t.\end{aligned}$$

9. Using $1-\cos^2\theta = \sin^2\theta$, we have

$$\frac{1-\cos^2\theta}{\sin\theta} = \frac{\sin^2\theta}{\sin\theta} = \sin\theta.$$

13. $8-5i$

Problems

17. We first solve for $\cos\alpha$,

$$\begin{aligned}2\cos\alpha &= 1 \\ \cos\alpha &= \frac{1}{2} \\ \alpha &= \frac{\pi}{3}, \frac{5\pi}{3}.\end{aligned}$$

21. **(a)** The graph resembles a cosine function with midline $k = 8$, amplitude $A = 10$, and period $p = 60$, so

$$y = 10\cos\left(\frac{2\pi}{60}x\right) + 8.$$

(b) We find the zeros at x_1 and x_2 by setting $y = 0$. Solving gives

$$\begin{aligned}10\cos\left(\frac{2\pi}{60}x\right) + 8 &= 0 \\ \cos\left(\frac{2\pi}{60}x\right) &= -\frac{8}{10} \\ \frac{2\pi}{60}x &= \cos^{-1}(-0.8) \\ x &= \frac{60}{2\pi}\cos^{-1}(-0.8) \\ &= 23.8550.\end{aligned}$$

Judging from the graph, this is the value of x_1. By symmetry, $x_2 = 60 - x_1 = 36.1445$.

25. Graphs of the four functions are in Figures 9.9 –9.12. The graphs in Figures 9.10 and 9.11 suggest that $(\tan^2 x)(\sin^2 x)$ and $\tan^2 x - \sin^2 x$ may be identical.

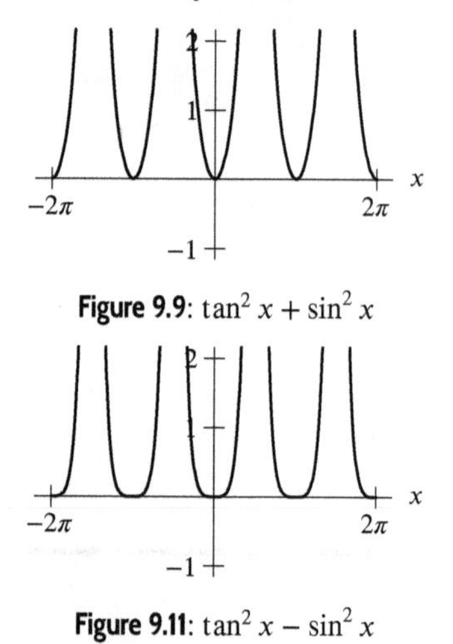

Figure 9.9: $\tan^2 x + \sin^2 x$

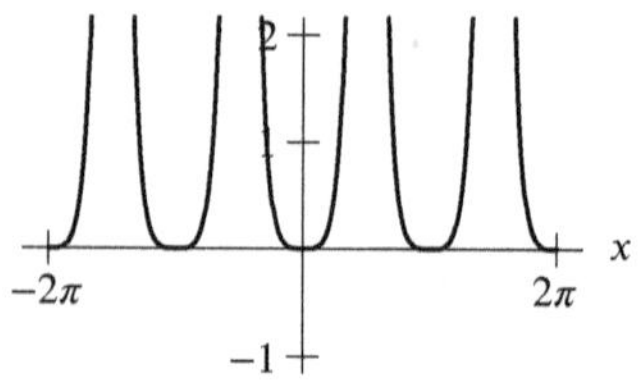

Figure 9.10: $(\tan^2 x)(\sin^2 x)$

Figure 9.11: $\tan^2 x - \sin^2 x$

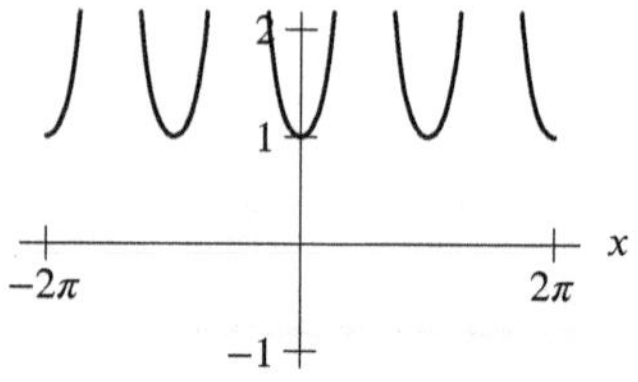

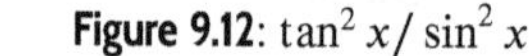

Figure 9.12: $\tan^2 x / \sin^2 x$

To prove the identity, use $\tan x = \sin x / \cos x$ to rewrite each side in terms of sine and cosine. We have

$$(\tan^2 x)(\sin^2 x) = \left(\frac{\sin x}{\cos x}\right)^2 (\sin^2 x) = \frac{\sin^4 x}{\cos^2 x}.$$

In addition,

$$\begin{aligned}
\tan^2 x - \sin^2 x &= \left(\frac{\sin x}{\cos x}\right)^2 - \sin^2 x \\
&= \frac{\sin^2 x}{\cos^2 x} - \sin^2 x \\
&= \sin^2 x\left(\frac{1}{\cos^2 x} - 1\right) \\
&= \frac{\sin^2 x(1-\cos^2 x)}{\cos^2 x} \\
&= \frac{\sin^2 x(\sin^2 x)}{\cos^2 x} \\
&= \frac{\sin^4 x}{\cos^2 x}.
\end{aligned}$$

Since both expressions equal $\frac{\sin^4 x}{\cos^2 x}$, they are identical.

29. By the Pythagorean identity, we know that $\cos^2\theta + \sin^2\theta = 1$, so

$$\begin{aligned}
0.27^2 + \sin^2\theta &= 1 \\
\sin^2\theta &= 1 - 0.27^2 \\
\sin\theta &= \pm\sqrt{0.927} \approx \pm 0.963.
\end{aligned}$$

However, we only need one value, so we take $\sin\theta = 0.963$, Since $\tan\theta = \sin\theta/\cos\theta$, we have $\tan\theta = \sqrt{0.927}/0.27 \approx 3.566$.

33.

$$\begin{aligned}
\cos(2\alpha) &= -\sin\alpha \\
1 - 2\sin^2\alpha + \sin\alpha &= 0 \\
-2\sin^2\alpha + \sin\alpha + 1 &= 0 \\
2\sin^2\alpha - \sin\alpha - 1 &= 0 \\
(2\sin\alpha + 1)(\sin\alpha - 1) &= 0
\end{aligned}$$

$$\begin{aligned} 2\sin\alpha + 1 &= 0 \\ 2\sin\alpha &= -1 \\ \sin\alpha &= -\frac{1}{2} \\ \alpha &= \frac{7\pi}{6}, \frac{11\pi}{6} \end{aligned} \qquad \begin{aligned} \sin\alpha - 1 &= 0 \\ \sin\alpha &= 1 \\ \alpha &= \frac{\pi}{2} \end{aligned}$$

37. Start with the expression on the left and factor it as the difference of two squares, and then apply the Pythagorean identity to one factor.

$$\begin{aligned} \sin^4 x - \cos^4 x &= (\sin^2 x)^2 - (\cos^2 x)^2 \\ &= (\sin^2 x - \cos^2 x)(\sin^2 x + \cos^2 x) \\ &= (\sin^2 x - \cos^2 x)(1) \\ &= (\sin^2 x - \cos^2 x). \end{aligned}$$

41. We first solve for $\sin\alpha$,

$$\begin{aligned} 3\sin^2\alpha + 3\sin\alpha + 4 &= 3 - 2\sin\alpha \\ 3\sin^2\alpha + 5\sin\alpha + 1 &= 0 \\ \sin\alpha &= \frac{-5 \pm \sqrt{25-12}}{6} \\ \sin\alpha &= \frac{-5 \pm \sqrt{13}}{6} \end{aligned}$$

$$\begin{aligned} \sin\alpha &= \tfrac{-5+\sqrt{13}}{6} \\ \alpha &= 3.376,\ 6.049 \end{aligned} \qquad \begin{aligned} \sin\alpha &= \tfrac{-5-\sqrt{13}}{6} \\ &\text{No solution } (-1 \le \sin\alpha \le 1) \end{aligned}$$

45. **(a)** The graph of $g(\theta) = \sin\theta - \cos\theta$ is shown in Figure 9.13.

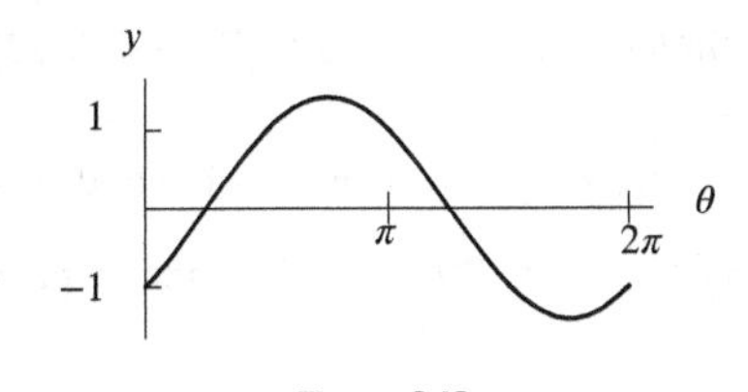

Figure 9.13

(b) We know $a_1 \sin t + a_2 \cos t = A\sin(t+\phi)$, where $A = \sqrt{a_1^2 + a_2^2}$ and $\tan\phi = a_2/a_1$. We let $a_1 = 1$ and $a_2 = -1$. This gives $A = \sqrt{2}$ and $\phi = \tan^{-1}(-1) = -\pi/4$, so

$$g(\theta) = \sqrt{2}\sin\left(\theta - \frac{\pi}{4}\right).$$

Note that we have chosen $B = 1$ and $k = 0$. We can check that this is correct by plotting the original function and $\sqrt{2}\sin(\theta - \pi/4)$ together.

We know that $\sin t = \cos(t - \pi/2)$, that is, the sine graph may be obtained by shifting the cosine graph $\pi/2$ units to the right. Thus,

$$g(\theta) = \sqrt{2}\cos\left(\theta - \frac{\pi}{4} - \frac{\pi}{2}\right) = \sqrt{2}\cos\left(\theta - \frac{3\pi}{4}\right).$$

49. **(a)** Since $z = 3 + 2i$, we have $iz = i(3 + 2i) = 3i + 2i^2 = -2 + 3i$. See Figure 9.14.

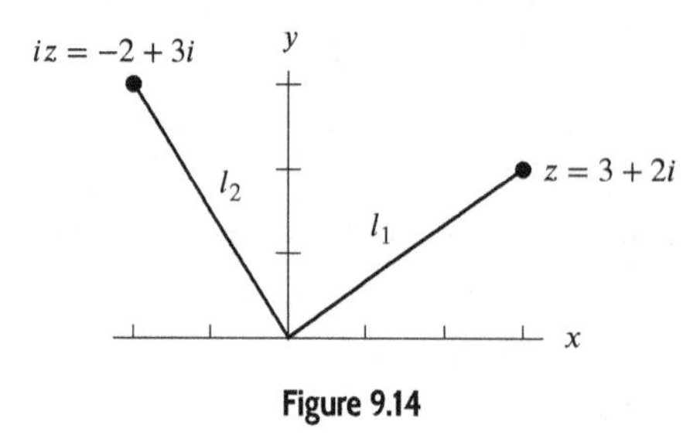

Figure 9.14

(b) The slope of the line l_1 in Figure 9.14 is 2/3, and the slope of the line l_2 is $-3/2$. Since the product $(2/3)(-3/2) = -1$ of these two slopes is -1, the lines are perpendicular.

STRENGTHEN YOUR UNDERSTANDING

1. True, because the factor e^{-t} decreases the oscillations of $\cos t$ as t grows.

5. True. This is an identity. Substitute using $\tan^2\theta = \sin^2\theta/\cos^2\theta$ and simplify to obtain $2 = 2\sin^2\theta + 2\cos^2\theta$. Divide by 2 to reach the Pythagorean identity.

9. True. There are many ways to prove this identity. We use the identity $\cos 2\theta = \cos^2\theta - \sin^2\theta$ to substitute in the right side of the equation. This becomes $\frac{1}{2}(1 - (\cos^2\theta - \sin^2\theta))$. Now substitute using $1 - \cos^2\theta = \sin^2\theta$ (a form of the Pythagorean identity.) The right side then simplifies to $\sin^2\theta$, which is the left side.

13. True. Start with the sine sum-of-angle identity:

$$\sin(\theta + \phi) = \sin\theta\cos\phi + \sin\phi\cos\theta$$

and let $\phi = \pi/2$, so

$$\sin(\theta + \pi/2) = \sin\theta\cos(\pi/2) + \sin(\pi/2)\cos\theta.$$

Simplify to

$$\sin(\theta + \pi/2) = \sin\theta \cdot 0 + 1 \cdot \cos\theta = \cos\theta.$$

17. True. Since $A\cos(Bt) = A\sin(Bt + \pi/2) = A\sin(B(t + \pi/(2B)))$, the graph of $A\cos(Bt)$ is a shift of $A\sin(Bt)$ to the left by $\pi/(2B)$.

21. True. We use the assumption that a_1 and a_2 are nonzero. The amplitude of the single sine function is $A = \sqrt{a_1^2 + a_2^2}$. Thus, A is greater than either a_1 or a_2.

25. True. Hertz is a measure of cycles per second and so a single cycle will take 1/60th of a second.

29. True. Since $0 \le \cos^{-1}(x) \le \pi$, we have $0 \le \sin(\cos^{-1}x) \le 1$. Thus, $\cos^{-1}(\sin(\cos^{-1}x))$ is an angle θ whose cosine is between 0 and 1. In addition, we have $0 \le \theta \le \pi$, as this is part of the definition of $\cos^{-1}x$. Hence $0 \le \theta \le \pi/2$.

33. False, since $(1 + i)^2 = 2i$ is not real.

37. True. This is Euler's formula, fundamental in higher mathematics.

CHAPTER TEN

Solutions for Section 10.1

Exercises

1. **(a)** We have $g(-1) = -1$, so $f(g(-1)) = f(-1) = 2$.
(b) We have $f(2) = 1$, so $g(f(2)) = g(1) = 1$.
(c) We have $g(-2) = -2$, so $f(g(-2)) = f(-2) = -1$.
(d) We have $f(0) = 0$, so $f(g(0)) = g(0) = 2$
(e) We have $f(1) = -2$, so $f(f(1)) = f(-2) = -1$
(f) Since $g(-1) = -1$, we have $g(g(g(-1))) = g(g(-1)) = g(-1) = -1$.

5. We have

$$g(f(x)) = \frac{2^x}{2^x + 1}.$$

9. Since $g(x) = 9x - 2$, we substitute $9x - 2$ for x in $r(x)$, giving us $r(g(x)) = \sqrt{3(9x-2)}$, which simplifies to $r(g(x)) = \sqrt{27x - 6}$.

13. Since $f(x) = 3x^2$, we substitute $3x^2$ for x in $m(x)$, giving us $m(f(x)) = 4(3x^2)$, which simplifies to $m(f(x)) = 12x^2$, which we then substitute for x in $g(x)$, giving $g(m(f(x))) = 9(12x^2) - 2$, which simplifies to $g(m(f(x))) = 108x^2 - 2$.

17. The inside function is $f(x) = \sin x$.

21. The inside function is $f(x) = 5 + 1/x$.

Problems

25. The function $t(f(H))$ gives the time of the trip as a function of temperature, H.

29. It is easiest to find values of h, because we can use the fact that $h(x) = g(f(x))$:

$$\begin{aligned} h(0) &= g(f(0)) \\ &= g(2) \qquad \text{because } f(0) = 2 \\ &= 3. \end{aligned}$$

Next, we will find values of f. To find $f(1)$, we know the output of $g(f(1))$ must be the same as $h(1)$, or 0. Since 0 is the output of g, we see from the table that its input must be 1. This means the value of $f(1)$ must be 1:

$$\begin{aligned} g(f(1)) = h(1) = 0 &\quad \text{because } h(1) = 0 \\ g(\underbrace{f(1)}_{1}) = 0 &\quad \text{because } g(1) = 0 \\ \text{so} \quad f(1) = 1. & \end{aligned}$$

Likewise, for $f(2)$, we see that

$$\begin{aligned} g(f(2)) = h(2) = 2 &\quad \text{because } h(2) = 2 \\ g(\underbrace{f(2)}_{4}) = 2 &\quad \text{because } g(4) = 2 \\ \text{so} \quad f(2) = 4. & \end{aligned}$$

Finally, we will find the values of g. To find $g(0)$, we know that the input of g is 0, so the output of f must be zero. This means that $x = 3$, because $f(3) = 0$. Thus, $g(0)$ is the same as $g(f(3))$, which equals $h(3)$ or 1:

$$\begin{aligned} g(0) &= g(\underbrace{f(3)}_{0}) && \text{because } f(3) = 0 \\ &= h(3) && \text{because } h(3) = g(f(3)) \\ &= 1. \end{aligned}$$

Likewise, to find $g(3)$, we have

$$\begin{aligned} g(3) &= g(\underbrace{f(4)}_{3}) && \text{because } f(4) = 3 \\ &= h(4) && \text{because } g(f(4)) = h(4) \\ &= 4. \end{aligned}$$

See Table 10.1.

Table 10.1

x	$f(x)$	$g(x)$	$h(x)$
0	2	1	3
1	1	0	0
2	4	3	2
3	0	4	1
4	3	2	4

33. First we find

$$f(x+h) - f(x) = (x+h)^2 + x + h - (x^2 + x) = (x^2 + 2xh + h^2 + x + h) - x^2 - x = 2xh + h^2 + h.$$

Then

$$\frac{f(x+h) - f(x)}{h} = \frac{2xh + h^2 + h}{h} = \frac{(2x + h + 1)h}{h} = 2x + h + 1.$$

37. Since $k(f(x)) = e^{f(x)} = e^{2x}$, we can let $f(x) = 2x$.

41. $g(x) = x^2$ and $h(x) = x + 3$

45. **(a)** Since $v(x) = x^2$ and y can be written $\dfrac{1 + (x^2)}{2 + (x^2)}$, we take

$$u(x) = \frac{1 + x}{2 + x}.$$

(b) Since $v(x) = x^2 + 1$ and y can be written $\dfrac{1 + x^2}{1 + 1 + x^2}$, we take

$$u(x) = \frac{x}{1 + x}.$$

49. **(a)** Since $u(x) = x^2$ and y can be written as $y = (\sin x)^2$, we take $v(x) = \sin x$.
(b) Since $v(x) = x^2$ and y can be written $y = \sin^2(\sqrt{x})^2$, we take $u(x) = \sin^2(\sqrt{x})$.

53. **(a)** From the graph of $f(x)$, we see that $f(g(x)) = 0$ when $g(x) = 0$ and when $g(x) = 4$. Since the solution to $g(x) = 0$ is $x = 4$, and the solution to $g(x) = 4$ is $x = 0$, we see that $x = 0$ and $x = 4$ are the only solutions to the equation $f(g(x)) = 0$.
(b) From the graph of $g(x)$, we see that $g(f(x)) = 0$ only when $f(x) = 4$, which occurs only when $x = 2$. Thus, $x = 2$ is the only solution to $g(f(x)) = 0$.

57. Reading values of the graph, we make an approximate table of values; we use these values to sketch Figure 10.1.

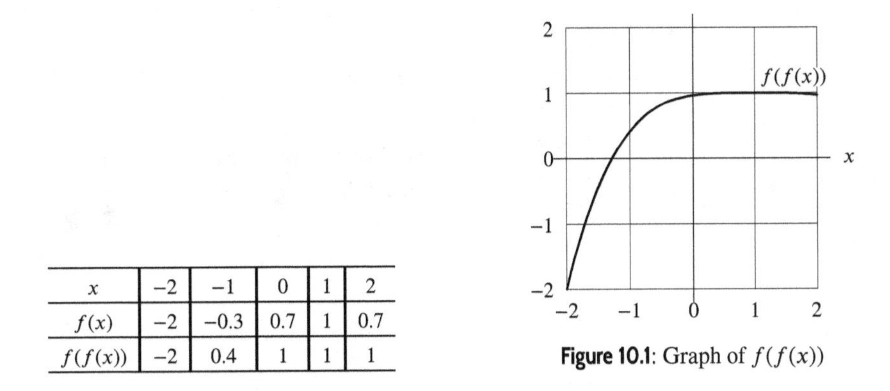

x	-2	-1	0	1	2
$f(x)$	-2	-0.3	0.7	1	0.7
$f(f(x))$	-2	0.4	1	1	1

Figure 10.1: Graph of $f(f(x))$

61. **(a)** For $f(g(x)) = [g(x)]^2$, the graph always lies on or above the x-axis. It touches the x-axis where $g(x) = 0$, which is the x-intercept of the graph of $g(x)$ and is increasing to the right and decreasing from the left. The graph is shown in Figure 10.2.

(b) For $g(f(x)) = g(x^2)$, we only need to consider the graph of $g(x)$ for positive input values, that is to the right of the y-axis. We see then that the graph of $g(f(x))$ is increasing to the right of the y-intercept and decreasing from the left. The graph is shown in Figure 10.3.

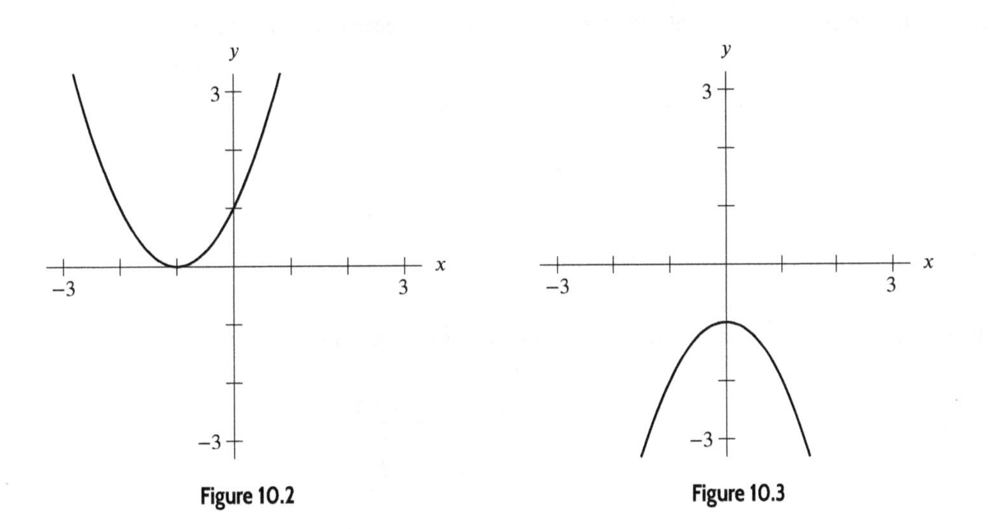

Figure 10.2 **Figure 10.3**

65. If $s(x) = 5 + \dfrac{1}{x+5} + x = x + 5 + \dfrac{1}{x+5}$ and $k(x) = x + 5$, then

$$s(x) = k(x) + \frac{1}{k(x)}.$$

However, $s(x) = v(k(x))$, so

$$v(k(x)) = k(x) + \frac{1}{k(x)}.$$

This is possible if

$$v(x) = x + \frac{1}{x}.$$

69. We have

$$\begin{aligned} q(r(x)) &= \frac{8^{x^3}}{16^{x^2}} \\ &= 8^{x^3} \cdot 16^{-x^2} && \text{exponent rule} \\ &= (2^3)^{x^3} \cdot (2^4)^{-x^2} \\ &= 2^{3x^3} \cdot 2^{-4x^2} && \text{exponent rule} \\ &= 2^{3x^3-4x^2} && \text{exponent rule} \\ &= 2^{r(x)}, \end{aligned}$$

so $q(x) = 2^x$.

73. We have

$$\begin{aligned} f(b) &= 1000\left(1 - 2^{-10/5}\right) && \text{because } b = 10 \\ &= 1000\left(1 - 2^{-2}\right) \\ &= 1000\left(1 - \frac{1}{4}\right) \\ &= 1000(3/4) \\ &= 750. \end{aligned}$$

This tells us that after 10 weeks, 750 out of every 1000 cars have been repaired. Likewise,

$$\begin{aligned} f(a) &= 1000\left(1 - 2^{-5/5}\right) && \text{because } a = 5 \\ &= 1000\left(1 - 2^{-1}\right) \\ &= 1000\left(1 - \frac{1}{2}\right) \\ &= 1000(1/2) \\ &= 500, \end{aligned}$$

telling us that after 5 weeks, 500 out of every 1000 cars have been repaired. Thus,

$$\begin{aligned} \frac{f(b) - f(a)}{b - a} &= \frac{750 - 500}{10 - 5} \\ &= \frac{250 \text{ cars}}{5 \text{ weeks}} \\ &= 50 \text{ cars/week}, \end{aligned}$$

telling us that an average of 50 cars per week are repaired between weeks 5 and 10.

77. **(a)** The function $y = f(x)$ has a y-intercept of 0, and a slope of $\frac{-10-0}{5-0} = -2$. So $f(x) = -2x$.

(b) The function is defined for the domain $0 \leq x \leq 5$, and it takes values in the range $-10 \leq y \leq 0$.

(c) The function $y = g(x)$ has a y-intercept of 1, and slope of $\frac{4-1}{1-0} = 3$. So $g(x) = 3x + 1$.

(d) The function $g(x)$ is defined on the domain $0 \leq x \leq 1$, and takes values in the range $1 \leq y \leq 4$.

(e) Since $f(x) = -2x$, and $g(x) = 3x + 1$,we know that $h(x) = f(g(x)) = -2(g(x)) = -2(3x + 1) = -6x - 2$.

(f) Since $g(x)$ is only defined for the domain $0 \leq x \leq 1$ and the range of $g(x)$ is contained in the domain of $f(x)$, $h(x)$ has the same domain as $g(x)$. Since $h(x)$ is a linear function, we can find its range by evaluating $h(x)$ for the extreme values of its domain, e.g. at $x = 0$ and at $x = 1$. We find that $h(0) = -6(0) - 2 = -2$ and $h(1) = -6(1) - 2 = -8$, so the range of $h(x)$ is $-8 \leq y \leq -2$.

(g) Since $h(x) = -6x - 2$, we know that its y-intercept is -2 and its slope is -6. Given a domain of $0 \leq x \leq 1$, $h(x)$ goes from $(0, -2)$ to $(1, -8)$.

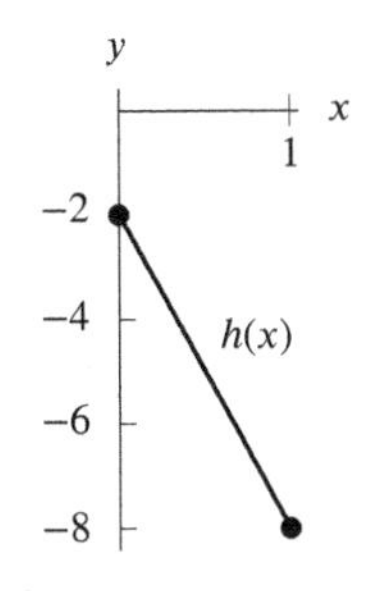

81. **(a)** We have $g(x) = g(-x)$, so $f(g(-x)) = f(g(x))$, making $f(g(x))$ even.
(b) We have $g(-x) = -g(x)$, so $f(g(-x)) = f(-g(x)) = f(g(x))$, making $f(g(x))$ even.
(c) We have $g(-x) = x$, so $f(g(-x)) = f(g(x))$, making $f(g(x))$ even.
(d) We have $g(-x) = -g(x)$, so

$$f(g(-x)) = f(-g(x)) = -f(g(x)),$$

which means $f(g(x))$ is odd.
(e) We have $g(-x) = g(x)$, so $f(g(-x)) = f(g(x))$, so $f(g(x))$ is even.
(f) Suppose that $g(x) = 5x$, which is odd, and $f(x) = 2x+1$, which is neither odd nor even. Then $f(g(x)) = 2(5x)+1 = 10x + 1$, which is neither even nor odd.
(g) Suppose that $f(x) = x^2$, which is even, and $g(x) = x + 1$, which is neither odd nor even. Then $f(g(x)) = (x + 1)^2$, which is neither odd nor even; for example $f(g(1)) = 4$ and $f(g(-1)) = 0$.

Solutions for Section 10.2

Exercises

1. A solution is $x = \tan^{-1}(-5) \approx -1.373$. Since $y = \tan x$ is a periodic function, there are multiple x-values for each of its y-values, so there are multiple solutions.

5. Let $y = x + 5$. Then $x = y - 5$, so $f^{-1}(x) = x - 5$.

9. Start with $x = f(f^{-1}(x))$ and substitute $y = f^{-1}(x)$. We have

$$\begin{aligned} x &= f(y) \\ x &= 3y - 7 \\ x + 7 &= 3y \\ \frac{x+7}{3} &= y \end{aligned}$$

Therefore,

$$f^{-1}(x) = \frac{x+7}{3}.$$

13. Since $f(x)$ passes the horizontal line test it is invertible.

17. It is not invertible.

21. It is not invertible.

25. The graph of a function $f(x)$ and its inverse are symmetric across the line $y = x$, so we first draw the line $y = x$ and then sketch the mirror image of $f(x)$ across this line to obtain the graph of $f^{-1}(x)$. See Figure 10.4. We sketch the graph of $f(x)$ as a dotted line to help distinguish between graphs.

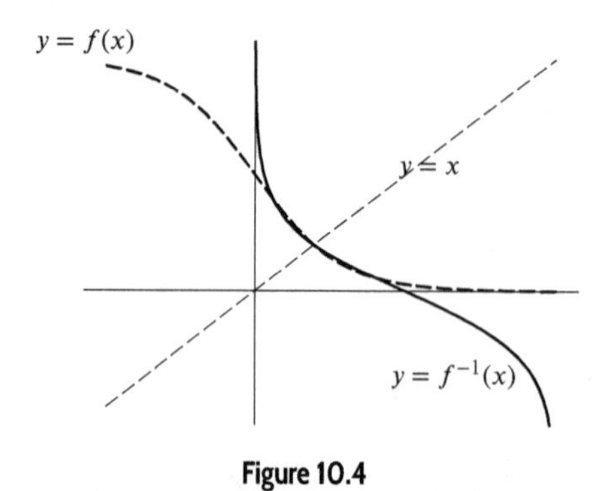

Figure 10.4

29. One way to check that these functions are inverses is to make sure they satisfy the identities $g(g^{-1}(x)) = x$ and $g^{-1}(g(x)) = x$.

$$\begin{aligned} g(g^{-1}(x)) &= 1 - \frac{1}{\left(1 + \frac{1}{1-x}\right) - 1} \\ &= 1 - \frac{1}{\left(\frac{1}{1-x}\right)} \\ &= 1 - (1 - x) \\ &= x. \end{aligned}$$

Also,

$$\begin{aligned} g^{-1}(g(x)) &= 1 + \frac{1}{1 - \left(1 - \frac{1}{x-1}\right)} \\ &= 1 + \frac{1}{\frac{1}{x-1}} \\ &= 1 + x - 1 = x. \end{aligned}$$

So the expression for g^{-1} is correct.

33. Check using the two compositions

$$f(f^{-1}(x)) = e^{f^{-1}(x)/2} = e^{(2\ln x)/2} = e^{\ln x} = x$$

and

$$f^{-1}(f(x)) = 2\ln f(x) = 2\ln e^{x/2} = 2(x/2) = x.$$

They are inverses of one another.

37. Let $y = 1/(1 + \frac{1}{x})$. Then

$$y = \frac{1}{\left(\frac{1+x}{x}\right)} = \frac{x}{x+1}.$$

This means that

$$\begin{aligned} y(x+1) &= x \\ yx + y &= x \\ x - yx &= y \\ x(1-y) &= y \\ x &= \frac{y}{1-y}. \end{aligned}$$

Thus, $o^{-1}(x) = x/(1 - x)$.

41. Start with $x = h(h^{-1}(x))$ and substitute $y = h^{-1}(x)$. We have

$$\begin{aligned}
x &= h(y)\\
x &= \log\frac{y+5}{y-4}\\
10^x &= \frac{y+5}{y-4}\\
10^x(y-4) &= y+5\\
10^x y - 4\cdot 10^x &= y+5\\
10^x y - y &= 5+4\cdot 10^x\\
y(10^x-1) &= 5+4\cdot 10^x\\
y &= \frac{5+4\cdot 10^x}{10^x-1}\\
h^{-1}(x) &= \frac{5+4\cdot 10^x}{10^x-1}.
\end{aligned}$$

Problems

45. The inverse function $f^{-1}(P)$ gives the time, t, in years at which the population is P thousand. Its units are years.

49. Solving for t gives

$$\begin{aligned}
P &= 10e^{0.02t}\\
\frac{P}{10} &= e^{0.02t}\\
0.02t &= \ln\left(\frac{P}{10}\right)\\
t &= \frac{\ln(P/10)}{0.02} = 50\ln(P/10).
\end{aligned}$$

The inverse function $t = f^{-1}(P) = 50\ln(P/10)$ gives the time, t, in years at which the population reaches P million.

53. **(a)** $f^{-1}(-1)$ is the value of x with $f(x) = -1$. Reading off the graph, we see that $f(0) = -1$ and so $f^{-1}(-1) = 0$.

(b) $f^{-1}(0)$ is the value of x with $f(x) = 0$. Reading off the graph, we see that $f(1) = 0$ and so $f^{-1}(0) = 1$.

(c) $f^{-1}(3)$ is the value of x with $f(x) = 3$. Reading off the graph, we see that $f(2) = 3$ and so $f^{-1}(3) = 2$.

(d) $f^{-1}(-2)$ is the value of x with $f(x) = -2$. Reading off the graph, we see that $f(-2) = -2$ and so $f^{-1}(-2) = -2$.

(e) Since $f^{-1}(f(x)) = x$ for any x, we have $f^{-1}(f(\frac{1}{2})) = \frac{1}{2}$. Alternatively, we could read values from the graph.

(f) Since $f(f^{-1}(y)) = y$ for any y, we have $f(f^{-1}(-0.05)) = -0.05$. Alternatively, we could read values from the graph.

57. **(a)** Definitely false. We know the graph of w lies entirely above the line $y = 3 + 2x$. This means

$$\begin{aligned}
w(x) &> 3+2x\\
w(2) &> 3+2\cdot 2\\
w(2) &> 7.
\end{aligned}$$

Since $w(2)$ must be greater than 7, $w(2)$ can't equal 6.

(b) Possibly true. If $w^{-1}(5) = 0$, this would mean $w(0) = 5$, so that the point $(0, 5)$ lies on the graph of w. This is possible, for this point falls above the line $y = 3 + 2x$, which contains the point $(0, 3)$.

(c) Definitely false. If this were true, then

$$\begin{aligned}
\left(w^{-1}(-6)\right)^{-1} &= -2\\
\frac{1}{w^{-1}(-6)} &= -2\\
w^{-1}(-6) &= -0.5\\
w(-0.5) &= -6,
\end{aligned}$$

which would mean the point $(-0.5, -6)$ lies on the graph of w. However, the graph must lie above $y = 3+2x$, so that

$$\begin{aligned} w(-0.5) &> 3 + 2(-0.5) \\ &> 2. \end{aligned}$$

Since $w(-0.5)$ must be greater than 2, $w(-0.5)$ cannot equal -6.

(d) Possibly true. This statement tells us the average rate of change of w on the interval $-2 \le x \le 3$ equals 4. This is consistent with the given information: it does not violate the conditions that w be invertible and that its graph lie above the line $y = 3 + 2x$.

61. Solving for r gives

$$\begin{aligned} r &= \log\left(\frac{I}{I_0}\right) \\ r &= \log\left(\frac{I}{I_0}\right) \\ \frac{I}{I_0} &= 10^r \\ I &= I_0 10^r. \end{aligned}$$

The inverse function $f^{-1}(r) = I_0 10^r$ gives the intensity of an earthquake with Richter rating r.

65. (a) Yes, given Table 10.2.

Table 10.2

x	-9	-8	-5	-4	6	7	9
$f^{-1}(x)$	3	2	1	0	-1	-2	-3

(b) No, because for example, $g(-3) = g(-1) = g(3) = 3$. Therefore, g^{-1} cannot exist, as we would be unable to determine whether $g^{-1}(3) = -1$ or $g^{-1}(3) = -3$ or $g^{-1}(3) = 3$.

(c)

Table 10.3

x	-3	-2	-1	0	1	2	3
$f(g(x))$	-9	-5	-9	-8	9	6	-9

(d) No element of the range of $f(x)$ is in the domain of $g(x)$. Therefore, $g(f(x))$ will be undefined for all values of x given by the above table.

69. We have

$$\begin{aligned} y &= 0.5(x^{-1} + A^{-1})^{-1} \\ 2y &= (x^{-1} + A^{-1})^{-1} \\ (2y)^{-1} &= x^{-1} + A^{-1} \\ x^{-1} &= 0.5y^{-1} - A^{-1} \\ x &= \left(0.5y^{-1} - A^{-1}\right)^{-1} \\ \text{so} \quad f^{-1}(x) &= \left(0.5x^{-1} - A^{-1}\right)^{-1}. \end{aligned}$$

73. (a) $C(0)$ is the concentration of alcohol in the 100 ml solution after 0 ml of alcohol is removed. Thus, $C(0) = 99\%$.

(b) Note that there are initially 99 ml of alcohol and 1 ml of water.

$$\begin{aligned} C(x) &= \begin{matrix}\text{Concentration of alcohol} \\ \text{after removing } x \text{ ml}\end{matrix} = \frac{\text{Amount of alcohol remaining}}{\text{Amount of solution remaining}} \\ &= \frac{\begin{matrix}\text{Original amount} \\ \text{of alcohol}\end{matrix} - \begin{matrix}\text{Amount of} \\ \text{alcohol removed}\end{matrix}}{\begin{matrix}\text{Original amount} \\ \text{of solution}\end{matrix} - \begin{matrix}\text{Amount of alcohol} \\ \text{removed}\end{matrix}} = \frac{99 - x}{100 - x}. \end{aligned}$$

(c) If $y = C(x)$, then $x = C^{-1}(y)$. We have

$$\begin{aligned} y &= \frac{99-x}{100-x} \\ y(100-x) &= 99-x \\ 100y - xy &= 99-x \\ x - xy &= 99-100y \\ x(1-y) &= 99-100y \\ x &= \frac{99-100y}{1-y}. \end{aligned}$$

Thus, $C^{-1}(y) = \dfrac{99-100y}{1-y}$.

(d) The function $C^{-1}(y)$ tells us how much alcohol we need to remove in order to obtain a solution whose concentration is y.

77. Since f is assumed to be an increasing function, its inverse is well defined. This is an amount of caffeine: the amount giving a pulse 20 bpm higher than r_c, that is, 20 bpm higher than the pulse of a person having 1 serving of coffee.

81. Since f is assumed to be an increasing function, its inverse is well defined. This is an amount of caffeine. We know that $1.1f(q_c) = 1.1r_c$ is 10% higher than the pulse of a person who has had 1 serving of coffee. This makes $f^{-1}(1.1f(q_c))$ is the amount of caffeine that will lead to a pulse 10% higher than will a serving of coffee.

Solutions for Section 10.3

Exercises

1. **(a)** We have $f(x) + g(x) = x + 5 + x - 5 = 2x$.
(b) We have $f(x) - g(x) = x + 5 - (x - 5) = 10$.
(c) We have $f(x)g(x) = (x + 5)(x - 5) = x^2 - 25$.
(d) We have $f(x)/g(x) = (x + 5)/(x - 5)$.

5. **(a)** We have $f(x) + g(x) = x^2 + 4 + x^2 + 2 = 2x^2 + 6$.
(b) We have $f(x) - g(x) = x^2 + 4 - (x^2 + 2) = 2$.
(c) We have $f(x)g(x) = (x^2 + 4)(x^2 + 2) = x^4 + 6x^2 + 8$.
(d) We have $f(x)/g(x) = (x^2 + 4)/(x^2 + 2)$.

9. $h(x) = 2(2x - 1) - 3(1 - x) = 4x - 2 - 3 + 3x = 7x - 5$.

13. We have $f(x) = e^x(2x + 1) = 2xe^x + e^x$.

17. Since $h(x) = f(x) + g(x)$, we know that $h(-1) = f(-1) + g(-1) = -4 + 4 = 0$. Similarly, $j(x) = 2f(x)$ tells us that $j(-1) = 2f(-1) = 2(-4) = -8$. Repeat this process for each entry in the table.

Table 10.4

x	$h(x)$	$j(x)$	$k(x)$	$m(x)$
−1	0	-8	16	-1
0	0	-2	1	-1
1	2	4	0	0
2	6	10	1	0.2
3	12	16	16	0.5
4	20	22	81	9/11

21. $\dfrac{f(x)}{g(x)} = \dfrac{\sin x}{x^2}$.

Problems

25. **(a)** A formula for $h(x)$ would be

$$h(x) = f(x) + g(x).$$

To evaluate $h(x)$ for $x = 3$, we use this equation:

$$h(3) = f(3) + g(3).$$

Since $f(x) = x + 1$, we know that

$$f(3) = 3 + 1 = 4.$$

Likewise, since $g(x) = x^2 - 1$, we know that

$$g(3) = 3^2 - 1 = 9 - 1 = 8.$$

Thus, we have

$$h(3) = 4 + 8 = 12.$$

To find a formula for $h(x)$ in terms of x, we substitute our formulas for $f(x)$ and $g(x)$ into the equation $h(x) = f(x) + g(x)$:

$$h(x) = \underbrace{f(x)}_{x+1} + \underbrace{g(x)}_{x^2-1}$$

$$h(x) = x + 1 + x^2 - 1 = x^2 + x.$$

To check this formula, we use it to evaluate $h(3)$, and see if it gives $h(3) = 12$, which is what we got before. The formula is $h(x) = x^2 + x$, so it gives

$$h(3) = 3^2 + 3 = 9 + 3 = 12.$$

This is the result that we expected.

(b) A formula for $j(x)$ would be

$$j(x) = g(x) - 2f(x).$$

To evaluate $j(x)$ for $x = 3$, we use this equation:

$$j(3) = g(3) - 2f(3).$$

We already know that $g(3) = 8$ and $f(3) = 4$. Thus,

$$j(3) = 8 - 2 \cdot 4 = 8 - 8 = 0.$$

To find a formula for $j(x)$ in terms of x, we again use the formulas for $f(x)$ and $g(x)$:

$$\begin{aligned} j(x) &= \underbrace{g(x)}_{x^2-1} - 2\underbrace{f(x)}_{x+1} \\ &= (x^2 - 1) - 2(x + 1) \\ &= x^2 - 1 - 2x - 2 \\ &= x^2 - 2x - 3. \end{aligned}$$

We check this formula using the fact that we already know $j(3) = 0$. Since we have $j(x) = x^2 - 2x - 3$,

$$j(3) = 3^2 - 2 \cdot 3 - 3 = 9 - 6 - 3 = 0.$$

This is the result that we expected.

(c) A formula for $k(x)$ would be

$$k(x) = f(x)g(x).$$

Evaluating $k(3)$, we have

$$k(3) = f(3)g(3) = 4 \cdot 8 = 32.$$

A formula in terms of x for $k(x)$ would be

$$\begin{aligned} k(x) &= \underbrace{f(x)}_{x+1} \cdot \underbrace{g(x)}_{x^2-1} \\ &= (x+1)(x^2-1) \\ &= x^3 - x + x^2 - 1 \\ &= x^3 + x^2 - x - 1. \end{aligned}$$

To check this formula,

$$k(3) = 3^3 + 3^2 - 3 - 1 = 27 + 9 - 3 - 1 = 32,$$

which agrees with what we already knew.

(d) A formula for $m(x)$ would be

$$m(x) = \frac{g(x)}{f(x)}.$$

Using this formula, we have

$$m(3) = \frac{g(3)}{f(3)} = \frac{8}{4} = 2.$$

To find a formula for $m(x)$ in terms of x, we write

$$\begin{aligned} m(x) = \frac{g(x)}{f(x)} &= \frac{x^2-1}{x+1} \\ &= \frac{(x+1)(x-1)}{(x+1)} \\ &= x - 1 \text{for } x \neq -1 \end{aligned}$$

We were able to simplify this formula by first factoring the numerator of the fraction $\frac{x^2-1}{x+1}$. To check this formula,

$$m(3) = 3 - 1 = 2,$$

which is what we were expecting.

(e) We have

$$n(x) = (f(x))^2 - g(x).$$

This means that

$$\begin{aligned} n(3) &= (f(3))^2 - g(3) \\ &= (4)^2 - 8 \\ &= 16 - 8 \\ &= 8. \end{aligned}$$

A formula for $n(x)$ in terms of x would be

$$\begin{aligned} n(x) &= (f(x))^2 - g(x) \\ &= (x+1)^2 - (x^2-1) \\ &= x^2 + 2x + 1 - x^2 + 1 \\ &= 2x + 2. \end{aligned}$$

To check this formula,

$$n(3) = 2 \cdot 3 + 2 = 8,$$

which is what we were expecting.

29. We have:

$$
\begin{aligned}
q(0) &= w^{-1}(v(0)) = w^{-1}(4) = 3 \quad &&\text{because } v(0) = 4 \text{ and } w(3) = 4\\
q(1) &= w^{-1}(v(1)) = w^{-1}(3) = 2 \quad &&\text{because } v(1) = 3 \text{ and } w(2) = 3\\
q(2) &= w^{-1}(v(2)) = w^{-1}(3) = 2 \quad &&\text{because } v(2) = 3 \text{ and } w(2) = 3\\
q(3) &= w^{-1}(v(3)) = w^{-1}(5) = 5 \quad &&\text{because } v(3) = 5 \text{ and } w(5) = 5\\
q(4) &= w^{-1}(v(4)) = w^{-1}(4) = 3 \quad &&\text{because } v(4) = 4 \text{ and } w(3) = 4\\
q(5) &= w^{-1}(v(5)) = w^{-1}(4) = 3. \quad &&\text{because } v(5) = 4 \text{ and } w(3) = 4
\end{aligned}
$$

See Table 10.5.

Table 10.5

x	0	1	2	3	4	5
$q(x)$	3	2	2	5	3	3

33. Where $g(x) = f(x)$, we see that $h(x) = g(x) - f(x) = 0$. Thus, the graph of h has an x-intercept wherever the graphs of f and g cross. See Figure 10.5.

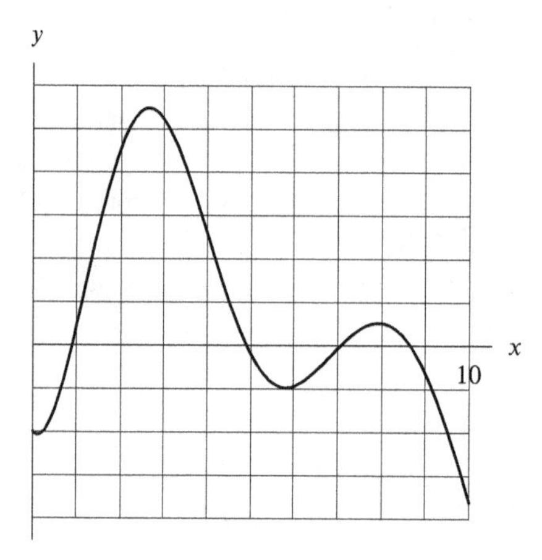

Figure 10.5

37. We can find the revenue function as a product:

$$\text{Revenue} = (\text{\# of customers}) \cdot (\text{price per customer}).$$

At the current price, 50,000 people attend every day. Since 2500 customers will be lost for each \$1 increase in price, the function $n(i)$ giving the number of customers who will attend given i one-dollar price increases, is given by $n(i) = 50{,}000 - 2500i$. The price function $p(i)$ giving the price after i one-dollar price increases is given by $p(i) = 15 + i$. The revenue function $r(i)$ is given by

$$
\begin{aligned}
r(i) &= n(i)p(i)\\
&= (50{,}000 - 2500i)(15 + i)\\
&= -2500i^2 + 12{,}500i + 750{,}000\\
&= -2500(i - 20)(i + 15).
\end{aligned}
$$

The graph $r(i)$ is a downward-facing parabola with zeros at $i = -15$ and $i = 20$, so the maximum revenue occurs at $i = 2.5$, which is halfway between the zeros. Thus, to maximize profits the ideal price is $\$15 + 2.5(\$1.00) = \$15 + \$2.50 = \$17.50$.

41. We have

$$\begin{aligned} H(x) &= F(G(x)) \\ &= F\left(\sqrt{x}\right) \\ &= \cos\left(\sqrt{x}\right) \\ h(x) &= f\,(G(x)) \cdot g(x) \\ &= f\left(\sqrt{x}\right) \cdot \frac{1}{2\sqrt{x}} \\ &= -\sin\left(\sqrt{x}\right) \cdot \frac{1}{2\sqrt{x}} \\ &= -\frac{\sin\left(\sqrt{x}\right)}{2\sqrt{x}}. \end{aligned}$$

45. **(a)** Since the initial amount was 316.75 and the growth factor is 1.004, we have $A(t) = 316.75(1.004)^t$.

(b) Since the CO_2 level oscillates once per year, the period is 1 year. The amplitude is 3.25 pm, so one possible answer is $V(t) = 3.25\sin(2\pi t)$. Any sinusoidal function with the same amplitude and period could describe the variation.

(c) The graph of $y = 316.75(1.004)^t + 3.25\sin(2\pi t)$ is in Figure 10.6.

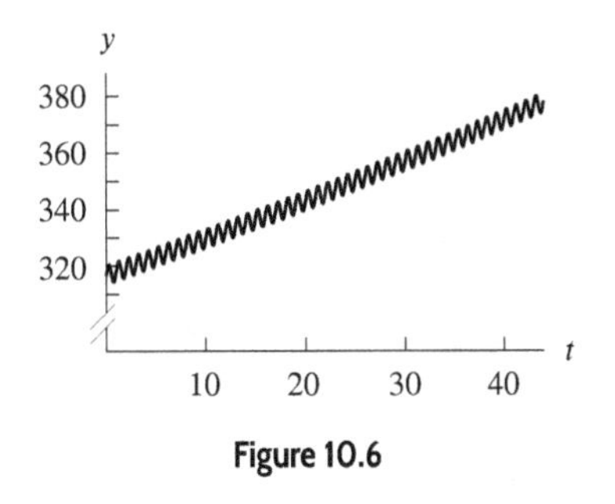

Figure 10.6

49. Since $2x$ represents twice as much office space as x, the cost of building twice as much space is $f(2x)$. The cost of building x amount of space is $f(x)$, so twice this cost is $2f(x)$. Thus, the contractor's statement is expressed

$$f(2x) < 2f(x).$$

53. The inequality $h(f(x)) < x$ tells us that Space can build fewer than x square feet of office space with the money Ace needs to build x square feet. You get more for your money with Ace.

Solutions for Chapter 10 Review

Exercises

1. Using substitution, we have $h(k(x)) = 2^{k(x)} = 2^{x^2}$ and $k(h(x)) = (h(x))^2 = (2^x)^2 = 2^{2x} = 4^x$.

5. Substituting the expression $x^2 + 1$ for the x term in the formula for $h(x)$ gives $\sqrt{x^2+1}$.

9. **(a)** A graph of this function on a window which contains both positive and negative values of x reveals that it fails the horizontal line test. Therefore, this function is not invertible.

(b) A graph reveals that the output values of this function oscillate back and forth between $y = -1$ and $y = 1$. It is therefore not invertible.

(c) A graph reveals that this function is always increasing. It passes the horizontal line test and is therefore invertible.

13. Start with $x = g(g^{-1}(x))$ and substitute $y = g^{-1}(x)$. We have

$$\begin{aligned} x &= g(y) \\ x &= e^{3y+1} \\ \ln x &= \ln e^{(3y+1)} \\ \ln x &= 3y + 1 \\ \ln x - 1 &= 3y \\ y &= \frac{1}{3}(\ln x - 1). \end{aligned}$$

Thus, $y = g^{-1}(x) = \frac{1}{3}(\ln x - 1)$.

17. Solving $y = g(x)$ for x gives

$$\begin{aligned} y &= \frac{x-2}{2x+3} \\ (2x+3)y &= x - 2 \\ 2xy + 3y &= x - 2 \\ 3y + 2 &= x - 2xy \\ x(1-2y) &= 3y + 2 \\ x &= \frac{3y+2}{1-2y}, \end{aligned}$$

so $g^{-1}(x) = \dfrac{3x+2}{1-2x}$.

21. Start with $x = s(s^{-1}(x))$ and substitute $y = s^{-1}(x)$. We have

$$\begin{aligned} x &= s(y) \\ x &= \frac{3}{2+\log y} \\ \frac{x}{3} &= \frac{1}{2+\log y} \\ \frac{3}{x} &= 2 + \log y \\ \frac{3}{x} - 2 &= \log y \\ 10^{\frac{3}{x}-2} &= 10^{\log y} \\ 10^{\frac{3}{x}-2} &= y \end{aligned}$$

So $s^{-1}(x) = 10^{(3/x)-2}$.

25. This function is not invertible. It does not pass the horizontal line test.

29. Since $h(x) = \sqrt{1-x}/x$ and $h^{-1}(x) = 1/(x^2+1)$, we have

$$
\begin{aligned}
h^{-1}(h(x)) &= h^{-1}\left(\sqrt{\frac{1-x}{x}}\right) \\
&= \frac{1}{\left(\sqrt{\frac{1-x}{x}}\right)^2 + 1} \\
&= \frac{1}{\frac{1-x}{x} + 1} \\
&= \frac{1}{\frac{1-x}{x} + \frac{x}{x}} \\
&= \frac{1}{\frac{1-x+x}{x}} = \frac{1}{\left(\frac{1}{x}\right)} = x.
\end{aligned}
$$

33. $g(g(x)) = g(2x-1) = 2(2x-1) - 1 = 4x - 3$

37. **(a)** $f(2x) = (2x)^2 + (2x) = 4x^2 + 2x$

(b) $g(x^2) = 2x^2 - 3$

(c) $h(1-x) = \dfrac{(1-x)}{1-(1-x)} = \dfrac{1-x}{x}$

(d) $(f(x))^2 = (x^2+x)^2$

(e) Since $g(g^{-1}(x)) = x$, we have

$$
\begin{aligned}
2g^{-1}(x) - 3 &= x \\
2g^{-1}(x) &= x + 3 \\
g^{-1}(x) &= \frac{x+3}{2}.
\end{aligned}
$$

(f) $(h(x))^{-1} = \left(\dfrac{x}{1-x}\right)^{-1} = \dfrac{1-x}{x}$

(g) $f(x)g(x) = (x^2+x)(2x-3)$

(h) $h(f(x)) = h(x^2+x) = \dfrac{x^2+x}{1-(x^2+x)} = \dfrac{x^2+x}{1-x^2-x}$

41. To find $f(x)$, we add $m(x)$ and $n(x)$ and simplify: $m(x) + n(x) = 3x^2 - x + 2x = 3x^2 + x = f(x)$.

45. To find $j(x)$, we divide $m(x)$by $n(x)$ and simplify: $(m(x))/n(x) = (3x^2 - x)/(2x) = 3x/2 - 1/2 = j(x)$.

49. Evaluate the two parts of the subtraction

$$
h(g(x)) = \tan\left(2\left(\frac{(3x-1)^2}{4}\right)\right) = \tan\frac{(3x-1)^2}{2} \quad \text{and} \quad f(9x) = (9x)^{3/2} = 9^{3/2}\cdot x^{3/2} = 27x^{3/2}
$$

and subtract

$$
h(g(x)) - f(9x) = \tan\left(\frac{(3x-1)^2}{2}\right) - 27x^{3/2}.
$$

Problems

53. If $f(x) = u(v(x))$, then one solution is $u(x) = \sqrt{x}$ and $v(x) = 3 - 5x$.

57. One possible solution is $F(x) = u(v(x))$ where $u(x) = x^3$ and $v(x) = 2x + 5$.

61. The troughs (where the graph is below the x-axis) are reflected about the horizontal axis to become humps. The humps (where the graph is above the x-axis) are unchanged.

65. First, we have $h(0) = f(g(0)) = f(1) = 0$, which completes the first row of the table. From the information in the second row of the table, we see that $h(1) = 1$. Therefore, since $h(1) = f(g(1))$, we conclude that $f(g(1)) = 1$, which is equivalent to $f(x) = 1$ if we let $x = g(1)$. Since f is invertible, our table indicates that $x = 2$ is the only solution to $f(x) = 1$. Therefore, $g(1) = x = 2$, which fills in the blank in the second row of the table. Finally, we have $h(2) = f(g(2)) = f(0) = 9$, which fills in the final entry in the table. See Table 10.6.

Table 10.6

x	$f(x)$	$g(x)$	$h(x)$
0	9	1	0
1	0	2	1
2	1	0	9

69. The inverse function $g^{-1}(t)$ represents the velocity needed for a trip of t hours. Its units are mph.

73. We take logarithms to help solve when x is in the exponent:

$$\begin{aligned} 2^{x+5} &= 3 \\ \ln(2^{x+5}) &= \ln 3 \\ (x+5)\ln 2 &= \ln 3 \\ x &= \frac{\ln 3}{\ln 2} - 5. \end{aligned}$$

77. Squaring eliminates square roots:

$$\begin{aligned} \sqrt{x+\sqrt{x}} &= 3 \\ x+\sqrt{x} &= 9 \\ \sqrt{x} &= 9 - x \quad (\text{so} \quad x \le 9) \\ x &= (9-x)^2 = 81 - 18x + x^2. \end{aligned}$$

So $x^2 - 19x + 81 = 0$. The quadratic formula gives the solutions

$$x = \frac{19 \pm \sqrt{37}}{2}.$$

The only solution is $x = \dfrac{19 - \sqrt{37}}{2}$. The other solution is too large to satisfy the original equation.

81. **(a)** $f(g(a)) = f(a) = a$
(b) $g(f(c)) = g(c) = b$
(c) $f^{-1}(b) - g^{-1}(b) = 0 - c = -c$
(d) $0 < x \le a$

85.

$$\begin{aligned} p(x) &= r(x) \\ 2x - 3 &= \frac{2x-1}{2x+1} \\ (2x-3)(2x+1) &= 2x - 1 \\ 4x^2 - 4x - 3 &= 2x - 1 \\ 4x^2 - 6x - 2 &= 0 \\ 2x^2 - 3x - 1 &= 0. \end{aligned}$$

Using the quadratic formula, we have:

$$x = \frac{-(-3) \pm \sqrt{(-3)^2 - 4 \cdot 2(-1)}}{2 \cdot 2}$$
$$= \frac{3 \pm \sqrt{17}}{4}.$$

89. **(a)** The function $f(x) = \sin^2 x$ is equal to $(u(x))^2$ but is not equal to $u(u(x))$. As an illustration of this, note that $f(\pi/2) = (\sin(\pi/2))^2 = 1$, but $u(u(\pi/2)) = \sin 1 \approx 0.84$. Since $f(1) \neq u(u(1))$, the functions $f(x)$ and $u(u(x))$ are not the same.

(b) First, we note that in the expression $p(x) = \sin(\cos^2 x)$, we are taking the composition of the sine function with $\cos^2 x$; we are not multiplying $\sin x$ by $\cos^2 x$. This tells us immediately that $u(x)(v(x))^2$ and $u(x)w(v(x))$ are not equal to $p(x)$. On the other hand, since $(v(x))^2 = w(v(x)) = \cos^2 x$, we see that $u((v(x))^2)$ and $u(w(v(x)))$ both equal $p(x)$, meaning that (ii) and (iii) are the only correct answers.

(c) (i) We have $(u(x) + v(x))^2 = \sin^2 x + 2\sin x \cos x + \cos^2 x$. Since $\sin^2 x + \cos^2 x = 1$ and $2\sin x \cos x = \sin 2x$, our answer simplifies to $1 + \sin 2x$.

(ii) We have $(u(x))^2 + (v(x))^2 = \sin^2 x + \cos^2 x = 1$.

(iii) We have $u(x^2) + v(x^2) = \cos(x^2) + \sin(x^2)$, which cannot be simplified.

93. See Figure 10.7.

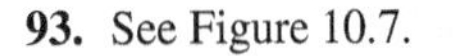

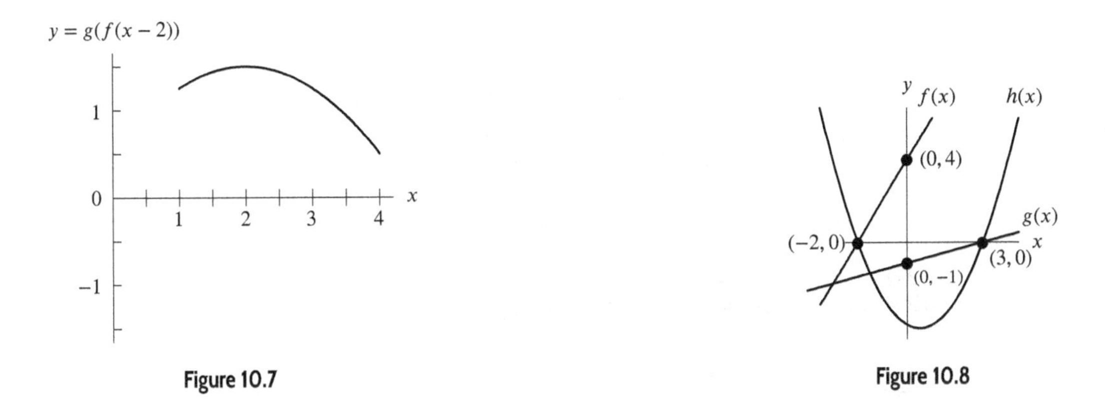

Figure 10.7

Figure 10.8

97. **(a)** Since $f(x)$ is a linear function, its formula can be written in the form $f(x) = mx + b$, where m represents the slope and b represents the y-intercept. According to the graph, the y-intercept is 4. Since $(-2, 0)$ and $(0, 4)$ both lie on the line, we know that

$$m = \frac{y_2 - y_1}{x_2 - x_1} = \frac{4 - 0}{0 - (-2)} = \frac{4}{2} = 2.$$

So we know that the formula is $f(x) = 2x + 4$. Similarly, we can find the slope of $g(x)$, $\frac{0-(-1)}{3-0} = \frac{1}{3}$, and the y-intercept, -1, so its formula is $g(x) = \frac{1}{3}x - 1$.

(b) To graph $h(x) = f(x) \cdot g(x)$, we first take note of where $f(x) = 0$ and $g(x) = 0$. At those places, $h(x) = 0$. Since the zero of $f(x)$ is -2 and the zero of $g(x)$ is 3, the zeros of $h(x)$ are -2 and 3. When $x < -2$, both $f(x)$ and $g(x)$ are negative, so we know that $h(x)$, their product, is positive. Similarly, when $x > 3$, both $f(x)$ and $g(x)$ are positive so $h(x)$ is positive. When $-2 < x < 3$, $f(x)$ is positive and $g(x)$ is negative, so $h(x)$ is negative. Also, since $h(x)$ is the product of two linear functions, we know that it is a quadratic function $h(x) = (2x + 4)(\frac{1}{3}x - 1) = \frac{2}{3}x^2 - \frac{2}{3}x - 4$. Putting these pieces of information together, we know that the graph of $h(x)$ is a parabola with zeros at -2 and 3 (and, therefore, an axis of symmetry at $x = \frac{1}{2}$) and that it is positive when $x < -2$ or $x > 3$ and negative when $-2 < x < 3$. [Note: since you know the axis of symmetry is $x = \frac{1}{2}$, you know that the x-coordinate of the vertex is $\frac{1}{2}$. You could find the y-coordinates of its vertex by finding $h(\frac{1}{2}) = (2(\frac{1}{2}) + 4)(\frac{1}{3}(\frac{1}{2}) - 1) = (1 + 4)(\frac{1}{6} - 1) = 5(-\frac{5}{6}) = -\frac{25}{6} = -4\frac{1}{6}$.] See Figure 10.8.

101. The statement is false. For example, if $f(x) = x$ and $g(x) = x^2$, then $f(x) \cdot g(x) = x^3$. In this case, $f(x) \cdot g(x)$ is an odd function, but $g(x)$ is an even function.

105. **(a)** One possible answer is:

$$y = \underbrace{6x}_{g(x)} \cdot \underbrace{e^{\overbrace{3x^2}^{G(x)}}}_{f(G(x))}$$

$$\begin{aligned} \text{so} \quad f(x) &= e^x \\ g(x) &= 6x \\ G(x) &= 3x^2. \end{aligned}$$

(b) One possible answer is:

$$y = -\frac{\sin\left(\sqrt{x}\right)}{2\sqrt{x}} = \underbrace{-\frac{1}{2\sqrt{x}}}_{g(x)} \cdot \underbrace{\sin\left(\overbrace{\sqrt{x}}^{G(x)}\right)}_{f(G(x))}$$

$$\begin{aligned} \text{so} \quad f(x) &= \sin x \\ g(x) &= -\frac{1}{2\sqrt{x}} \\ G(x) &= \sqrt{x}. \end{aligned}$$

109. This is an increasing function, because as x increases, $f(x)$ increases, and as $f(x)$ increases, $f(f(x))$ increases.

113. **(a)** $f(8) = 2$, because 8 divided by 3 equals 2 with a remainder of 2. Similarly, $f(17) = 2$, $f(29) = 2$, and $f(99) = 0$.
(b) $f(3x) = 0$ because, no matter what x is, $3x$ will be divisible by 3.
(c) No. Knowing, for example, that $f(x) = 0$ tells us that x is evenly divisible by 3, but gives us no other information regarding x.
(d) $f(f(x)) = f(x)$, because $f(x)$ equals either 0, 1, or 2, and $f(0) = 0$, $f(1) = 1$, and $f(2) = 2$.
(e) No. For example, $f(1) + f(2) = 1 + 2 = 3$, but $f(1 + 2) = f(3) = 0$.

STRENGTHEN YOUR UNDERSTANDING

1. False, since $f(4) + g(4) = \frac{1}{4} + \sqrt{4}$ but $(f + g)(8) = \frac{1}{8} + \sqrt{8}$.

5. True, since $g(f(x)) = g\left(\frac{1}{x}\right) = \sqrt{\frac{1}{x}}$.

9. True. Evaluate $\dfrac{f(3) + g(3)}{h(3)} = \dfrac{\frac{1}{3} + \sqrt{3}}{3 - 5}$ and simplify.

13. False. As a counterexample, let $f(x) = x^2$ and $g(x) = x + 1$. Then $f(g(x)) = (x + 1)^2 = x^2 + 2x + 1$, but $g(f(x)) = x^2 + 1$.

17. False. $f(x + h) = \dfrac{1}{x + h} \neq \dfrac{1}{x} + \dfrac{1}{h}$.

21. False. If $f(x) = ax^2 + bx + c$ and $g(x) = px^2 + qx + r$, then

$$f(g(x)) = f(px^2 + qx + r) = a(px^2 + qx + r)^2 + b(px^2 + qx + r) + c.$$

Expanding shows that $f(g(x))$ has an x^4 term.

25. True, since $g(f(2)) = g(1) = 3$ and $f(g(3)) = f(1) = 3$.

29. True. The function g is not invertible if two different points in the domain have the same function value.

33. True. Each x value has only one y value.

37. True. The inverse of a function reverses the action of the function and returns the original value of the independent variable x.

CHAPTER ELEVEN

Solutions for Section 11.1

Skill Refresher

S1. $\sqrt{36t^2} = (36t^2)^{1/2} = 36^{1/2} \cdot (t^2)^{1/2} = 6|t^1| = 6|t|$

S5. We have

$$\begin{aligned} 10x^{5-2} &= 2 \\ 10x^3 &= 2 \\ x^3 &= 0.2 \\ x &= (0.2)^{1/3} = 0.585. \end{aligned}$$

S9. False

Exercises

1. This is a power function in the form $y = ax^p$:

$$y = \frac{48}{30625} \cdot x^{-2}, \qquad a = \frac{48}{30625}, \qquad p = -2.$$

We have

$$\begin{aligned} y &= 3\left(\frac{2}{5\sqrt{7x}}\right)^4 \\ &= 3 \cdot \frac{2^4}{5^4\left(\sqrt{7}\sqrt{x}\right)^4} \\ &= \frac{48}{625 \cdot 49x^2} \\ &= \frac{48}{30625} \cdot x^{-2}. \end{aligned}$$

5. Although y is a power function of $(x+7)$, it is not a power function of x and cannot be written in the form $f(x) = kx^p$.

9. Since the graph is symmetric about the y-axis, the power function is even.

13. Since $f(1) = k \cdot 1^p = k$, we know $k = f(1) = \frac{3}{2}$.

Since $f(2) = k \cdot 2^p = \frac{3}{8}$, and since $k = \frac{3}{2}$, we know

$$\left(\frac{3}{2}\right) \cdot 2^p = \frac{3}{8}$$

which implies

$$2^p = \frac{3}{8} \cdot \frac{2}{3} = \frac{1}{4}.$$

Thus $p = -2$, and $f(x) = \frac{3}{2} \cdot x^{-2}$.

17. We need to solve $j(x) = kx^p$ for p and k. We know that $j(x) = 2$ when $x = 1$. Since $j(1) = k \cdot 1^p = k$, we have $k = 2$. To solve for p, use the fact that $j(2) = 16$ and also $j(2) = 2 \cdot 2^p$, so

$$2 \cdot 2^p = 16,$$

giving $2^p = 8$, so $p = 3$. Thus, $j(x) = 2x^3$.

21. Substituting into the general formula $c = kd^2$, we have $45 = k(3)^2$ or $k = 45/9 = 5$. So the formula for c is

$$c = 5d^2.$$

When $d = 5$, we get $c = 5(5)^2 = 125$.

25. **(a)** $\lim_{x\to\infty} x^{-4} = \lim_{x\to\infty}(1/x^4) = 0$.

(b) $\lim_{x\to-\infty} 2x^{-1} = \lim_{x\to-\infty}(2/x) = 0$.

Problems

29. **(a)** For $f(x) = x^{1/2}$, between $x = 0$ and $x = 2$, we have

$$\text{Average rate of change } = \frac{f(2) - f(0)}{2 - 0} = \frac{\sqrt{2} - \sqrt{0}}{2 - 0} = 0.707.$$

Similar calculations show the rate of change of $f(x)$ between $x = 2$ and $x = 4$ is 0.293 and give all the values in Table 11.1.

Table 11.1

Interval	$0-2$	$2-4$	$4-6$	$6-8$
Rate of change of $f(x) = x^{1/2}$	0.707	0.293	0.225	0.189
Rate of change of $g(x) = x^2$	2	6	10	14

(b) For $f(x) = x^{1/2}$, as x increases, the rate of change decreases (from 0.707 to 0.293 to 0.225 to 0.189). This reflects the fact that the graph of $f(x) = x^{1/2}$ is concave down.

For $g(x) = x^2$, as x increases, the rate of change increases. This reflects the fact that the graph of $g(x) = x^2$ is concave up.

33. $c(t) = \frac{1}{t}$ is indeed one possible formula. It is not, however, the only one. Because the vertical and horizontal axes are asymptotes for this function, we know that the power p is a negative number and

$$c(t) = kt^p.$$

If $p = -3$ then $c(t) = kt^{-3}$. Since $(2, \frac{1}{2})$ lies on the curve, $\frac{1}{2} = k(2)^{-3}$ or $k = 4$. So, $c(t) = 4t^{-3}$ could describe this function. Similarly, so could $c(t) = 16x^{-5}$ or $c(t) = 64x^{-7}$...

37. Since $s < r < 0$, we know that these are negative-powered power functions, so their graphs have horizontal asymptotes at $y = 0$ and vertical asymptotes at $x = 0$.

As $x \to \infty$, the graph of g approaches its horizontal asymptote more rapidly than the graph of f, because $s < r$; that is, s is "more negative" than r. Thus, the graph of g should lie below the graph of f for large x-values, as it does at $x = x_1$. Thus, $x_1 \geq x_0$.

41. We have

$$\begin{aligned} F(x) &= \frac{1}{\sqrt[3]{7x}} \\ &= \frac{1}{\sqrt[3]{7}\sqrt[3]{x}} \\ &= \underbrace{\frac{1}{\sqrt[3]{7}}}_{k} \cdot x^{\overbrace{-1/3}^{n}} \end{aligned}$$

$$\text{so} \quad k = \frac{1}{\sqrt[3]{7}}$$
$$n = -\frac{1}{3}$$
$$\text{which means} \quad f(x) = nkx^{n-1}$$
$$= -\frac{1}{3} \cdot \frac{1}{\sqrt[3]{7}} \cdot x^{-\frac{1}{3}-1}$$
$$= -\frac{1}{3\sqrt[3]{7}} \cdot x^{-\frac{4}{3}}.$$

45. The number of calories is directly proportional to the number of ounces because as the amount of beef, x, increases, the number of calories, c, also increases.

Substituting into the formula $c = kx$, we have $245 = k(3)$ or $k = 81.667$. So the formula is

$$c = 81.667x.$$

When $x = 4$, we have $c = 81.667(4) = 326.667$. Therefore, 4 ounces of hamburger contain 326.667 calories.

49. The number of hours is inversely proportional to the speed because as the speed, v, increases, the number of hours, h, decreases.

Substituting into the formula $h = k/v$, we get $3.5 = k/55$, so $k = 192.5$, giving

$$h = \frac{192.5}{v}.$$

(The value of k is the distance to Albany, 192.5 miles.) When $h = 3$, we have $3 = 192.5/v$, so $v = 64.167$. Thus, getting to Albany in 3 hours requires a speed of 64.167 mph.

53. **(a)** Since the speed of sound is 340 meters/sec = 0.34 km/sec, if t is in seconds,

$$d = f(t) = 0.34t.$$

Thus, when $t = 5$ sec, $d = 0.34 \cdot 5 = 1.7$ km. Other values in the second row of Table 11.2 are calculated in a similar manner.

Table 11.2

Time, t	5 sec	10 sec	1 min	5 min
Distance, d (km)	1.7	3.4	20.4	102
Area, A (km^2)	9.1	36.3	1307	32685

(b) Using $d = 0.34t$, we want to calculate t when $d = 200$, so

$$200 = 0.34t$$
$$t = \frac{200}{0.34} = 588.24 \text{ sec} = 9.8 \text{ mins}.$$

(c) The values for A are listed in Table 11.2. These were calculated using the fact that the area of a circle of radius r is $A = \pi r^2$. At time t, the radius of the circle of people who have heard the explosion is $d = 0.34t$. Thus

$$A = \pi d^2 = \pi(0.34t)^2 = \pi(0.34)^2 t^2 = 0.363t^2.$$

This formula was used to calculate the values of A in Table 11.2.

(d) Since the population density is 31 people/km^2, the population, P, who have heard the explosion is given by

$$P = 31A.$$

Since $A = 0.363t^2$, we have

$$P = 31 \cdot 0.363t^2 = 11.25t^2.$$

So $P = f(t) = 11.25t^2$.

(e) The graph of $P = f(t) = 11.25t^2$ is in Figure 11.1.

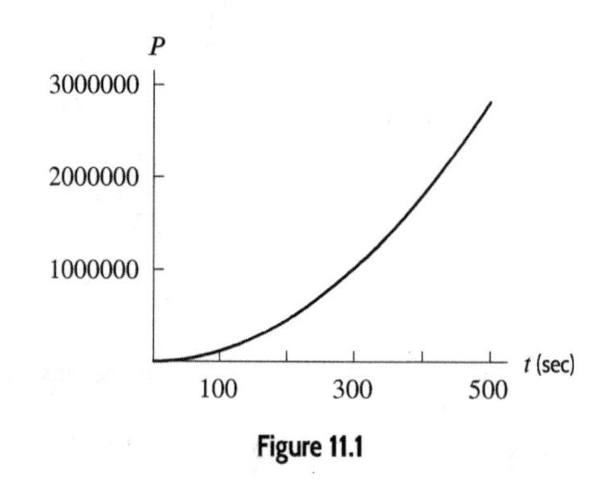

Figure 11.1

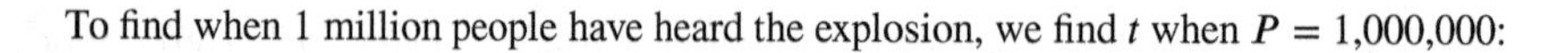

To find when 1 million people have heard the explosion, we find t when $P = 1{,}000{,}000$:

$$\begin{aligned} 1000000 &= 11.25t^2 \\ t^2 &= \frac{1000000}{11.25} \\ t &= \sqrt{\frac{1000000}{11.25}} = 298 \text{ sec} \approx 5 \text{ min.} \end{aligned}$$

57. (a) We have

$$f(x) = g(h(x)) = 16x^4.$$

Since $g(x) = 4x^2$, we know that

$$\begin{aligned} g(h(x)) = 4(h(x))^2 &= 16x^4 \\ (h(x))^2 &= 4x^4. \\ \text{Thus,} \qquad h(x) &= 2x^2 \text{ or } -2x^2. \end{aligned}$$

Since $h(x) \leq 0$ for all x, we know that

$$h(x) = -2x^2.$$

(b) We have

$$f(x) = j(2g(x)) = 16x^4, \qquad j(x) \text{ a power function.}$$

Since $g(x) = 4x^2$, we know that

$$j(2g(x)) = j(8x^2) = 16x^4.$$

Since $j(x)$ is a power function, $j(x) = kx^p$. Thus,

$$\begin{aligned} j(8x^2) = k(8x^2)^p &= 16x^4 \\ k \cdot 8^p x^{2p} &= 16x^4. \end{aligned}$$

Since $x^{2p} = x^4$ if $p = 2$, letting $p = 2$, we have

$$\begin{aligned} k \cdot 64x^4 &= 16 \cdot x^4 \\ 64k &= 16 \\ k &= \frac{1}{4} \end{aligned}$$

Thus, $j(x) = \frac{1}{4}x^2$.

Solutions for Section 11.2

Skill Refresher

S1. $x^2 + 2x - x - 2 = x^2 + x - 2$

S5. $\frac{1}{3}$

Exercises

1. Since 5^x is not a power function, this is not a polynomial.

5. This is not a polynomial because $2e^x$ is not a power function.

9. Rewriting in standard form, we get

$$\begin{aligned} 5x^4 - 3x^2 - 2x^4 + 1 - 3x^4 &= 5x^4 - 2x^4 - 3x^4 - 3x^2 + 1 \\ &= -3x^2 + 1, \end{aligned}$$

so we see that the leading term is $-3x^2$.

13. Since n is a positive integer, we see that $n + 3$ is larger than the other powers. This means the leading term is $4x^{n+3}$.

17. Here, the leading term is $x^2 \cdot 2x^3 = 2x^5$, and the contributing terms are $x^2, 2x^3$.

21. **(a)** $\lim_{x\to\infty}(3x^2 - 5x + 7) = \lim_{x\to\infty}(3x^2) = \infty$.

(b) $\lim_{x\to-\infty}(7x^2 - 9x^3) = \lim_{x\to-\infty}(-9x^3) = \infty$.

Problems

25. **(a)** Note that the leading terms of both $u(x)$ and $v(x)$ are $-\frac{1}{5}x^3$ so the graphs of u and v have the same end behavior. As $x \to -\infty$, both $u(x)$ and $v(x) \to \infty$, and as $x \to \infty$, both $u(x)$ and $v(x) \to -\infty$.

The graphs have different y-intercepts, and u has three distinct zeros. The function v has a multiple zero at $x = 0$. See Figure 11.2.

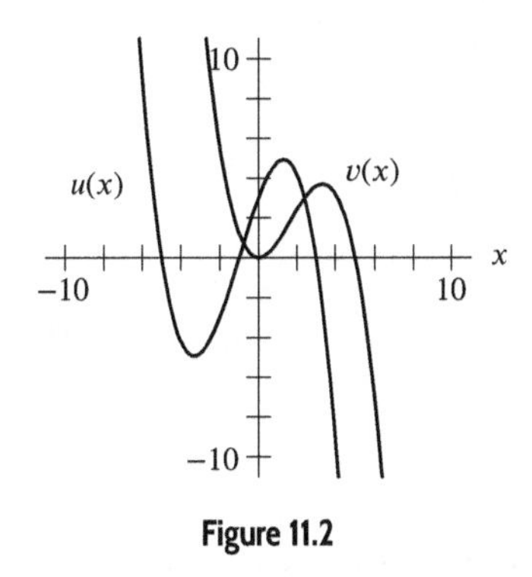

Figure 11.2

(b) On the window $-20 \le x \le 20$ by $-1600 \le y \le 1600$, the peaks and valleys of both functions are not distinguishable. Near the origin, the behavior of both functions looks the same. The functions are still distinguishable from one another on the ends.

On the window $-50 \le x \le 50$ by $-25{,}000 \le y \le 25{,}000$, the functions are still slightly distinct from one another on the ends—but barely.

On the last window the graphs of both functions appear identical. Both functions look like the function $y = -\frac{1}{5}x^3$.

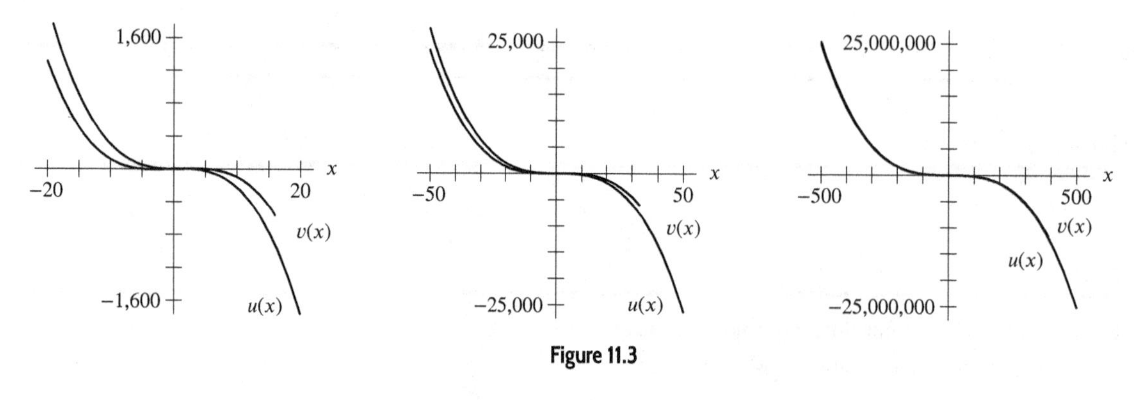

Figure 11.3

29. We have $-1.1 \leq x \leq -0.9$, $f(-1.1) \leq y \leq f(-0.9)$ or $-0.121 \leq y \leq 0.081$.

33. This is a fourth-degree polynomial whose leading coefficient is positive. Thus, in the long run its graph resembles Figure (II).

37. **(a)** Using a computer or a graphing calculator, we can get a picture of $f(x)$ like the one in Figure 11.4. On this window f appears to be invertible because it passes the horizontal line test.

(b) Substituting gives

$$f(0.5) = (0.5)^3 + 0.5 + 1 = 1.625.$$

To find $f^{-1}(0.5)$, we solve $f(x) = 0.5$. With a computer or graphing calculator, we trace along the graph of f in Figure 11.5 to find

$$f^{-1}(0.5) \approx -0.424.$$

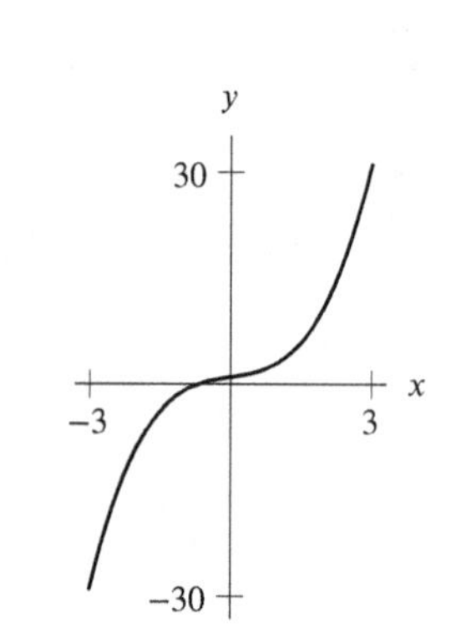

Figure 11.4: $f(x) = x^3 + x + 1$

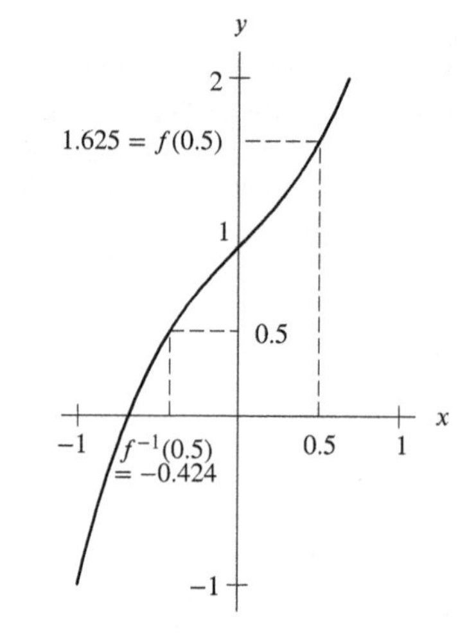

Figure 11.5

41. Yes. For the sake of illustration, suppose $f(x) = x^2 + x + 1$, a second-degree polynomial. Then

$$\begin{aligned} f(g(x)) &= (g(x))^2 + g(x) + 1 \\ &= g(x) \cdot g(x) + g(x) + 1. \end{aligned}$$

Since $f(g(x))$ is formed from products and sums involving the polynomial g, the composition $f(g(x))$ is also a polynomial. In general, $f(g(x))$ will be a sum of powers of $g(x)$, and thus $f(g(x))$ will be formed from sums and products involving the polynomial $g(x)$. A similar situation holds for $g(f(x))$, which will be formed from sums and products involving the polynomial $f(x)$. Thus, either expression will yield a polynomial.

45. **(a)** A graph of V is shown in Figure 11.6 for $0 \le t \le 5, 0 \le V \le 1$.

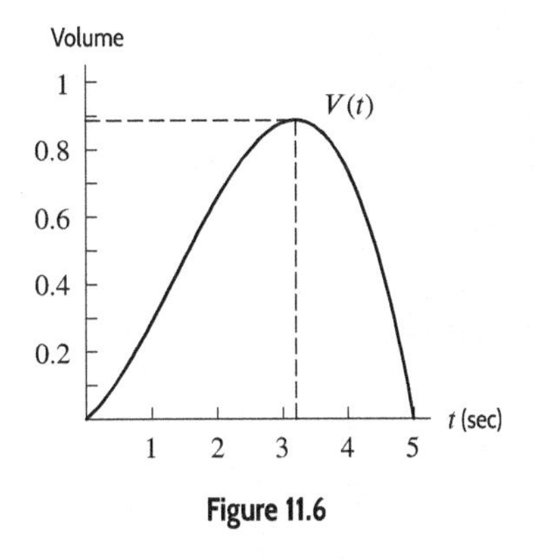

Figure 11.6

(b) The maximum value of V for $0 \le t \le 5$ occurs when $t \approx 3.195$, $V \approx .886$. Thus, at just over 3 seconds into the cycle, the lungs contain ≈ 0.88 liters of air.

(c) The volume is zero at $t = 0$ and again at $t \approx 5$. This indicates that at the beginning and end of the 5-second cycle the lungs are empty.

49. **(a)** Substituting $x = 0.5$ into p, we have

$$p(0.5) = 1 - 0.5 + 0.5^2 - 0.5^3 + 0.5^4 - 0.5^5 \approx 0.65625.$$

Since $f(0.5) = 2/3 = 0.6666....$, the approximation is accurate to 2 decimal places.

(b) We have $p(1) = 1 - 1 + 1 - 1 + 1 - 1 = 0$, but $f(1) = 0.5$. Thus $p(1)$ is a poor approximation to $f(1)$.

(c) See Figure 11.7. The two graphs are difficult to tell apart for $-0.5 \le x \le 0.5$, but for x outside this region the fit is not good.

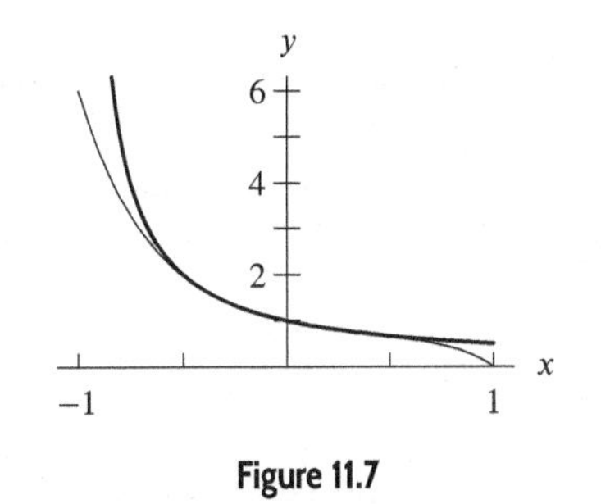

Figure 11.7

Solutions for Section 11.3

Skill Refresher

S1. $x^4(x^2 - 1) = x^4(x - 1)(x + 1)$

S5. We must have either $2x + 4 = 0$ or $(x - 1)^2 = 0$. So $x = -2$ or $x = 1$.

Exercises

1. Zeros occur where $y = 0$, at $x = -3$, $x = 2$, and $x = -7$.

5. The function has a common factor of $4x$ which gives

$$f(x) = 4x(2x^2 - x - 15),$$

and the quadratic factor reduces further, giving

$$f(x) = 4x(2x + 5)(x - 3).$$

Thus, the zeros of f are $x = 0$, $x = \frac{-5}{2}$, and $x = 3$.

9. This polynomial must be of fourth (or higher even-powered) degree, so either (but not both) of the zeros at $x = 2$ or $x = 5$ could be doubled. One possible formula is $y = k(x + 2)(x - 2)^2(x - 5)$. Solving for k gives

$$\begin{aligned} k(0 + 2)(0 - 2)^2(0 - 5) &= 5 \\ -40k &= 5 \\ k &= -\frac{1}{8}, \end{aligned}$$

so

$$y = -\frac{1}{8}(x + 2)(x - 2)^2(x - 5).$$

Another possible formula is $y = k(x + 2)(x - 2)(x - 5)^2$. Solving for k gives

$$\begin{aligned} k(0 + 2)(0 - 2)(0 - 5)^2 &= 5 \\ -100k &= 5 \\ k &= -\frac{1}{20}, \end{aligned}$$

so

$$y = -\frac{1}{20}(x + 2)(x - 2)(x - 5)^2.$$

There are other possible polynomials, but all are of degree higher than 4, so these are the simplest.

Problems

13. The graph represents a polynomial of even degree, at least fourth. Zeros are shown at $x = -2$, $x = -1$, $x = 2$, and $x = 3$. Since $f(x) \to -\infty$ as $x \to \infty$ or $x \to -\infty$, the leading coefficient must be negative. Thus, of the choices in the table, only C and E are possibilities. When $x = 0$, function C gives

$$y = -\frac{1}{2}(2)(1)(-2)(-3) = -\frac{1}{2}(12) = -6,$$

and function E gives

$$y = -(2)(1)(-2)(-3) = -12.$$

Since the y-intercept appears to be $(0, -6)$ rather than $(0, -12)$, function C best fits the polynomial shown.

17. Clearly $f(x) = x$ works. However, the solution is not unique. If f is of the form $f(x) = ax^2 + bx + c$, then $f(0) = 0$ gives $c = 0$, and $f(1) = 1$ gives

$$a(1)^2 + b(1) + 0 = 1,$$

so

$$a + b = 1,$$

or

$$b = 1 - a.$$

Since these are the only conditions which must be satisfied, any polynomial of the form

$$f(x) = ax^2 + (1 - a)x$$

will work. If $a = 0$, we get $f(x) = x$.

21. The function g has zeros at $x = -1$ and $x = 5$, and a double zero at $x = 3$. Thus, let $g(x) = k(x-3)^2(x-5)(x+1)$. Use $g(0) = 3$ to solve for k; $g(0) = k(-3)^2(-5)(1) = -45k$. Thus $-45k = 3$ and $k = -\frac{1}{15}$. So

$$g(x) = -\frac{1}{15}(x-3)^2(x-5)(x+1).$$

25. Note that $y = x^4 + 6x^2 + 9 = (x^2+3)^2$. This implies that $y = 0$ if $x^2 = -3$, but $x^2 = -3$ has no real solutions. Thus, there are no zeros.

29. Zeros occur where $y = 0$, at $x = 0$ and $x = -3$. Since x^2 is never less than zero, $x^2 + 4$ is never less than 4, so $x^2 + 4$ has no zeros.

33. We know that $g(-2) = 0$, $g(-1) = -3$, $g(2) = 0$, and $g(3) = 0$. We also know that $x = -2$ is a multiple zero. Thus, let

$$g(x) = k(x+2)^2(x-2)(x-3).$$

Then, using $g(-1) = -3$, gives

$$g(-1) = k(-1+2)^2(-1-2)(-1-3) = k(1)^2(-3)(-4) = 12k,$$

so $12k = -3$, and $k = -\frac{1}{4}$. Thus,

$$g(x) = -\frac{1}{4}(x+2)^2(x-2)(x-3)$$

is a possible formula for g.

37. Notice that we can think of g as a vertically shifted polynomial. That is, if we let $g(x) = h(x)+4$, then $h(x)$ is a polynomial with zeros at $x = -1$, $x = 2$, and $x = 4$; furthermore, since $g(-2) = 0$, $h(-2) = 0 - 4 = -4$. Thus,

$$h(x) = k(x+1)(x-2)(x-4).$$

To find k, note that $h(-2) = k(-2+1)(-2-2)(-2-4) = k(-1)(-4)(-6) = -24k$. Since $h(-2) = -24k = -4$, we have $k = \frac{1}{6}$, which gives

$$h(x) = \frac{1}{6}(x+1)(x-2)(x-4).$$

Thus since $g(x) = h(x) + 4$, we have

$$g(x) = \frac{1}{6}(x+1)(x-2)(x-4) + 4.$$

41. From its formula, we know that f has double zeros at $x = 5$ and $x = 3$ and a single zero at $x = 1$. Since it is an even function, we know the graph of f has even symmetry—that is, its graph is symmetrical across the y-axis. So we can use the unknowns r, s, and $g(x)$ to balance the zeros we do know. Here is one possibility:

- To balance the double zero at $x = 3$ with a double zero at $x = -3$, we can let $s = 2$.
- To balance the single zero at $x = 1$ with a single zero at $x = -1$, we can let $r = -1$.
- To balance the double zero at $x = 5$, we can let the second-degree polynomial $g(x) = k(x+5)^2$.

Putting this together gives

$$f(x) = (x-5)^2(x-3)^2(x-1)\underbrace{(x+1)}_{(x-r)}\underbrace{(x+3)^2}_{(x+3)^s}\cdot\underbrace{k(x+5)^2}_{g(x)}$$

Another possibility:

- To balance the double zero at $x = 3$ with a double zero at $x = -3$, we again let $s = 2$.
- If we let $g(x) = k(x+1)(x+5)$, then g is a second-degree polynomial that balances the single zero at $x = 1$ and one of the repeated zeros at $x = 5$.
- We can balance the other of the repeated zeros at $x = 5$ by letting $r = -5$.

Putting this together gives

$$f(x) = (x-5)^2(x-3)^2(x-1)\underbrace{(x+5)}_{(x-r)}\underbrace{(x+3)^2}_{(x+3)^s}\cdot\underbrace{k(x+5)(x+1)}_{g(x)}$$

Note that k is an arbitrary nonzero constant. In Figure 11.8, we assume $k = 1$.

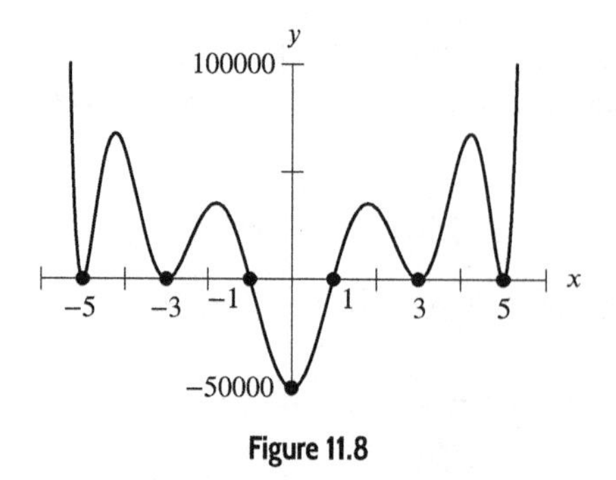

Figure 11.8

45. We express the volume as a function of the length, x, of the square's side that is cut out in Figure 11.9.

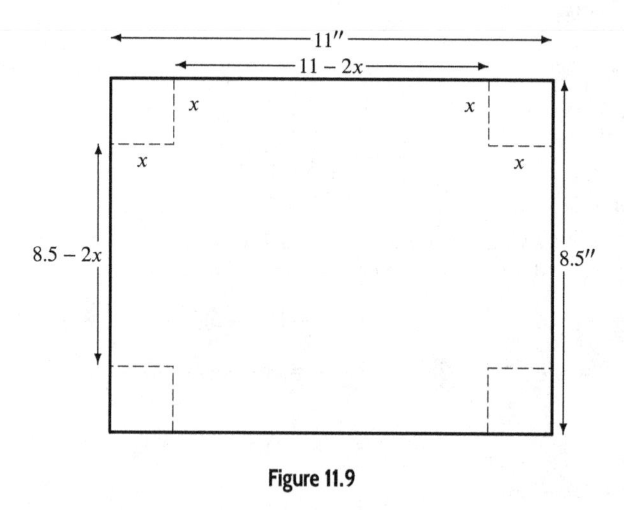

Figure 11.9

Since the sides of the base are $(11 - 2x)$ and $(8.5 - 2x)$ inches and the depth is x inches, the volume, $V(x)$, is given by

$$V(x) = x(11 - 2x)(8.5 - 2x).$$

The graph of $V(x)$ in Figure 11.10 suggest that the maximum volume occurs when $x \approx 1.585$ inches. (A good viewing window is $0 \le x \le 5$ and $0 \le y \le 70$.) So one side is $x = 1.585$, and therefore the others are $11 - 2(1.585) = 7.83$ and $8.5 - 2(1.585) = 5.33$.

The dimensions of the box are 7.83 by 5.33 by 1.585 inches.

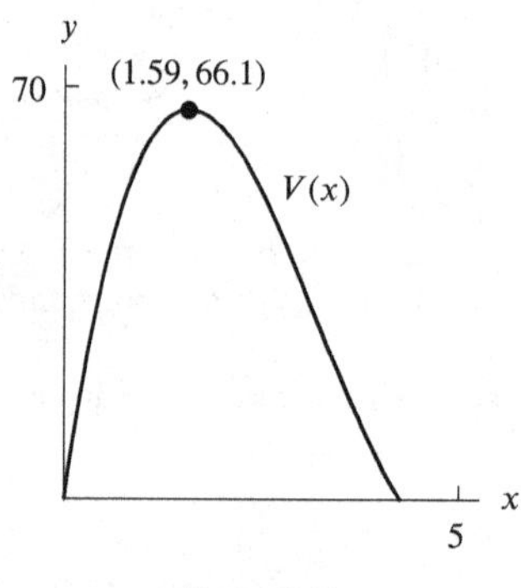

Figure 11.10

49. **(a)** Never true, because $f(x) \to -\infty$ as $x \to \pm\infty$, which means $f(x)$ must be of even degree.
(b) Sometimes true, since f could be an even-degree polynomial without being symmetric to the y-axis.
(c) Sometimes true, since f could have a multiple zero.
(d) Never true, because f must be of even degree.
(e) True, because, since f is of even degree, $f(-x)$ must have the same long-run behavior as $f(x)$.
(f) Never true, because, since $f(x) \to -\infty$ as $x \to \pm\infty$, f will fail the horizontal line test.

Solutions for Section 11.4

Skill Refresher

S1. $\dfrac{6}{y} + \dfrac{7}{y^3} = \dfrac{6y^2+7}{y^3}$

S5.

$$\begin{aligned}\frac{5}{(x-2)^2(x+1)} - \frac{18}{(x-2)} &= \frac{5-18(x-2)(x+1)}{(x-2)^2(x+1)}\\ &= \frac{5-18x^2+18x+36}{(x-2)^2(x+1)0}\\ &= \frac{-18x^2+18x+41}{(x-2)^2(x+1)}\end{aligned}$$

S9. $$\frac{x^{-1}+x^{-2}}{1-x^{-2}} = \frac{\frac{1}{x}+\frac{1}{x^2}}{1-\frac{1}{x^2}} = \frac{\frac{x+1}{x^2}}{\frac{x^2-1}{x^2}} = \frac{x+1}{x^2}\cdot\frac{x^2}{x^2-1} = \frac{x+1}{(x+1)(x-1)} = \frac{1}{x-1}.$$

Exercises

1. This is a rational function, and it is already in the form of one polynomial divided by another.

5. This is not a rational function, as we cannot put it in the form of one polynomial divided by another, since $\sqrt{x}+1$ is not a polynomial.

9. Rewriting this in the form $p(x)/q(x)$ where $p(x)$ and $q(x)$ are polynomials in standard form, we have:

$$\begin{aligned}\frac{1}{x+2} + \frac{x}{x+1} &= \frac{1}{x+2}\cdot\frac{x+1}{x+1} + \frac{x}{x+1}\cdot\frac{x+2}{x+2}\\ &= \frac{x+1}{(x+1)(x+2)} + \frac{x(x+2)}{(x+1)(x+2)}\\ &= \frac{x^2+3x+1}{x^2+3x+2}.\end{aligned}$$

The leading term in the numerator is x^2 and the leading term in the denominator is x^2, so for large enough values of x (either positive or negative),

$$\frac{x^2+3x+1}{x^2+3x+2} \approx \frac{x^2}{x^2}.$$

Since $\dfrac{x^2}{x^2} \to 1$ as $x \to \pm\infty$ we see that

$$\frac{1}{x+2} + \frac{x}{x+1} = \frac{x^2+3x+1}{x^2+3x+2} \to 1$$

as $x \to \pm\infty$.

13. As $x \to \infty$, we see that $y \to -\dfrac{3x^2}{x^2}$, so the long-run behavior is $y \to -3$. This matches graph (III).

17. We have

$$\lim_{x\to\infty}(2x^{-3}+4) = \lim_{x\to\infty}(2/x^3+4) = 0+4 = 4.$$

21. As $x \to \pm\infty$, $1/x \to 0$ and $x/(x+1) \to 1$, so $h(x)$ approaches $3 - 0 + 1 = 4$.
Therefore $y = 4$ is the horizontal asymptote.

Problems

25. Let $y = f(x)$. Then $x = f^{-1}(y)$. Solving for x,

$$\begin{aligned} y &= \frac{4-3x}{5x-4} \\ y(5x-4) &= 4-3x \\ 5xy - 4y &= 4 - 3x \\ 5xy + 3x &= 4y + 4 \\ x(5y+3) &= 4y+4 \quad \text{(factor out an } x\text{)} \\ x &= \frac{4y+4}{5y+3} \end{aligned}$$

Therefore,

$$f^{-1}(x) = \frac{4x+4}{5x+3}.$$

29. **(a)** The time to travel the first 10 miles is $\frac{10}{40} = 0.25$ hours. The time for the remaining 50 miles in $50/V$ hours so the total journey time is $T = 0.25 + 50/V$. Thus, the average speed is

$$\text{Average speed} = \frac{60}{T} = \frac{60}{\left(0.25 + \frac{50}{V}\right)} = \frac{240V}{V+200}.$$

(b) If you want to average 60 mph for the trip then you need

$$\frac{240V}{V+200} = 60.$$

Solving this equation gives $V = 200/3$ mph, nearly 70 mph.

33. **(a)** $f(x) = \dfrac{\text{Amount of Alcohol}}{\text{Amount of Liquid}} = \dfrac{x}{x+5}$

(b) $f(7) = \frac{7}{7+5} = \frac{7}{12} \approx 58.333\%$. Also, $f(7)$ is the concentration of alcohol in a solution consisting of 5 gallons of water and 7 gallons of alcohol.

(c) $f(x) = 0$ implies that $\dfrac{x}{x+5} = 0$ and so $x = 0$. The concentration of alcohol is 0% when there is no alcohol in the solution, that is, when $x = 0$.

(d) The horizontal asymptote is given by the ratio of the highest-power terms of the numerator and denominator:

$$y = \frac{x}{x} = 1 = 100\%$$

This means that as the amount of alcohol added, x, grows large, the concentration of alcohol in the solution approaches 100%.

37. Line l_1 has a smaller slope than line l_2. We know the slope of line l_1 represents the average cost of producing n_1 units, and the slope of l_2 represents the average cost of producing n_2 units. Thus, the average cost of producing n_2 units is more than that of producing n_1 units. For these goods, the average cost actually goes up between n_1 and n_2 units.

Solutions for Section 11.5

Exercises

1. **(a)** $y = -\dfrac{1}{(x-5)^2} - 1$ has a vertical asymptote at $x = 5$, no x intercept, horizontal asymptote $y = -1$: (iii)

(b) $y = \dfrac{x-2}{(x+1)(x-3)}$ has vertical asymptotes at $x = -1, 3$, x intercept at 2, horizontal asymptote $y = 0$: (i)

(c) $y = \dfrac{2x+4}{x-1}$ has a vertical asymptote at $x = 1$, x intercept at $x = -2$, horizontal asymptote $y = 2$: (ii)

(d) $y = \dfrac{x-3+x+1}{(x+1)(x-3)} = \dfrac{2x-2}{(x+1)(x-3)}$ has vertical asymptotes at $x = -1, 3$, x intercept at 1, horizontal asymptote at $y = 0$: (iv)

(e) $y = \dfrac{(1+x)(1-x)}{x-2}$ has vertical asymptote at $x = 2$, two x intercepts at ± 1: (vi)

(f) $y = \dfrac{1-4x}{2x+2}$ has a vertical asymptote at $x = -1$, x intercept at $x = \frac{1}{4}$, horizontal asymptote at $y = -2$: (v)

5. The zeros of this function are at $x = \pm 2$. It has a vertical asymptote at $x = 9$. Its long-run behavior is that it looks like the line $y = x$. See Figure 11.11.

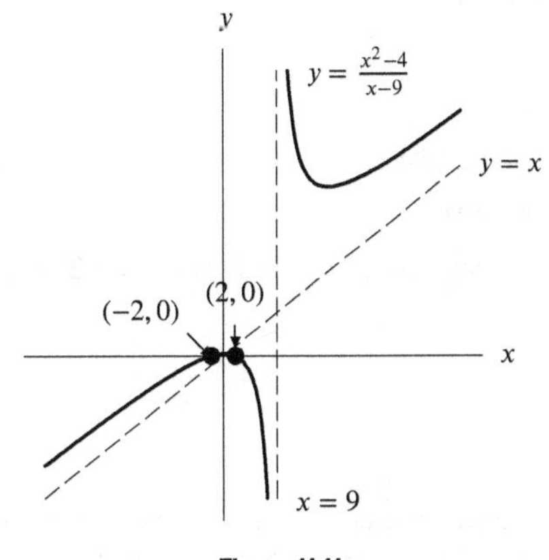

Figure 11.11

9. Since

$$k(x) = \frac{x(4-x)}{x^2-6x+5} = \frac{x(4-x)}{(x-1)(x-5)},$$

the x-intercepts are $x = 0, x = 4$; the y-intercept is $y = 0$; the vertical asymptotes are $x = 1, x = 5$. Since we can also write

$$k(x) = \frac{4x-x^2}{x^2-6x+5} = \frac{-x^2+4x}{x^2-6x+5},$$

the horizontal asymptote is $y = -1$.

Problems

13. (a) To estimate

$$\lim_{x\to 2^+} \frac{5-x}{(x-2)^2},$$

we consider what happens to the function when x is slightly larger than 2. The numerator is positive and the denominator is positive and is approaching 0 as x approaches 2. We suspect that $\frac{5-x}{(x-2)^2}$ gets larger and larger as x approaches 2 from the right. We can also use either a graph or a table of values as in Table 11.3 to estimate this limit. We see that

$$\lim_{x\to 2^+} \frac{5-x}{(x-2)^2} = +\infty.$$

Table 11.3

x	2.1	2.01	2.001	2.0001
$f(x)$	290	29900	2999000	299990000

(b) To estimate

$$\lim_{x\to 2^-} \frac{5-x}{(x-2)^2},$$

we consider what happens to the function when x is slightly smaller than 2. The numerator is positive and the denominator is positive and is approaching 0 as x approaches 2. We suspect that $\frac{5-x}{(x-2)^2}$ gets larger and larger as x approaches 2 from the left. We can also use either a graph or a table of values to estimate this limit. We see that

$$\lim_{x\to 2^-} \frac{5-x}{(x-2)^2} = +\infty.$$

17. **(a)** If $f(n)$ is large, then $\frac{1}{f(n)}$ is small.
(b) If $f(n)$ is small, then $\frac{1}{f(n)}$ is large.
(c) If $f(n) = 0$, then $\frac{1}{f(n)}$ is undefined.
(d) If $f(n)$ is positive, then $\frac{1}{f(n)}$ is also positive.
(e) If $f(n)$ is negative, then $\frac{1}{f(n)}$ is negative.

21. **(a)** The graph shows $y = 1/x$ flipped across the x-axis and shifted left 2 units. Therefore

$$y = -\frac{1}{x+2}$$

is a choice for a formula.
(b) The formula $y = -1/(x+2)$ is already written as a ratio of two linear functions.
(c) The graph has a y-intercept if $x = 0$. Thus, $y = -\frac{1}{2}$. Since y cannot be zero if $y = -1/(x+2)$, there is no x-intercept. The only intercept is $(0, -\frac{1}{2})$.

25. First, we can simplify the formula for $h(x)$:

$$\begin{aligned} h(x) &= \frac{1}{x-1} + \frac{2}{1-x} + 2 \\ &= \frac{1}{x-1} - \frac{2}{x-1} + 2 = -\frac{1}{x-1} + 2. \end{aligned}$$

Thus h is a transformation of $y = \frac{1}{x}$ with $p = 1$. The graph of $y = \frac{1}{x}$ has been shifted one unit to the right, flipped over the x-axis, and shifted up 2 units. To find the y-intercept, we need to evaluate $h(0)$:

$$h(0) = -\frac{1}{-1} + 2 = 3.$$

To find the x-intercepts, we need to solve $h(x) = 0$ for x:

$$\begin{aligned} 0 &= -\frac{1}{x-1} + 2 \\ -2 &= -\frac{1}{x-1} \\ -2x + 2 &= -1 \\ -2x &= -3, \\ \text{so,}\quad x &= \frac{3}{2} \quad \text{is the only } x\text{-intercept.} \end{aligned}$$

The graph of h is shown in Figure 11.12.

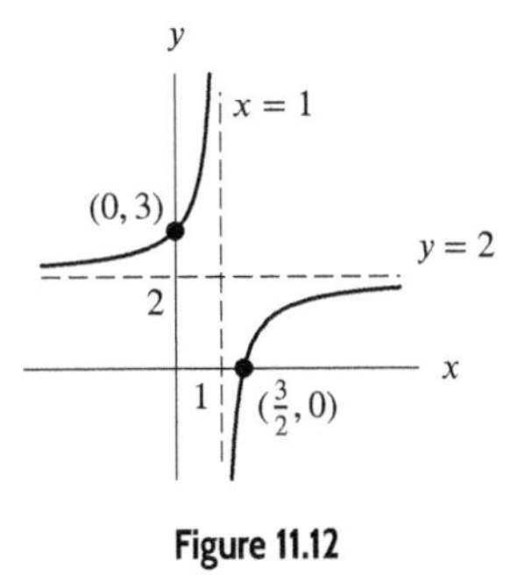

Figure 11.12

29. **(a)** The table indicates translation of $y = 1/x$ because the values of the function are headed in opposite directions near the vertical asymptote.

(b) The data points in the table indicate that $y \to \frac{1}{2}$ as $x \to \pm\infty$. The vertical asymptote does not appear to have been shifted. Thus, we might try

$$y = \frac{1}{x} + \frac{1}{2}.$$

A check of x-values shows that this formula works. To express as a ratio of polynomials, we get a common denominator. Then

$$y = \frac{1(2)}{x(2)} + \frac{1(x)}{2(x)}$$
$$y = \frac{2+x}{2x}$$

33. We try $(x+3)(x-1)$ in the numerator in order to get zeros at $x = -3$ and $x = 1$. There is only one vertical asymptote at $x = -2$, but in order to have the horizontal asymptote of $y = 1$, the numerator and denominator must be of same degree. Thus, try

$$y = \frac{(x+3)(x-1)}{(x+2)^2}.$$

Note that this answer gives the correct y-intercept of $(0, -\frac{3}{4})$ and $y \to 1$ as $x \to \pm\infty$.

37. The graph of $y = \dfrac{x}{(x+2)(x-3)}$ fits.

41. Writing

$$\begin{aligned} y &= \frac{x-1}{x^2-4x+3} \\ &= \frac{x-1}{(x-1)(x-3)} \\ &= \frac{1}{x-3} \qquad \text{for } x \neq 1, \end{aligned}$$

we see that y is undefined at $x = 1$. The behavior near $x = 1$ is like $y = 1/(x-3)$ and this graph passes through $y = -1/2$ at $x = 1$. So the graph of the original function has a hole at $x = 1$, not an asymptote. See Figure 11.13.

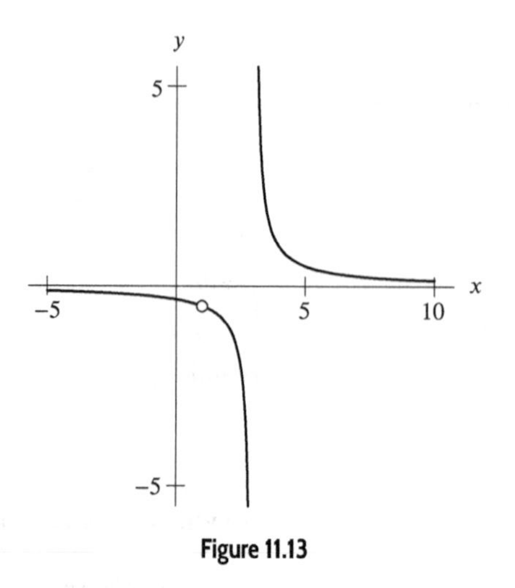

Figure 11.13

45.
$$f(x) = \frac{p(x)}{q(x)} = \frac{-3(x-2)(x-3)}{(x-5)^2}$$

We need the factor of -3 in the numerator and the exponent of 2 in the denominator, because we have a horizontal asymptote of $y = -3$. The ratio of highest term of $p(x)$ to highest term of $q(x)$ will be $\frac{-3x^2}{x^2} = -3$.

Solutions for Section 11.6

Exercises

1. Larger powers of x give smaller values for $0 < x < 1$.
   A - (iii)
   B - (ii)
   C - (iv)
   D - (i)

5. The function fits neither form. If the expression in the parentheses expanded, then $m(x) = 3(9x^2+6x+1) = 27x^2+18x+3$.

9. **(a)**

Table 11.4

x	$f(x)$	$g(x)$
−3	1/27	−27
−2	1/9	−8
−1	1/3	−1
0	1	0
1	3	1
2	9	8
3	27	27

(b) As $x \to -\infty$, $f(x) \to 0$. For f, large negative values of x result in small $f(x)$ values because a large negative power of 3 is very close to zero. For g, large negative values of x result in large negative values of $g(x)$, because the cube of a large negative number is a larger negative number. Therefore, as $x \to -\infty$, $g(x) \to -\infty$.

As $x \to \infty$, $f(x) \to \infty$ and $g(x) \to \infty$. For $f(x)$, large x-values result in large powers of 3; for $g(x)$, large x values yield the cubes of large x-values. f and g both climb *fast*, but f climbs faster than g (for $x > 3$).

13. As $x \to \infty$, the higher power dominates, so $x^{1.1}$ dominates $x^{1.08}$. The coefficients 1000 and 50 do not change this, so $y = 50x^{1.1}$ dominates.

Problems

17. Table 11.5 shows that 3^{-x} approaches zero faster than x^{-3} as $x \to \infty$.

Table 11.5

x	2	10	100
3^{-x}	1/9	0.000017	1.94×10^{-48}
x^{-3}	1/8	0.001	10^{-6}

21. **(a)** If f is linear,

$$m = \frac{128-16}{2-1} = 112,$$

and

$$16 = 112(1) + b, \qquad \text{so} \qquad b = -96.$$

Thus,

$$f(x) = 112x - 96.$$

(b) If f is exponential, then

$$\frac{128}{16} = \frac{a(b)^2}{a(b)} = b, \qquad \text{so} \qquad b = 8$$

and

$$16 = a(8), \qquad \text{so} \qquad a = 2.$$

Therefore

$$f(x) = 2(8)^x.$$

(c) If f is a power function, $f(x) = k(x)^p$. Then

$$\frac{f(2)}{f(1)} = \frac{k(2)^p}{k(1)^p} = (2)^p = \frac{128}{16} = 8,$$

so $p = 3$. Using $f(1) = 16$ to solve for k, we have

$$16 = k(1^3), \qquad \text{so} \qquad k = 16.$$

Thus,

$$f(x) = 16x^3.$$

25. **(a)** We are given $t(v) = v^{-2} = 1/v^2$ and $r(v) = 40v^{-3} = 40/v^3$; therefore

$$\frac{1}{v^2} = \frac{40}{v^3}.$$

Multiplying by v^3, we get $v = 40$.

(b) We found in part (a) that $v = 40$. By graphing or substituting values of x between 0 and 40, we see that $r(x) > t(x)$ for $0 < x < 40$.

(c) For values of $x > 40$ we see by graphing or substituting values that $t(x) > r(x)$.

29. For large positive t, the value of $3^{-t} \to 0$ and $4^t \to \infty$. Thus, $y \to 0$ as $t \to \infty$.

For large negative t, the value of $3^{-t} \to \infty$ and $4^t \to 0$. Thus,

$$y \to \frac{\text{Very large positive number}}{0+7} \qquad \text{as } t \to -\infty.$$

So $y \to \infty$ as $t \to -\infty$.

33. Since e^t and t^2 both dominate $\ln|t|$, we have $y \to \infty$ as $t \to \infty$.

For large negative t, the value of $e^t \to 0$, but t^2 is large and dominates $\ln|t|$. Thus, $y \to \infty$ as $t \to -\infty$,

37. Since e^{3t} dominates e^{2t}, the value of y is very small for large positive t. Thus, $y \to 0$ as $t \to \infty$.

For large negative t, the value of $e^{2t} \to 0$, so $y \to 0$ as $t \to -\infty$.

Solutions for Section 11.7

Exercises

1. Judging from the figure, an exponential function might best model this data.

5. $f(x) = ax^p$ for some constants a and c. Since $f(1) = 1 = a(1)^p$, it follows that $a = 1$. Also, $f(2) = 2^p = c$. Solving for p, we have $p = \ln c / \ln 2$. Thus, $f(x) = x^{\ln c/\ln 2}$.

9. **(a)** (i) Calculator result: $y = 46.79t^{0.301}$. Answers may vary.

(ii) Calculator result: $y = 0.822t^2 - 6.282t + 76.53$. Answers may vary.

(b) For the time period 2002–2012, the quadratic function is the better fit. The power function is approximately the square root function, so it is concave down, but the catch values increase rapidly toward the end of this period. Notice that the power function goes through $(0, 0)$, meaning that the predicted value of the 2000 catch is zero, which is not a realistic prediction. The quadratic function is shifted and stretched, so is the better fit. See Figure 11.14. However, outside of the interval 2002–2012, there is no reason to suppose that either function is a good fit. Our results hold only for this time period.

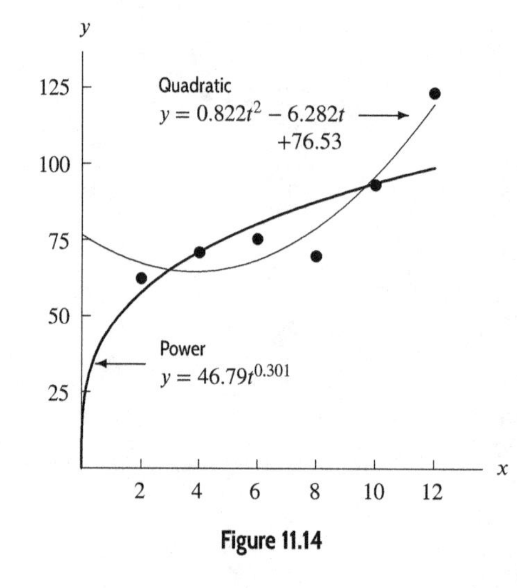

Figure 11.14

13. The slope of this line is $m = \frac{y_2 - y_1}{x_2 - x_1} = \frac{3}{2}$. The vertical intercept is 0, thus $y = \frac{3}{2}x$.

Problems

17. **(a)** The function $y = -83.039 + 61.514x$ gives a superb fit, with correlation coefficient $r = 0.99997$.

(b) When the power function is plotted for $2 \le x \le 2.05$, it resembles a line. This is true for most of the functions we have studied. If you zoom in close enough on any given point, the function begins to resemble a line. However, for other values of x (say, $x = 3, 4, 5 \ldots$), the fit no longer holds.

21. **(a)** A computer or calculator gives

$$N = -14t^4 + 433t^3 - 2255t^2 + 5634t - 4397.$$

(b) The graph of the data and the quartic in Figure 11.15 shows a good fit between 1980 and 1996.

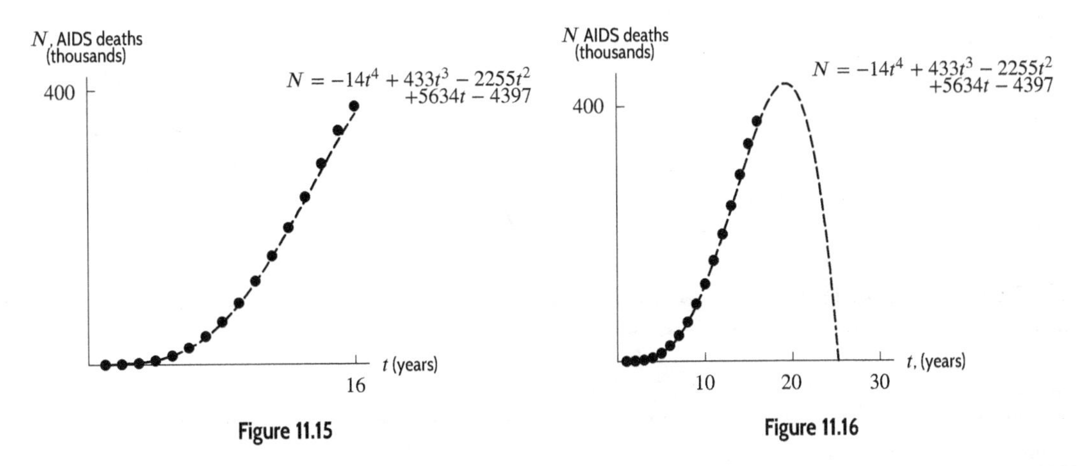

Figure 11.15

Figure 11.16

(c) Figure 11.15 shows that the quartic model fits the 1980-1996 data well. However, this model predicts that in 2000, the number of deaths decreases. See Figure 11.16. Since N is the total number of deaths since 1980, this is impossible. Therefore the quartic is definitely not a good model for $t > 20$.

25. (a) See Figure 11.17.

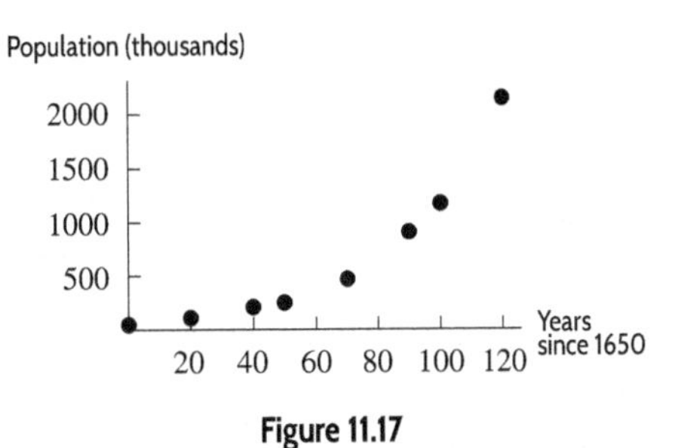

Figure 11.17

(b) Using a calculator or computer, we get $P(t) = 56.108(1.031)^t$. Answers may vary.
(c) The 56.108 represents a population of 56,108 people in 1650. Note that this is more than the actual population of 50,400. The growth factor of 1.031 means the rate of growth is approximately 3.1% per year.
(d) We find $P(100) = 1194.308$, which is slightly higher than the given data value of 1,170.8.
(e) The estimated population, $P(150) = 5510.118$, is higher than the given census population.

29. (a) The formula is $N = 1148.55e^{0.3617t}$. See Figure 11.18.

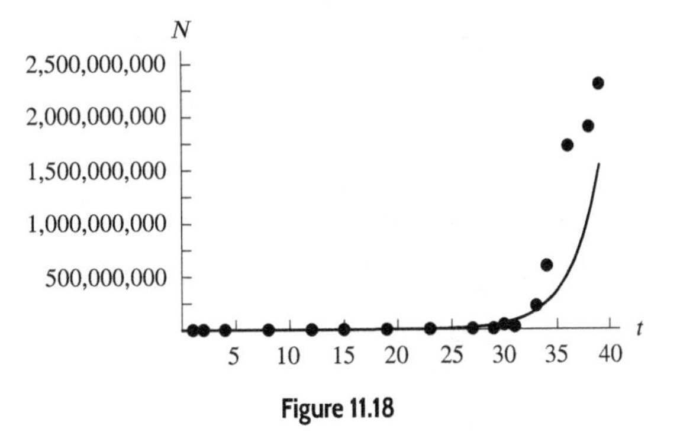

Figure 11.18

(b) The doubling time is given by $\ln 2/0.3617 \approx 1.916$. This is consistent with *Moore's Law*, which states that the number of transistors doubles about once every two years. Dr. Gordon E. Moore is Chairman Emeritus of Intel Corporation According to the Intel Corporation, "Gordon Moore made his famous observation in 1965, just four years after the first planar integrated circuit was discovered. The press called it 'Moore's Law' and the name has stuck. In his original paper, Moore observed an exponential growth in the number of transistors per integrated circuit and predicted that this trend would continue."

33. **(a)** Quadratic is the only choice that increases and then decreases to match the data.

(b) Using ages of $x = 20, 30, \ldots, 80$, a quadratic function is $y = -34.136x^2 + 3497.733x - 39{,}949.714$. Answers may vary.

(c) The value of the function at 37 is $y = -34.136 \cdot 37^2 + 3497.733 \cdot 37 - 39{,}949.714 = \$42{,}734$.

(d) The value of the function for age 10 is $y = -34.136 \cdot 10^2 + 3497.733 \cdot 10 - 39{,}949.714 = -\$8386$. Answers may vary. Not reasonable, as income is positive. In addition, 10-year-olds do not usually work.

Solutions for Chapter 11 Review

Exercises

1. This function represents proportionality to a power.

$$y = \frac{\frac{1}{3}}{2x^7} = \frac{1}{6x^7} = \left(\frac{1}{6}\right)x^{(-7)}.$$

Thus $k = 1/6$ and $p = -7$.

5. While $y = 6x^3$ is a power function, when we add two to it, it can no longer be written in the form $y = kx^p$, so this is not a power function.

9. Since the graph is symmetric about the y-axis, the power function is even.

13. Since the graph is symmetric about the origin, the power function has an odd power.

17. By multiplying out the expression $(x^2 - 4)(x^2 - 2x - 3)$ and then simplifying the result, we see that

$$y = x^4 - 2x^3 - 7x^2 + 8x + 12$$

which is a fourth-degree polynomial.

21. Since $2x^2/x^{-7} = 2x^9$, we can rewrite this polynomial as $y = 2x^9 - 7x^5 + 3x^3 + 2$. Since the leading term of the polynomial is $2x^9$, the value of y goes to infinity as $x \to \infty$. The graph resembles $y = 2x^9$.

25. This is not a rational function, as we cannot put it in the form of one polynomial divided by another, since e^x is an exponential function, not a polynomial.

29. **(a)** Since the long-run behavior of $r(x) = p(x)/q(x)$ is given by the ratio of the leading terms of p and q, we have

$$\lim_{x\to\infty} \frac{2x+1}{x-5} = \lim_{x\to\infty} \frac{2x}{x} = 2.$$

(b) We have

$$\lim_{x\to-\infty} \frac{2+5x}{6x+3} = \lim_{x\to-\infty} \frac{5x}{6x} = \frac{5}{6}.$$

Problems

33. **(a)** $x^{1/n}$ is concave down: its values increase quickly at first and then more slowly as x gets larger. The function x^n, on the other hand, is concave up. Its values increase at an increasing rate as x gets larger. Thus

$$f(x) = x^{1/n}$$

and

$$g(x) = x^n.$$

(b) Since the point A is the intersection of $f(x)$ and $g(x)$, we want the solution of the equation $x^n = x^{1/n}$. Raising both sides to the power of n, we get

$$(x^n)^{n} = (x^{1/n})^{n}$$

or in other words

$$x^{n^2} = x.$$

Since $x \neq 0$ at the point A, we can divide both sides by x, giving

$$x^{n^2-1} = 1.$$

Since $n^2 - 1$ is just some integer we rewrite the equation as

$$x^p = 1$$

where $p = n^2 - 1$. If p is even, $x = \pm 1$, if p is odd $x = 1$. By looking at the graph we can tell that we are not interested in the situation when $x = -1$. When $x = 1$, the quantities x^n and $x^{1/n}$ both equal 1. Thus the coordinates of point A are $(1, 1)$.

37. **(a)** The function w is linear because equal spacing between successive input values ($\Delta x = 1$) results in equal spacing between successive output values ($\Delta y = 0.4$). The function z is exponential because ratios of successive output values are equal:

$$\frac{16}{64} = \frac{4}{16} = \frac{1}{4} = \frac{0.25}{1} = \frac{0.06}{0.25} = 0.25$$

The function v has to be the trigonometric function because it increases, then decreases, and then increases again, and none of the other three types of functions can exhibit this behavior. This means that u must be the power function by the process of elimination.

(b) Since z is exponential, we know that $z(x) = ab^x$, and we are given that $z(1) = 64$ and $z(2) = 16$. This means that $ab = 64$ and $ab^2 = 16$. Solving for a in the first equation yields $a = 64/b$, which we substitute into the second equation to solve for b:

$$\begin{aligned} ab^2 &= 16 \\ \frac{64}{b} \cdot b^2 &= 16 \\ 64b &= 16 \\ b &= \frac{1}{4}. \end{aligned}$$

Therefore, we have $a = 64/b = 256$, which yields a final answer of

$$z(x) = 256 \left(\frac{1}{4}\right)^x.$$

(c) Since w is linear, we know that $w(x) = b + mx$. The slope of the line is given by

$$m = \frac{-0.94 - (-1.34)}{2 - 1} = 0.4.$$

To find b, we note that $w(1) = -1.34$, so we have

$$\begin{aligned} w(x) &= 0.4x + b \\ -1.34 &= 0.4(1) + b \\ -1.74 &= b. \end{aligned}$$

Therefore, our final answer is $w(x) = -1.74 + 0.4x$.

(d) Since u is a power function, we know that $u(x) = kx^p$, and we are given that $u(1) = 0.5$ and $u(2) = 1.41$. Therefore, we have $k \cdot 1^p = 0.5$ and $k \cdot 2^p = 1.41$. The first equation yields $k = 0.5$, and substituting this value into the second equation yields

$$\begin{aligned} 0.5 \cdot 2^p &= 1.41 \\ 2^p &= 2.82 \\ p &= \frac{\ln 2.82}{\ln 2} \\ &= 1.50. \end{aligned}$$

Therefore, our answer is $u(x) = 0.5x^{1.5}$.

(e) Since g is a trigonometric function, we know that it must have the form $v(x) = A\sin(B(x-h)) + k$ or $v(x) = A\cos(B(x-h)) + k$. We can see from the data that $k = 0.5$ is the midline of the function, the amplitude is given by

$$A = \frac{0.75 - 0.25}{2} = 0.25,$$

and that the period is given by $6 - 2 = 4$ units. Therefore, we have

$$\frac{2\pi}{B} = 4$$
$$B = \frac{\pi}{2}$$

If we start reading the table of data for v from the point $x = 2$ and read from left to right, we notice that the values of v start from the midline and then increase, matching the behavior of the standard sine function. Therefore, we can choose to represent v with a sine function and a horizontal shift of $h = 2$, yielding a final answer of

$$v(x) = 0.25\sin\left(\frac{\pi}{2}(x-2)\right) + 0.5.$$

41. The shape of the graph suggests an odd-degree polynomial with a positive leading term. Since the graph crosses the x-axis at 0, -2, and 2, there are factors of x, $(x+2)$ and $(x-2)$, giving $y = ax(x+2)(x-2)$. To find a we use the fact that at $x = 1$, $y = -6$. Substituting gives:

$$-6 = a(1)(1+2)(1-2)$$
$$-6 = -3a$$
$$2 = a.$$

Thus, a possible polynomial is $y = 2x(x+2)(x-2)$.

45. We use the position of the "bounce" on the x-axis to indicate a multiple zero at that point. Since there is not a sign change at those points, the zero occurs an even number of times. Letting

$$y = k(x+2)^2(x)(x-2)^2$$

represent (c), we use the point $(1, -3)$ to get

$$-3 = f(1) = k(3)^2(1)(-1)^2,$$

so

$$-3 = 9k,$$
$$k = -\frac{1}{3}.$$

Thus, a possible formula is

$$y = -\frac{1}{3}(x+2)^2(x)(x-2)^2.$$

49. **(a)** The graph shows the graph of $y = 1/x^2$ shifted up 2 units. Therefore, a formula is

$$y = \frac{1}{x^2} + 2.$$

(b) The equation $y = (1/x^2) + 2$ can be written as

$$y = \frac{2x^2 + 1}{x^2}.$$

(c) We see that the graph has no intercepts on either axis. Algebraically this is seen by the fact that $x = 0$ is not in the domain of the function, and there are no real solutions to $2x^2 + 1 = 0$.

53. Notice that $f(x) = (x-3)^2$ has one zero (at $x = 3$), $g(x) = (x-2)(x+2)$ has 2 zeros (at $x = 2$ and $x = -2$), $h(x) = x+1$ has one zero (at $x = -1$), and $j(x)$ has no zeros.

(a) $s(x) = \dfrac{x^2-4}{x^2+1}$ has 2 zeros, no vertical asymptote, and a horizontal asymptote at $y = 1$.

(b) $r(x) = (x-3)^2(x+1)$ has 2 zeros, no vertical asymptote, and no horizontal asymptote.

(c) $h(x)/f(x)$ fits this description, but is not among the functions above.

(d) $p(x) = \dfrac{(x-3)^2}{x^2-4}$ and $q(x) = \dfrac{x+1}{x^2-4}$ each have 1 zero, 2 vertical asymptotes, and a horizontal asymptote. p has a horizontal asymptote at $y = 1$, and q has a horizontal asymptote at $y = 0$.

(e) $v(x) = \dfrac{x^2+1}{(x-3)^2}$ has no zeros, 1 vertical asymptote, and a horizontal asymptote at $y = 1$.

(f) $t(x) = \dfrac{1}{h(x)} = \dfrac{1}{x+1}$ has no zeros, 1 vertical asymptote, and a horizontal asymptote at $y = 0$.

57. A denominator of $(x+1)$ will give the vertical asymptote at $x = -1$. The numerator will have a highest-powered term of $1 \cdot x^1$ to give a horizontal asymptote of $y = 1$. If there is a zero at $x = -3$, try

$$f(x) = \frac{(x+3)}{(x+1)}.$$

Note, this agrees with the y-intercept at $y = 3$.

61.

$$\begin{aligned} p(x) &= k(x+3)(x-2)(x-5)(x-6)^2 \\ 7 = p(0) &= k(3)(-2)(-5)(-6)^2 \\ &= k(1080) \\ k &= \frac{7}{1080} \\ p(x) &= \frac{7}{1080}(x+3)(x-2)(x-5)(x-6)^2 \end{aligned}$$

65. Since the frequency, f, is inversely proportional to the length, L, we have

$$f = k\frac{1}{L},$$

so

$$L = k\frac{1}{f}.$$

Thus, the length of the string is inversely proportional to the frequency.

69. **(a)** The graph of $C(x) = (x-1)^3 + 1$ is the graph of $y = x^3$ shifted right one unit and up one unit. The graph is shown in Figure 11.19.

(b) The price is \$1000 per unit, since $R(1) = 1$ means selling 1000 units yields \$1,000,000.

(c)

$$\begin{aligned} \text{Profit} &= R(x) - C(x) \\ &= x - [(x-1)^3 + 1] \\ &= x - (x-1)^3 - 1. \end{aligned}$$

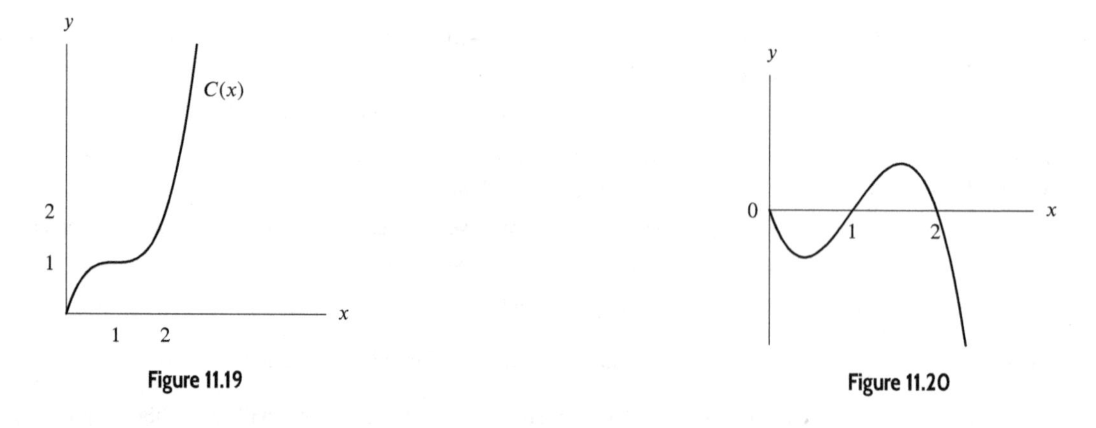

Figure 11.19

Figure 11.20

The graph of $R(x) - C(x)$ is shown in Figure 11.20. Profit is negative for $x < 1$ and for $x > 2$. Profit $= 0$ at $x = 1$ and $x = 2$. Thus, the firm will break even with either 1000 or 2000 units, make a profit for $1000 < x < 2000$ units, and lose money for any number of units between 0 and 1000 or greater than 2000.

73. **(a)** See Figure 11.21.

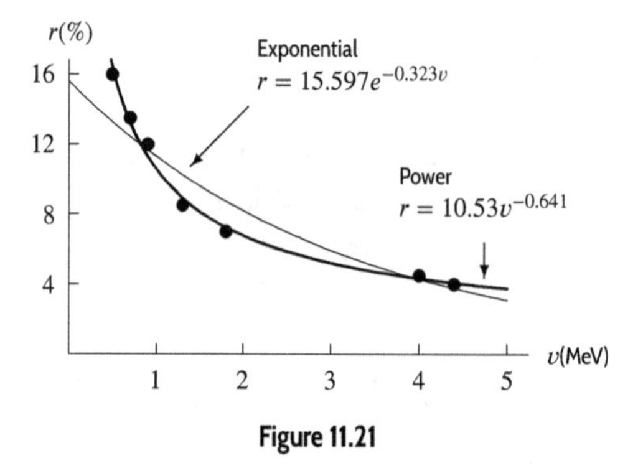

Figure 11.21

(b) The value of r tends to go down as v increases. This means that the telescope is better able to distinguish between high-energy gamma ray photons than low-energy ones.

(c) The first curve is a power function given by $r = 10.53v^{-0.641}$. The second is an exponential function given by $r = 15.597e^{-0.323v}$. The power function appears to give a better fit.

(d) The power function predicts that $r \to \infty$ as $v \to 0$, and so is most consistent with the prediction that the telescope gets rapidly worse and worse at low energies. In contrast, exponential function predicts that r gets close to 15.6% as E gets close to 0.

STRENGTHEN YOUR UNDERSTANDING

1. False. The quadratic function $y = 3x^2 + 5$ is not of the form $y = kx^n$, so it is not a power function.

5. True. All positive even power functions have an upward opening U shape.

9. True. The x-axis is an asymptote for $f(x) = x^{-1}$, so the values approach zero.

13. False. As x grows very large the exponential decay function g approaches the x-axis faster than any power function with a negative power.

17. False. For example, the polynomial $x^2 + x^3$ has degree 3 because the degree is the highest power, not the first power, in the formula for the polynomial.

21. True. The graph crosses the y-axis at the point $(0, p(0))$.

25. True. We can write $p(x) = (x - a) \cdot C(x)$. Evaluating at $x = a$, we get $p(a) = (a - a) \cdot C(a) = 0 \cdot C(a) = 0$.

29. True. This is the definition of a rational function.

33. True. The ratio of the highest-degree terms in the numerator and denominator is $2x/x^2 = 2/x$, so for large positive x-values, y approaches 0.

37. False. The ratio of the highest-degree terms in the numerator and denominator is $3x^4/x^2 = 3x^2$. So for large positive x-values, y behaves like $y = 3x^2$.

41. True. At $x = -4$, we have $f(-4) = (-4+4)/(-4-3) = 0/(-7) = 0$, so $x = -4$ is a zero.

45. False. If $p(x)$ has no zeros, then $r(x)$ has no zeros. For example, if $p(x)$ is a nonzero constant or $p(x) = x^2 + 1$, then $r(x)$ has no zeros.

Solutions to Skills Review for Chapter 11

1. $\dfrac{3}{5} + \dfrac{4}{7} = \dfrac{3\cdot 7 + 4\cdot 5}{35} = \dfrac{21+20}{35} = \dfrac{41}{35}$

5. $\dfrac{-2}{yz} + \dfrac{4}{z} = \dfrac{-2z + 4yz}{yz^2} = \dfrac{-2+4y}{yz} = \dfrac{-2(1-2y)}{yz}$

9. $\dfrac{\frac{5}{6}}{15} = \dfrac{5}{6}\cdot\dfrac{1}{15} = \dfrac{1}{18}$

13. $\dfrac{4z}{x^2y} - \dfrac{3w}{xy^4} = \dfrac{4zxy^4 - 3wx^2y}{x^3y^5} = \dfrac{xy(4zy^3 - 3wx)}{x^3y^5} = \dfrac{4y3z - 3wx}{x^2y^4}$

17.

$$\begin{aligned}\frac{8}{3x^2 - x - 4} - \frac{9}{x+1} &= \frac{8}{(x+1)(3x-4)} - \frac{9}{x+1}\\ &= \frac{8 - 9(3x-4)}{(x+1)(3x-4)}\\ &= \frac{-27x+44}{(x+1)(3x-4)}\end{aligned}$$

21. The second denominator $4r^2 + 6r = 2r(2r+3)$, while the first denominator is $2r+3$. Therefore the common denominator is $2r(2r+3)$. We have:

$$\begin{aligned}\frac{1}{2r+3} + \frac{3}{4r^2+6r} &= \frac{1}{2r+3} + \frac{3}{2r(2r+3)}\\ &= \frac{1\cdot 2r}{2r(2r+3)} + \frac{3}{2r(2r+3)}\\ &= \frac{2r+3}{2r(2r+3)} = \frac{1}{2r}.\end{aligned}$$

25. If we factor the number and denominator of the second fraction, we can cancel some terms with the first,

$$\frac{a+b}{2}\cdot\frac{8x+2}{b^2-a^2} = \frac{a+b}{2}\cdot\frac{2(4x+1)}{(b+a)(b-a)} = \frac{4x+1}{b-a}.$$

29. $\dfrac{2a+3}{(a+3)(a-3)}$

33. We expand within the first brackets first. Therefore,

$$\begin{aligned}\frac{[4-(x+h)^2] - [4-x^2]}{h} &= \frac{[4-(x^2+2xh+h^2)] - [4-x^2]}{h}\\ &= \frac{[4-x^2-2xh-h^2] - 4 + x^2}{h} = \frac{-2xh - h^2}{h}\\ &= -2x - h.\end{aligned}$$

37. $$\frac{\frac{3}{xy} - \frac{5}{x^2y}}{\frac{6x^2 - 7x - 5}{x^4y^2}} = \frac{\frac{3x - 5}{x^2y}}{\frac{(3x - 5)(2x + 1)}{y^4y^2}} = \frac{3x - 5}{x^2y} \cdot \frac{x^4y^2}{(3x - 5)(2x + 1)} = \frac{x^2y}{2x + 1}$$

41. Dividing $2x^3$ into each term in the numerator yields:

$$\frac{26x + 1}{2x^3} = \frac{26x}{2x^3} + \frac{1}{2x^3} = \frac{13}{x^2} + \frac{1}{2x^3}.$$

45.

$$\frac{\frac{1}{3}x - \frac{1}{2}}{2x} = \frac{\frac{x}{3}}{2x} - \frac{\frac{1}{2}}{2x} = \frac{x}{3} \cdot \frac{1}{2x} - \frac{1}{2} \cdot \frac{1}{2x} = \frac{1}{6} - \frac{1}{4x}$$

49. Dividing the denominator R into each term in the numerator yields

$$\frac{R+1}{R} = \frac{R}{R} + \frac{1}{R} = 1 + \frac{1}{R}.$$

53. False

57. True

CHAPTER TWELVE

Solutions for Section 12.1

Exercises

1. We can describe an elevation with one number, so this is a scalar.

5. Scalar.

9.

$$\vec{p} = 2\vec{w}, \quad \vec{q} = -\vec{u}, \quad \vec{r} = \vec{w} + \vec{u} = \vec{u} + \vec{w},$$
$$\vec{s} = \vec{p} + \vec{q} = 2\vec{w} - \vec{u}, \quad \vec{t} = \vec{u} - \vec{w}$$

Problems

13. See Figure 12.1.

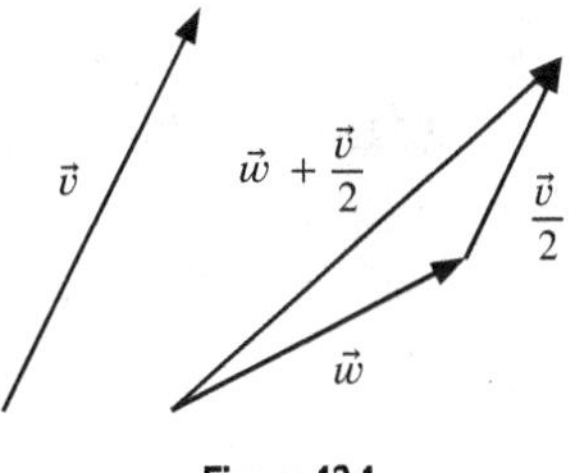

Figure 12.1

17. **(a)** From Figure 12.2,

$$\text{Distance from Oracle Road} = 5\sin 20^\circ = 1.710 \text{ miles}.$$

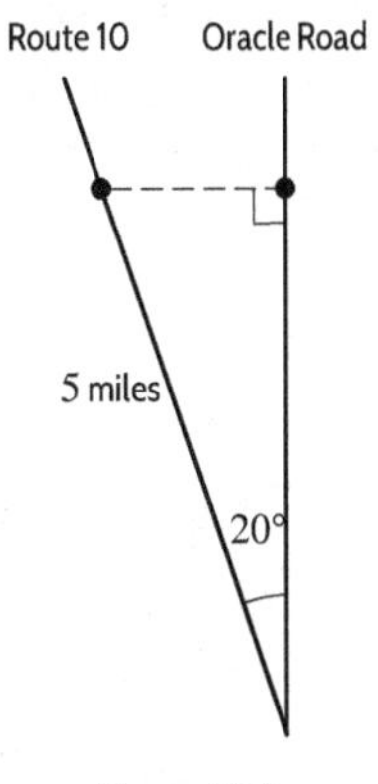

Figure 12.2

(b) If the distance along Route 10 is x miles, we have

$$x\sin 20^\circ = 2 \text{ miles}$$
$$x = \frac{2}{\sin 20^\circ} = 5.848 \text{ miles}.$$

21. 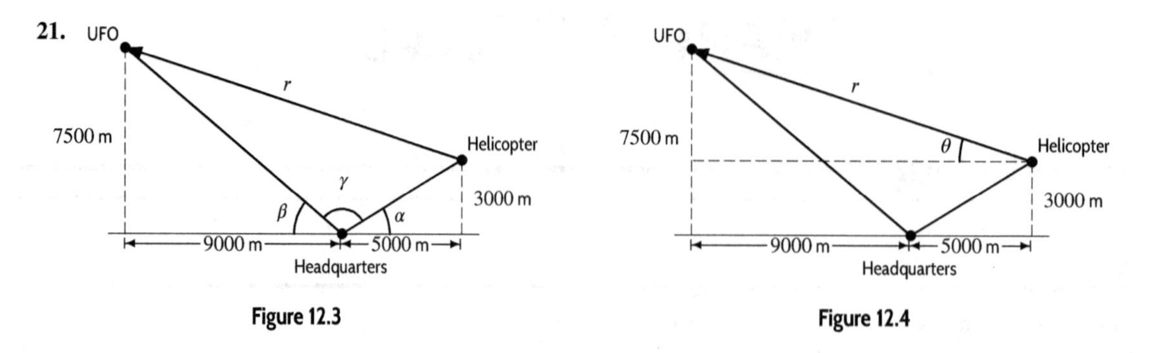

Figure 12.3 Figure 12.4

Figure 12.3 shows the headquarters at the origin, and a positive y-value as up, and a positive x-value as east. To solve for r, we must first find γ:

$$\begin{aligned}\gamma &= 180^\circ - \alpha - \beta \\ &= 180^\circ - \arctan\frac{3000}{5000} - \arctan\frac{7500}{9000} \\ &= 109.231^\circ.\end{aligned}$$

We now can find r using the Law of Cosines in the triangle formed by the position of the headquarters, the helicopter, and the UFO.

In kilometers:

$$\begin{aligned}r^2 &= 34 + 137.250 - 2 \cdot \sqrt{34} \cdot \sqrt{137.250} \cdot \cos\gamma \\ r^2 &= 216.250 \\ r &= 14.705 \text{ km} \\ &= 14{,}705 \text{ m}.\end{aligned}$$

From Figure 12.4 we see:

$$\begin{aligned}\tan\theta &= \frac{4500}{14{,}000} \\ \theta &= 17.819^\circ.\end{aligned}$$

Therefore, the helicopter must fly 14,705 meters with an angle of 17.819° from the horizontal.

25. The vector $\vec{v} + \vec{w}$ is equivalent to putting the vectors $\overrightarrow{OA}$ and $\overrightarrow{AB}$ end-to-end as shown in Figure 12.5; the vector $\vec{w} + \vec{v}$ is equivalent to putting the vectors $\overrightarrow{OC}$ and $\overrightarrow{CB}$ end-to-end. Since they form a parallelogram, $\vec{v} + \vec{w}$ and $\vec{w} + \vec{v}$ are both equal to the vector $\overrightarrow{OB}$, we have $\vec{v} + \vec{w} = \vec{w} + \vec{v}$.

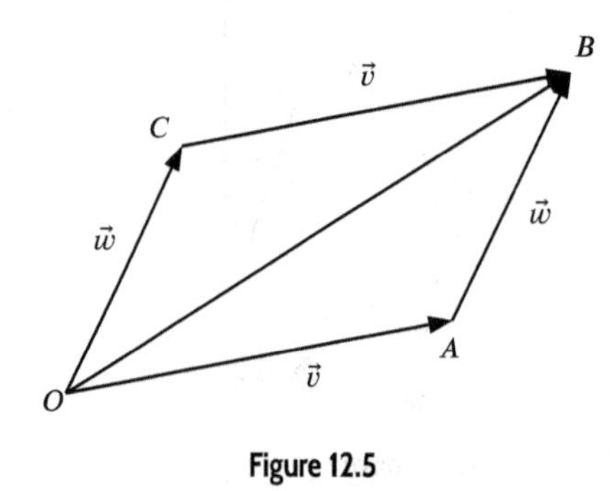

Figure 12.5

29. According to the definition of scalar multiplication, $1 \cdot \vec{v}$ has the same direction and magnitude as $\vec{v}$, so it is the same as $\vec{v}$.

Solutions for Section 12.2

Exercises

1. The vector we want is the displacement from Q to P, which is given by

$$\overrightarrow{QP} = (1-4)\vec{i} + (2-6)\vec{j} = -3\vec{i} - 4\vec{j}.$$

5. $4\vec{i} + 2\vec{j} - 3\vec{i} + \vec{j} = \vec{i} + 3\vec{j}$.

9. $\|\vec{v}\| = \sqrt{1^2 + (-1)^2 + 3^2} = \sqrt{11} \approx 3.317$

Problems

13. The velocity of the ship in still water is $10\vec{j}$ knots and the velocity of the current is $-5\vec{i}$ since the current is east to west. The velocity of the ship is $-5\vec{i} + 10\vec{j}$ knots.

17. Figure 12.6 shows the vector $\vec{w}$ redrawn to show that it is perpendicular to the displacement vector $\overrightarrow{PQ}$, which lies along the dotted line. Thus, the angle is $90°$ or $\pi/2$.

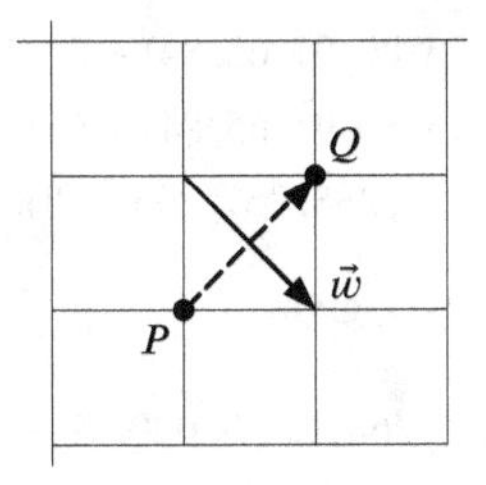

Figure 12.6

21. We need to calculate the length of each vector.

$$\|21\vec{i} + 35\vec{j}\| = \sqrt{21^2 + 35^2} = \sqrt{1666} \approx 40.8,$$
$$\|40\vec{i}\| = \sqrt{40^2} = 40.$$

So the first car is faster.

25. **(a)** The velocity, $\vec{v}$, is represented by a vector of length 5 in a northeasterly direction. The vector $\vec{i} + \vec{j}$ points northeast, but has length $\sqrt{1^2 + 1^2} = \sqrt{2}$. Thus,

$$\vec{v} = \frac{5}{\sqrt{2}}(\vec{i} + \vec{j}) = 3.536(\vec{i} + \vec{j})$$

(b) The current flows northward, so it is represented by $\vec{c} = 1.2\vec{j}$. The swimmer's velocity relative to the riverbed is

$$\vec{s} = \vec{c} + \vec{v} = 1.2\vec{j} + 3.536(\vec{i} + \vec{j}) = 3.536\vec{i} + 4.736\vec{j}.$$

29. We get displacement by subtracting the coordinates of the bottom of the tree, $(2, 4, 0)$, from the coordinates of the squirrel, $(2, 4, 1)$, giving:

$$\text{Displacement} = (2-2)\vec{i} + (4-4)\vec{j} + (1-0)\vec{k} = \vec{k}.$$

Solutions for Section 12.3

Exercises

1. $\vec{B} = 2\vec{M} = 2(1,1,2,3,5,8) = (2,2,4,6,10,16)$.

5. $\vec{K} = \frac{\vec{N}}{3} + \frac{2\vec{N}}{3} = \frac{3\vec{N}}{3} = \vec{N} = (5,6,7,8,9,10)$.

9. Since the components of $\vec{Q}$ represent millions of people, an increase of 120,000 people will increase each component by 0.12. Therefore,

$$\begin{aligned}\vec{S} &= \vec{Q} + (0.12, 0.12, 0.12, 0.12, 0.12, 0.12)\\ &= (3.57, 1.33, 6.55, 1.32, 1.05, 0.63) + (0.12, 0.12, 0.12, 0.12, 0.12, 0.12)\\ &= (3.69, 1.45, 6.67, 1.44, 1.17, 0.75).\end{aligned}$$

Problems

13. The total scores are out of 300 and are given by the total score vector $\vec{v} + 2\vec{w}$:

$$\begin{aligned}\vec{v} + 2\vec{w} &= (73, 80, 91, 65, 84) + 2(82, 79, 88, 70, 92)\\ &= (73, 80, 91, 65, 84) + (164, 158, 176, 140, 184)\\ &= (237, 238, 267, 205, 268).\end{aligned}$$

To get the scores as a percentage, we divide by 3, giving

$$\frac{1}{3}(237, 238, 267, 205, 268) \approx (79.000, 79.333, 89.000, 68.333, 89.333).$$

17. **(a)** See the sketch in Figure 12.7, where $\vec{v}$ represents the first part of the man's walk, and $\vec{w}$ represents the second part. Since the man first walks 5 miles, we know $\|\vec{v}\| = 5$. Since he walks 30° north of east, resolving gives

$$\vec{v} = 5\cos 30°\vec{i} + 5\sin 30°\vec{j} = 4.330\vec{i} + 2.500\vec{j}.$$

For the second leg of his journey, the man walks a distance x miles due east, so $\vec{w} = x\vec{i}$.

(b) The vector from finish to start is $-(\vec{v} + \vec{w}) = -(4.330 + x)\vec{i} - 2.500\vec{j}$. This vector is at an angle of 10° south of west. So, using the magnitudes of the sides in the triangle in Figure 12.8:

$$\begin{aligned}\frac{2.500}{4.330 + x} &= \tan(10°) = 0.176\\ 2.500 &= 0.176(4.330 + x)\\ x &= \frac{2.500 - 0.176 \cdot 4.330}{0.176} = 9.848.\end{aligned}$$

This means that $x = 9.848$.

Figure 12.7

Figure 12.8

(c) The distance from the starting point is $\| -(4.330 + 9.848)\vec{i} - (2.500)\vec{j}\| = \sqrt{14.178^2 + 2.500^2} = 14.397$ miles.

21. In an actual video game, our rectangle would be replaced with a more sophisticated graphic (perhaps an airplane or an animated figure). But the principles involved in rotation about the origin are the same, and it will be easier to think about them using rectangles instead of fancy graphics.

We can represent the four corners of the rectangle (before rotation) using the position vectors $\vec{p}_a, \vec{p}_b, \vec{p}_c$, and $\vec{p}_d$. For instance, the components of $\vec{p}_a$ are $\vec{p}_a = 2\vec{i} + \vec{j}$.

After the rectangle has been rotated, its four corners are given by the position vectors $\vec{q}_a, \vec{q}_b, \vec{q}_c$, and $\vec{q}_d$. Notice that the lengths of these vectors have not changed; in other words,

$$||\vec{p}_a|| = ||\vec{q}_a||, \qquad ||\vec{p}_b|| = ||\vec{q}_b||, \qquad ||\vec{p}_c|| = ||\vec{q}_c||, \qquad \text{and} \qquad ||\vec{p}_d|| = ||\vec{q}_d||.$$

This is because in a rotation the only thing that changes is orientation, not length.

When the rectangle is rotated through a 35° angle, the angle made by corner a increases by 35°. So do the angles made by the other three corners. Letting θ be the angle made by corner a, we have

$$\begin{aligned} \tan\theta &= \frac{1}{2} \\ \theta &= \arctan 0.5 = 26.565^\circ. \end{aligned}$$

This is the direction of the position vector $\vec{p}_a$. After rotation, the angle θ is given by

$$\theta = 26.565^\circ + 35^\circ = 61.565^\circ.$$

This is the direction of the new position vector $\vec{q}_a$. The length of $\vec{q}_a$ is the same as the length of $\vec{p}_a$ and is given by

$$||\vec{q}_a||^2 = ||\vec{p}_a||^2 = 2^2 + 1^2 = 5,$$

and so $||\vec{q}_a|| = \sqrt{5}$. Thus, the components of $\vec{q}_a$ are given by

$$\begin{aligned} \vec{q}_a &= (\sqrt{5}\cos 61.565^\circ)\vec{i} + (\sqrt{5}\sin 61.565^\circ)\vec{j} \\ &= 1.065\vec{i} + 1.966\vec{j}. \end{aligned}$$

This process can be repeated for the other three corners. You can see for yourself that the angles made with the origin by the corners a, b, and c, respectively, are 14.036°, 26.565°, and 45°. After rotation, these angles are 49.036°, 61.565°, and 80°. Similarly, the lengths of the position vectors for these three points (both before and after rotation) are $\sqrt{17}$, $\sqrt{20}$, and $\sqrt{8}$. Thus, the final positions of these three points are

$$\begin{aligned} \vec{q}_b &= 2.703\vec{i} + 3.113\vec{j}, \\ \vec{q}_c &= 2.129\vec{i} + 3.933\vec{j}, \\ \vec{q}_d &= 0.491\vec{i} + 2.785\vec{j}. \end{aligned}$$

Solutions for Section 12.4

Exercises

1. $\vec{z} \cdot \vec{a} = (\vec{i} - 3\vec{j} - \vec{k}) \cdot (2\vec{j} + \vec{k}) = 1 \cdot 0 + (-3)2 + (-1)1 = 0 - 6 - 1 = -7.$

5. $\vec{a} \cdot \vec{b} = (2\vec{j} + \vec{k}) \cdot (-3\vec{i} + 5\vec{j} + 4\vec{k}) = 0(-3) + 2 \cdot 5 + 1 \cdot 4 = 0 + 10 + 4 = 14.$

9. Since $\vec{a} \cdot \vec{b}$ is a scalar and $\vec{a}$ is a vector, the expression is a vector parallel to $\vec{a}$. We have

$$\vec{a} \cdot \vec{b} = (2\vec{j} + \vec{k}) \cdot (-3\vec{i} + 5\vec{j} + 4\vec{k}) = 0(-3) + 2(5) + 1(4) = 14.$$

Thus,

$$(\vec{a} \cdot \vec{b}) \cdot \vec{a} = 14\vec{a} = 14(2\vec{j} + \vec{k}) = 28\vec{j} + 14\vec{k}.$$

Problems

13. We use the dot product to find this angle.

We have $\vec{w} = \vec{i} - \vec{j}$ and $\vec{v} = \vec{i} + 2\vec{j}$ so

$$\vec{w} \cdot \vec{v} = (\vec{i} - \vec{j}) \cdot (\vec{i} + 2\vec{j}) = -1,$$

therefore

$$\vec{w} \cdot \vec{v} = ||\vec{w}|| \cdot ||\vec{v}|| \cos\theta.$$

Since $||\vec{w}|| = \sqrt{5}$ and $||\vec{v}|| = \sqrt{2}$, and $\vec{w} \cdot \vec{v} = -1$, we have

$$\begin{aligned} -1 &= \sqrt{2}\sqrt{5}\cos\theta \\ \cos\theta &= -\frac{1}{\sqrt{10}} \\ \theta &= \arccos -\frac{1}{\sqrt{10}} = 108.435°. \end{aligned}$$

17. Using the dot product, the angle is given by

$$\cos\theta = \frac{(\vec{i} + \vec{j} + \vec{k}) \cdot (\vec{i} - \vec{j} - \vec{k})}{||\vec{i} + \vec{j} + \vec{k}||\,||\vec{i} - \vec{j} - \vec{k}||} = \frac{1 \cdot 1 + 1(-1) + 1(-1)}{\sqrt{1^1 + 1^2 + 1^2}\sqrt{1^2 + (-1)^2 + (-1)^2}} = -\frac{1}{3}.$$

So, $\theta = \arccos(-\frac{1}{3}) \approx 1.911$ radians, or $\approx 109.471°$.

21. It is clear from the Figure 12.9 that only angle $\angle CAB$ could possibly be a right angle. Subtraction of x, y values for the points gives $\overrightarrow{AB} = 3\vec{i} - \vec{j}$ and $\overrightarrow{AC} = 1\vec{i} + 2\vec{j}$. Taking the dot product yields $\overrightarrow{AB} \cdot \overrightarrow{AC} = (3)(1) + (-1)(2) = 1$. Since this is nonzero, the angle cannot be a right angle.

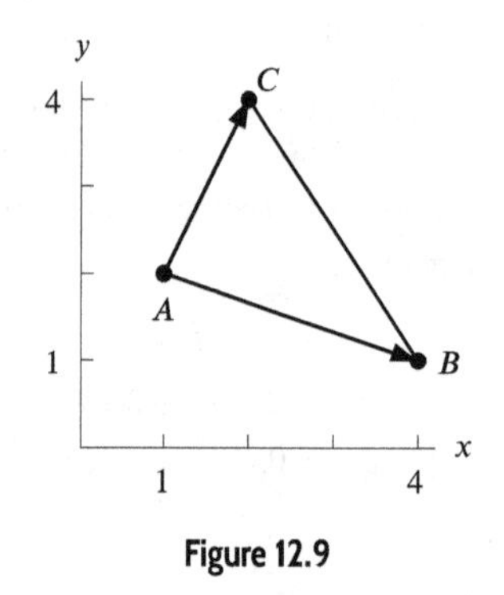

Figure 12.9

25. **(a)** We have the price vector $\vec{a} = (3, 2, 4)$. Let the consumption vector $\vec{c} = (c_b, c_e, c_m)$, then $3c_b + 2c_e + 4c_m = 40$ or $\vec{a} \cdot \vec{c} = 40$.

(b) Note $\vec{a} \cdot \vec{c}$ is the cost of consuming $\vec{c}$ groceries at Acme Store, so $\vec{b} \cdot \vec{c}$ is the cost of consuming $\vec{c}$ groceries at Beta Mart. Thus $\vec{b} \cdot \vec{c} - \vec{a} \cdot \vec{c} = (\vec{b} - \vec{a}) \cdot \vec{c}$ is the difference in costs between Beta and Acme for the same $\vec{c}$ groceries.

For $\vec{b} - \vec{a}$ to be perpendicular to $\vec{c}$, we must have $(\vec{b} - \vec{a}) \cdot \vec{c} = 0$. Since $\vec{b} - \vec{a} = (0.20, -0.20, 0.50)$, the vector $\vec{b} - \vec{a}$ is perpendicular to $\vec{c}$ if $0.20c_b - 0.20c_e + 0.50c_m = 0$. For example, this occurs when we consume the same number of loaves of bread as dozens of eggs, but no milk.

(c) Since $\vec{b} \cdot \vec{c}$ is the cost of groceries at Beta, you might think of $(1/1.1)\vec{b} \cdot \vec{c}$ as the "freshness-adjusted" cost at Beta. Then $(1/1.1)\vec{b} \cdot \vec{c} < \vec{a} \cdot \vec{c}$ means the "freshness-adjusted" cost is lower at Beta.

29. **(a)** We have

$$\begin{aligned} \vec{r} \cdot \vec{w} &= (x_1, y_1, x_2, y_2) \cdot (-1, 0, 1, 0) \\ &= -x_1 + x_2 = x_2 - x_1, \end{aligned}$$

so, since $x_2 > x_1$, this quantity represents the width w.

(b) We have

$$\vec{r}\cdot\vec{h} = (x_1, y_1, x_2, y_2)\cdot(0,-1,0,1) = -y_1 + y_2 = y_2 - y_1,$$

so, since $y_2 > y_1$, this quantity represents the height h.

(c) We have

$$2\vec{r}\cdot(\vec{w}+\vec{h}) = 2\vec{r}\cdot\vec{w} + 2\vec{r}\cdot\vec{h} = 2w + 2h,$$

where from part (a) w is the width and from part (b) h is the height. Thus, this quantity represents the perimeter p.

Solutions for Section 12.5

Exercises

1. (a) We have

$$5\mathbf{R} = 5\begin{pmatrix} 3 & 7 \\ 2 & -1 \end{pmatrix} = \begin{pmatrix} 5\cdot 3 & 5\cdot 7 \\ 5\cdot 2 & 5\cdot -1 \end{pmatrix} = \begin{pmatrix} 15 & 35 \\ 10 & -5 \end{pmatrix}.$$

(b) We have

$$-2\mathbf{S} = -2\begin{pmatrix} 1 & -5 \\ 0 & 8 \end{pmatrix} = \begin{pmatrix} -2\cdot 1 & -2\cdot -5 \\ -2\cdot 0 & -2\cdot 8 \end{pmatrix} = \begin{pmatrix} -2 & 10 \\ 0 & -16 \end{pmatrix}.$$

(c) We have

$$\mathbf{R}+\mathbf{S} = \begin{pmatrix} 3 & 7 \\ 2 & -1 \end{pmatrix} + \begin{pmatrix} 1 & -5 \\ 0 & 8 \end{pmatrix} = \begin{pmatrix} 3+1 & 7-5 \\ 2+0 & -1+8 \end{pmatrix} = \begin{pmatrix} 4 & 2 \\ 2 & 7 \end{pmatrix}.$$

(d) Writing $\mathbf{S} - 3\mathbf{R} = \mathbf{S} + (-3)\mathbf{R}$, we first find $-3\mathbf{R}$:

$$-3\mathbf{R} = -3\begin{pmatrix} 3 & 7 \\ 2 & -1 \end{pmatrix} = \begin{pmatrix} -3\cdot 3 & -3\cdot 7 \\ -3\cdot 2 & -3\cdot -1 \end{pmatrix} = \begin{pmatrix} -9 & -21 \\ -6 & 3 \end{pmatrix}.$$

This gives

$$\mathbf{S}+(-3)\mathbf{R} = \begin{pmatrix} 1 & -5 \\ 0 & 8 \end{pmatrix} + \begin{pmatrix} -9 & -21 \\ -6 & 3 \end{pmatrix} = \begin{pmatrix} 1-9 & -5-21 \\ 0-6 & 8+3 \end{pmatrix} = \begin{pmatrix} -8 & -26 \\ -6 & 11 \end{pmatrix}.$$

(e) Writing $\mathbf{R} + 2\mathbf{R} + 2(\mathbf{R} - \mathbf{S}) = 5\mathbf{R} + (-2)\mathbf{S}$, we use our answers to parts (a) and (b):

$$5\mathbf{R}+(-2)\mathbf{S} = \begin{pmatrix} 15 & 35 \\ 10 & -5 \end{pmatrix} + \begin{pmatrix} -2 & 10 \\ 0 & -16 \end{pmatrix} = \begin{pmatrix} 15-2 & 35+10 \\ 10+0 & -5-16 \end{pmatrix} = \begin{pmatrix} 13 & 45 \\ 10 & -21 \end{pmatrix}.$$

(f) We have

$$k\mathbf{S} = \begin{pmatrix} k\cdot 1 & k\cdot -5 \\ k\cdot 0 & k\cdot 8 \end{pmatrix} = \begin{pmatrix} k & -5k \\ 0 & 8k \end{pmatrix}.$$

5. **(a)** We have

$$\mathbf{A}\vec{u} = \begin{pmatrix} 2 & 5 & 7 \\ 4 & -6 & 3 \\ 16 & -5 & 0 \end{pmatrix}\begin{pmatrix} 3 \\ 2 \\ 5 \end{pmatrix}$$

$$= \begin{pmatrix} 2\cdot 3 + 5\cdot 2 + 7\cdot 5 \\ 4\cdot 3 - 6\cdot 2 + 3\cdot 5 \\ 16\cdot 3 - 5\cdot 2 + 0\cdot 5 \end{pmatrix} = \begin{pmatrix} 51 \\ 15 \\ 38 \end{pmatrix}.$$

(b) We have

$$\mathbf{B}\vec{v} = \begin{pmatrix} 8 & -6 & 0 \\ 5 & 3 & -2 \\ 3 & 7 & 12 \end{pmatrix}\begin{pmatrix} -1 \\ 0 \\ 3 \end{pmatrix}$$

$$= \begin{pmatrix} 8\cdot -1 - 6\cdot 0 + 0\cdot 3 \\ 5\cdot -1 + 3\cdot 0 - 2\cdot 3 \\ 3\cdot -1 + 7\cdot 0 + 12\cdot 3 \end{pmatrix} = \begin{pmatrix} -8 \\ -11 \\ 33 \end{pmatrix}.$$

(c) Letting $\vec{w} = \vec{u} + \vec{v} = (2, 2, 8)$, we have:

$$\mathbf{A}\vec{w} = \begin{pmatrix} 2 & 5 & 7 \\ 4 & -6 & 3 \\ 16 & -5 & 0 \end{pmatrix}\begin{pmatrix} 2 \\ 2 \\ 8 \end{pmatrix}$$

$$= \begin{pmatrix} 2\cdot 2 + 5\cdot 2 + 7\cdot 8 \\ 4\cdot 2 - 6\cdot 2 + 3\cdot 8 \\ 16\cdot 2 - 5\cdot 2 + 0\cdot 8 \end{pmatrix} = \begin{pmatrix} 70 \\ 20 \\ 22 \end{pmatrix}.$$

Another to work this problem would be to write $\mathbf{A}(\vec{u} + \vec{v})$ as $\mathbf{A}\vec{u} + \mathbf{A}\vec{v}$ and proceed accordingly.

(d) Letting $\mathbf{C} = \mathbf{A} + \mathbf{B}$, we have

$$\mathbf{C} = \begin{pmatrix} 2 & 5 & 7 \\ 4 & -6 & 3 \\ 16 & -5 & 0 \end{pmatrix} + \begin{pmatrix} 8 & -6 & 0 \\ 5 & 3 & -2 \\ 3 & 7 & 12 \end{pmatrix} = \begin{pmatrix} 10 & -1 & 7 \\ 9 & -3 & 1 \\ 19 & 2 & 12 \end{pmatrix}.$$

We can now write $(\mathbf{A} + \mathbf{B})\vec{v}$ as $\mathbf{C}\vec{v}$, and so:

$$\mathbf{C}\vec{v} = \begin{pmatrix} 10 & -1 & 7 \\ 9 & -3 & 1 \\ 19 & 2 & 12 \end{pmatrix}\begin{pmatrix} -1 \\ 0 \\ 3 \end{pmatrix}$$

$$= \begin{pmatrix} 10\cdot -1 - 1\cdot 0 + 7\cdot 3 \\ 9\cdot -1 - 3\cdot 0 + 1\cdot 3 \\ 19\cdot -1 + 2\cdot 0 + 12\cdot 3 \end{pmatrix} = \begin{pmatrix} 11 \\ -6 \\ 17 \end{pmatrix}.$$

(e) From part (a) we have $\mathbf{A}\vec{u} = (51, 15, 38)$, and from part (b) we have $\mathbf{B}\vec{v} = (-8, -11, 33)$. This gives

$$\begin{aligned} \mathbf{A}\vec{u} \cdot \mathbf{B}\vec{v} &= (51, 15, 38) \cdot (-8, -11, 33) \\ &= 51 \cdot -8 + 15 \cdot -11 + 38 \cdot 33 = 681. \end{aligned}$$

(f) We have $\vec{u}\cdot\vec{v} = 3\cdot -1 + 2\cdot 0 + 5\cdot 3 = 12$, and so

$$(\vec{u}\cdot\vec{v})\mathbf{A} = 12\mathbf{A} = 12\begin{pmatrix} 2 & 5 & 7 \\ 4 & -6 & 3 \\ 16 & -5 & 0 \end{pmatrix} = \begin{pmatrix} 24 & 60 & 84 \\ 48 & -72 & 36 \\ 192 & -60 & 0 \end{pmatrix}.$$

Problems

9. **(a)** We have

$$\begin{aligned} s_{\text{new}} &= s_{\text{old}} - \underbrace{0.10s_{\text{old}}}_{\text{10\% infected}} \\ &= 0.90s_{\text{old}} \\ i_{\text{new}} &= i_{\text{old}} + \underbrace{0.10s_{\text{old}}}_{\text{10\% infected}} - \underbrace{0.50i_{\text{old}}}_{\text{50\% recover}} + \underbrace{0.02r_{\text{old}}}_{\text{2\% reinfected}} \\ &= 0.10s_{\text{old}} + 0.50i_{\text{old}} + 0.02r_{\text{old}} \\ r_{\text{new}} &= r_{\text{old}} + \underbrace{0.50i_{\text{old}}}_{\text{50\% recover}} - \underbrace{0.02r_{\text{old}}}_{\text{2\% reinfected}} \\ &= 0.50i_{\text{old}} + 0.98r_{\text{old}}. \end{aligned}$$

Using matrix multiplication, we can rewrite these three equations as

$$\begin{pmatrix} s_{\text{new}} \\ i_{\text{new}} \\ r_{\text{new}} \end{pmatrix} = \begin{pmatrix} 0.90 & 0 & 0 \\ 0.10 & 0.50 & 0.02 \\ 0 & 0.50 & 0.98 \end{pmatrix}\begin{pmatrix} s_{\text{old}} \\ i_{\text{old}} \\ r_{\text{old}} \end{pmatrix},$$

and so $\vec{p}_{\text{new}} = \mathbf{T}\vec{p}_{\text{old}}$ where $\mathbf{T} = \begin{pmatrix} 0.90 & 0 & 0 \\ 0.10 & 0.50 & 0.02 \\ 0 & 0.50 & 0.98 \end{pmatrix}.$

(b) We have

$$\begin{aligned} \vec{p}_1 = \mathbf{T}\vec{p}_0 &= \begin{pmatrix} 0.90 & 0 & 0 \\ 0.10 & 0.50 & 0.02 \\ 0 & 0.50 & 0.98 \end{pmatrix}\begin{pmatrix} 2.0 \\ 0.0 \\ 0.0 \end{pmatrix} \\ &= \begin{pmatrix} 0.9(2) + 0(0) + 0(0) \\ 0.1(2) + 0.5(0) + 0.02(0) \\ 0(2) + 0.5(0) + 0.98(0) \end{pmatrix} = \begin{pmatrix} 1.8 \\ 0.2 \\ 0.0 \end{pmatrix} \\ \vec{p}_2 = \mathbf{T}\vec{p}_1 &= \begin{pmatrix} 0.90 & 0 & 0 \\ 0.10 & 0.50 & 0.02 \\ 0 & 0.50 & 0.98 \end{pmatrix}\begin{pmatrix} 1.8 \\ 0.2 \\ 0.0 \end{pmatrix} \\ &= \begin{pmatrix} 0.9(1.8) + 0(0.2) + 0(0) \\ 0.1(1.8) + 0.5(0.2) + 0.02(0) \\ 0(1.8) + 0.5(0.2) + 0.98(0) \end{pmatrix} = \begin{pmatrix} 1.62 \\ 0.28 \\ 0.10 \end{pmatrix} \\ \vec{p}_3 = \mathbf{T}\vec{p}_2 &= \begin{pmatrix} 0.90 & 0 & 0 \\ 0.10 & 0.50 & 0.02 \\ 0 & 0.50 & 0.98 \end{pmatrix}\begin{pmatrix} 1.62 \\ 0.28 \\ 0.10 \end{pmatrix} \end{aligned}$$

$$= \begin{pmatrix} 0.9(1.62) + 0(0.28) + 0(0.1) \\ 0.1(1.62) + 0.5(0.28) + 0.02(0.1) \\ 0(1.62) + 0.5(0.28) + 0.98(0.1) \end{pmatrix} = \begin{pmatrix} 1.458 \\ 0.304 \\ 0.238 \end{pmatrix}.$$

13. **(a)** We first find $\vec{v}$:

$$\vec{v} = \mathbf{A}\vec{u} = \begin{pmatrix} 2 & 1 \\ 3 & 2 \end{pmatrix}\begin{pmatrix} 3 \\ 5 \end{pmatrix} = \begin{pmatrix} 11 \\ 19 \end{pmatrix}.$$

Now, we show that $\vec{u} = \mathbf{A}^{-1}\vec{v}$:

$$\mathbf{A}^{-1}\vec{v} = \begin{pmatrix} 2 & -1 \\ -3 & 2 \end{pmatrix}\begin{pmatrix} 11 \\ 19 \end{pmatrix} = \begin{pmatrix} 3 \\ 5 \end{pmatrix} = \vec{u}.$$

(b) We first find $\vec{v}$:

$$\vec{v} = \mathbf{A}\vec{u} = \begin{pmatrix} 2 & 1 \\ 3 & 2 \end{pmatrix}\begin{pmatrix} -1 \\ 7 \end{pmatrix} = \begin{pmatrix} 5 \\ 11 \end{pmatrix}.$$

Now, we show that $\vec{u} = \mathbf{A}^{-1}\vec{v}$:

$$\mathbf{A}^{-1}\vec{v} = \begin{pmatrix} 2 & -1 \\ -3 & 2 \end{pmatrix}\begin{pmatrix} 5 \\ 11 \end{pmatrix} = \begin{pmatrix} -1 \\ 7 \end{pmatrix} = \vec{u}.$$

(c) We first find $\vec{v}$:

$$\vec{v} = \mathbf{A}\vec{u} = \begin{pmatrix} 2 & 1 \\ 3 & 2 \end{pmatrix}\begin{pmatrix} a \\ b \end{pmatrix} = \begin{pmatrix} 2a+b \\ 3a+2b \end{pmatrix}.$$

Now, we show that $\vec{u} = \mathbf{A}^{-1}\vec{v}$:

$$\begin{aligned} \mathbf{A}^{-1}\vec{v} &= \begin{pmatrix} 2 & -1 \\ -3 & 2 \end{pmatrix}\begin{pmatrix} 2a+b \\ 3a+2b \end{pmatrix} \\ &= \begin{pmatrix} 2(2a+b) - (3a+2b) \\ -3(2a+b) + 2(3a+2b) \end{pmatrix} \\ &= \begin{pmatrix} 4a+2b-3a-2b \\ -6a-3b+6a+4b \end{pmatrix} \\ &= \begin{pmatrix} a \\ b \end{pmatrix} = \vec{u}. \end{aligned}$$

17. **(a)** We have $\mathbf{C} = \begin{pmatrix} 3 & 5 \\ 2 & 4 \end{pmatrix}$ and so

$$\begin{aligned} \mathbf{C}\vec{u} &= \begin{pmatrix} 3 & 5 \\ 2 & 4 \end{pmatrix}\begin{pmatrix} a \\ b \end{pmatrix} \\ &= \begin{pmatrix} 3a+5b \\ 2a+4b \end{pmatrix} \\ &= \begin{pmatrix} 3a \\ 2a \end{pmatrix} + \begin{pmatrix} 5b \\ 4b \end{pmatrix} \\ &= a\begin{pmatrix} 3 \\ 2 \end{pmatrix} + b\begin{pmatrix} 5 \\ 4 \end{pmatrix} \\ &= a\vec{c}_1 + b\vec{c}_2. \end{aligned}$$

(b) Provided $\mathbf{C}^{-1}$ exists, we can write $\vec{u} = \mathbf{C}^{-1}v$. We can find $\mathbf{C}^{-1}$ using the approach from Problem 14: we have $a = 3$, $b = 5$, $c = 2$, $d = 4$, and so $D = ad - bc = 3(4) - 5(2) = 2$. We have

$$\mathbf{C}^{-1} = \frac{1}{D}\begin{pmatrix} d & -b \\ -c & a \end{pmatrix} = \frac{1}{2}\begin{pmatrix} 4 & -5 \\ -2 & 3 \end{pmatrix} = \begin{pmatrix} 2 & -2.5 \\ -1 & 1.5 \end{pmatrix}.$$

This means that

$$\begin{aligned} \vec{u} &= \mathbf{C}^{-1}\vec{v} \\ &= \begin{pmatrix} 2 & -2.5 \\ -1 & 1.5 \end{pmatrix}\begin{pmatrix} 2 \\ 5 \end{pmatrix} \\ &= \begin{pmatrix} 2(2) - 2.5(5) \\ -1(2) + 1.5(5) \end{pmatrix} \\ &= \begin{pmatrix} -8.5 \\ 5.5 \end{pmatrix}. \end{aligned}$$

(c) From parts (a) and (b), we see that

$$\vec{v} = \begin{pmatrix} 2 \\ 5 \end{pmatrix} = \mathbf{C}\vec{u} = \begin{pmatrix} 3 & 2 \\ 5 & 4 \end{pmatrix}\begin{pmatrix} -8.5 \\ 5.5 \end{pmatrix}.$$

Referring to Problem 16, we see that

$$\begin{aligned} \vec{v} = \begin{pmatrix} 2 \\ 5 \end{pmatrix} &= \Bigg(\underbrace{\boxed{\begin{matrix} 3 \\ 2 \end{matrix}}}_{\vec{c}_1}\ \underbrace{\boxed{\begin{matrix} 5 \\ 4 \end{matrix}}}_{\vec{c}_2}\Bigg)\underbrace{\begin{pmatrix} -8.5 \\ 5.5 \end{pmatrix}}_{\vec{u}} \\ &= -8.5\underbrace{(3,2)}_{\vec{c}_1} + 5.5\underbrace{(5,4)}_{\vec{c}_2} \\ &= -8.5\vec{c}_1 + 5.5\vec{c}_2, \end{aligned}$$

which gives $\vec{v}$ as a combination of $\vec{c}_1$ and $\vec{c}_2$.

Solutions for Chapter 12 Review

Exercises

1. $3\vec{c} = 3(1,1,2) = (3,3,6)$.

5. $2\vec{a} - 3(\vec{b} - \vec{c}) = 2(5,1,0) - 3((2,-1,9) - (1,1,2)) = (10,2,0) - 3(1,-2,7) = (7,8,-21)$.

9. $\vec{a} = \vec{b} = \vec{c} = 3\vec{k}$, $\vec{d} = 2\vec{i} + 3\vec{k}$, $\vec{e} = \vec{j}$, $\vec{f} = -2\vec{i}$.

13. $(3\vec{i} + \vec{j}) \cdot (5\vec{i} - 2\vec{j}) = 15 - 2 = 13$.

17. Since the dot product is a scalar, $(2\vec{i} + 5\vec{j}) \cdot 3\vec{i} = 6$, we have $\vec{i} + \vec{j} + \vec{k}$ multiplied by 6, that is

$$(2\vec{i} + 5\vec{j}) \cdot 3\vec{i}\,(\vec{i} + \vec{j} + \vec{k}) = 6(\vec{i} + \vec{j} + \vec{k}) = 6\vec{i} + 6\vec{j} + 6\vec{k}.$$

Problems

21. **(a)** To be parallel, vectors must be scalar multiples. The $\vec{k}$ component of the first vector is 2 times the $\vec{k}$ component of the second vector. So the $\vec{i}$ components of the two vectors must be in a 2:1 ratio, and the same is true for the $\vec{j}$ components. Thus, $4 = 2a$ and $a = 2(a-1)$. These equations have the solution $a = 2$, and for that value, the vectors are parallel.

(b) Perpendicular means a zero dot product. So $4a + a(a-1) + 18 = 0$, or $a^2 + 3a + 18 = 0$. Since $b^2 - 4ac = 9 - 4 \cdot 1 \cdot 18 = -63 < 0$, there are no real solutions. This means the vectors are never perpendicular.

25. See Figure 12.10, where $\vec{g}$ is the acceleration due to gravity and $g = \|\vec{g}\|$.
If $\theta = 0$ (the plank is at ground level), the sliding force is $F = 0$.
If $\theta = \pi/2$ (the plank is vertical), the sliding force equals g, the force due to gravity.
Therefore, we can guess that F is proportional with $\sin\theta$:

$$F = g\sin\theta.$$

This agrees with the bounds at $\theta = 0$ and $\theta = \pi/2$, and with the fact that the sliding force is smaller than g between 0 and $\pi/2$.

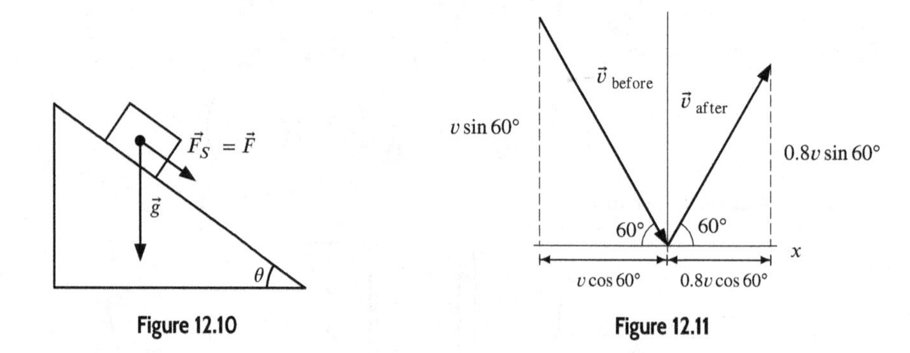

Figure 12.10 **Figure 12.11**

29. The speed of the particle before impact is v, so the speed after impact is $0.8v$. If we consider the barrier as being along the x-axis (see Figure 12.11), then the $\vec{i}$-component is $0.8v\cos 60° = 0.8v(0.5) = 0.4v$.
Similarly, the $\vec{j}$-component is $0.8v\sin 60° = 0.8v(0.8660) \approx 0.693v$. Thus

$$\vec{v}_{\text{after}} = 0.4v\vec{i} + 0.693v\vec{j}.$$

33. If $\vec{x}$ and $\vec{y}$ are two consumption vectors corresponding to points satisfying the same budget constraint, then

$$\vec{p} \cdot \vec{x} = k = \vec{p} \cdot \vec{y}.$$

Therefore we have

$$\vec{p} \cdot (\vec{x} - \vec{y}) = \vec{p} \cdot \vec{x} - \vec{p} \cdot \vec{y} = 0.$$

Thus $\vec{p}$ and $\vec{x} - \vec{y}$ are perpendicular; that is, the difference between two consumption vectors on the same budget constraint is perpendicular to the price vector.

37. Using the result of Problem 36, we have $\overrightarrow{AC} = \vec{w} + \vec{n} - \vec{m} = 3\vec{n} - 3\vec{m}$; $\overrightarrow{AB} = \vec{v} + \vec{m} + \vec{n} = 3\vec{m} + \vec{n}$; $\overrightarrow{AD} = \vec{v} + \vec{m} - (\vec{n} - \vec{m}) = 4\vec{m} - \vec{n}$; $\overrightarrow{BD} = (-\vec{n}) - (\vec{n} - \vec{m}) = \vec{m} - 2\vec{n}$.

STRENGTHEN YOUR UNDERSTANDING

1. False. The length $\|0.5\vec{i} + 0.5\vec{j}\| = \sqrt{(0.5)^2 + (0.5)^2} = \sqrt{0.5} \neq 1$.

5. False. The dot product $(2\vec{i} + \vec{j}) \cdot (2\vec{i} - \vec{j}) = 3$ is not zero.

9. True. If $\vec{u} = (u_1, u_2)$ and $\vec{v} = (v_1, v_2)$, then $\vec{u} + \vec{v}$ and $\vec{v} + \vec{u}$ both equal $(u_1 + v_1, u_2 + v_2)$.

13. False. Distance is not a vector. The vector $\vec{i} + \vec{j}$ is the displacement vector from P to Q. The distance between the points is the length of the displacement vector: $\|\vec{i} + \vec{j}\| = \sqrt{2}$.

17. True. Both vectors have length $\sqrt{13}$.

21. True. In the subscript, the first number gives the row and the second number gives the column.

CHAPTER THIRTEEN

Solutions for Section 13.1

Exercises

1. Not arithmetic. The differences are 5, 4, 3.

5. Arithmetic, with $a = 6$, $d = 3$, so $a_n = 6 + (n-1)3 = 3 + 3n$.

9. Not geometric. The ratios of successive terms are $2, 2, \frac{3}{2}$.

13. Not geometric, since the ratios of successive terms are $1/4, 1/4, 1/2$.

17. Geometric, since the ratios of successive terms are all $1/1.2$. Thus, $a = 1$ and $r = 1/1.2$, so $a_n = 1(1/1.2)^{n-1} = 1/(1.2)^{n-1}$.

Problems

21. $a_1 = \cos(\pi) = -1$, $a_2 = \cos(2\pi) = 1$, $a_3 = \cos(3\pi) = -1$, $a_4 = \cos(4\pi) = 1$. This is a geometric sequence.

25. Since the first term is 5 and the difference is 10, the arithmetic sequence is given by

$$a_n = 5 + (n-1)10.$$

For $a_n = 5 + (n-1) \cdot 10 > 1000$, we must have

$$\begin{aligned} 10(n-1) &> 995 \\ n - 1 &> 99.5 \\ n &> 100.5. \end{aligned}$$

The terms of the sequence exceed 1000 when $n \geq 101$.

29. Since $a_6 - a_3 = 3d = 9 - 5.7 = 3.3$, we have $d = 1.1$, so $a_1 = a_3 - 2d = 5.7 - 2 \cdot 1.1 = 3.5$. Thus

$$\begin{aligned} a_5 &= 3.5 + (5-1)1.1 = 7.9 \\ a_{50} &= 3.5 + (50-1)1.1 = 57.4 \\ a_n &= 3.5 + (n-1)1.1 = 2.4 + 1.1n. \end{aligned}$$

33. Since $a_2 = ar = 6$ and $a_4 = ar^3 = 54$, we have

$$\frac{a_4}{a_2} = \frac{ar^3}{ar} = r^2 = \frac{54}{6} = 9,$$

so

$$r = 3.$$

We have $a = 6/r = 6/3 = 2$. Thus

$$\begin{aligned} a_6 &= ar^5 = 2 \cdot (3)^5 = 486 \\ a_n &= ar^{n-1} = 2 \cdot 3^{n-1}. \end{aligned}$$

37. **(a)** The growth factor of the population is

$$r = \frac{19.042}{18.934} = 1.0057.$$

Thus, one and two years after 2012, the population is

$$\begin{aligned} a_1 &= 19.042(1.0057) = 19.151 \\ a_2 &= 19.042(1.0057)^2 = 19.260. \end{aligned}$$

(b) Using the growth factor in part (a), n years after 2012, the population is

$$a_n = 19.042(1.0057)^n.$$

(c) The doubling time is the value of n for which

$$\begin{aligned} a_n &= 2 \cdot 19.042 \\ 19.042(1.0057)^n &= 2 \cdot 19.042 \\ (1.0057)^n &= 2 \\ n \ln(1.0057) &= \ln 2 \\ n &= \frac{\ln 2}{\ln(1.0057)} = 121.951. \end{aligned}$$

Thus the doubling time is about 121.951 years

41. Arithmetic, because points lie on a line. The sequence is decreasing, so $d < 0$.

45. Since $a_1 = 3$ and $a_n = 2a_{n-1} + 1$, we have $a_2 = 2a_1 + 1 = 7$, $a_3 = 2a_2 + 1 = 15$, $a_4 = 2a_3 + 1 = 31$. Writing out the terms without simplification to try and guess the pattern, we have

$$\begin{aligned} a_1 &= 3 \\ a_2 &= 2 \cdot 3 + 1 \\ a_3 &= 2(2 \cdot 3 + 1) + 1 = 2^2 \cdot 3 + 2 + 1 \\ a_4 &= 2(2^2 \cdot 3 + 2 + 1) + 1 = 2^3 \cdot 3 + 2^2 + 2 + 1. \end{aligned}$$

Thus

$$a_n = 2^{n-1} \cdot 3 + 2^{n-2} + 2^{n-3} + \cdots + 2 + 1.$$

49. **(a)** You send the letter to 4 friends; at that stage your name is on the bottom of the list. The 4 friends send it to 4 each, so 4^2 people have the letter when you are third on the list. These 4^2 people send it to 4 each, so 4^3 people have the letter when you are second on the list. By similar reasoning, 4^4 people have the letter when you are at the top of the list. Thus, you should receive $\$4^4 = \$256$. (The catch is, of course, that someone usually breaks the chain.)

(b) By the same reasoning as in part (a), we see that $d_n = 4^n$.

Solutions for Section 13.2

Exercises

1. This series is not arithmetic, as each term is twice the previous one.

5. $\sum_{i=-1}^{5} i^2 = (-1)^2 + 0^2 + 1^2 + 2^2 + 3^2 + 4^2 + 5^2.$

9. $\sum_{j=2}^{10} (-1)^j = (-1)^2 + (-1)^3 + (-1)^4 + \cdots + (-1)^{10}.$

13. The pattern is $1/2, 2/2, 3/2, \ldots, 8/2$. Thus, a possible solution is

$$\sum_{n=1}^{8} \frac{1}{2} n.$$

17. One way to work this problem is to complete the first few values of a_n, and then continue the pattern across the row. Then, the values of S_n can be found by summing the values of a_n from left to right. See Table 13.1. We see that $a_1 = 3$ and $d = a_2 - a_1 = 7 - 3 = 4$.

Table 13.1

n	1	2	3	4	5	6	7	8
a_n	3	7	11	15	19	23	27	31
S_n	3	10	21	36	55	78	105	136

21. We use the formula $S_n = 1 + 2 + 3 + \cdots + n = \frac{1}{2}n(n+1)$ with $n = 1000$:

$$S_{1000} = \frac{1}{2} \cdot 1000 \cdot 1001 = 500{,}500.$$

25. $\sum_{n=0}^{10}(8 - 4n) = 8 + 4 + 0 + (-4) + \cdots$. This is an arithmetic series with 11 terms and $d = -4$.

$$S_{11} = \frac{1}{2} \cdot 11(2 \cdot 8 + 10(-4)) = -132.$$

29. This is an arithmetic series with $a_1 = -3.01$, $n = 35$, and $d = -0.01$. Thus

$$S_{35} = \frac{1}{2} \cdot 35(2(-3.01) + 34(-0.01)) = -111.3.$$

Problems

33. **(a)** (i) From the table, we have $S_4 = 248.7$, the population of the US in millions 4 decades after 1950, that is, in 1990. Similarly, $S_5 = 281.4$, the population in millions in 2000, and $S_6 = 308.7$, the population in 2010.

(ii) We have $a_2 = S_2 - S_1 = 203.3 - 179.3 = 24$; that is, the increase in the US population in millions in the 1960s. Similarly, $a_5 = S_5 - S_4 = 281.4 - 248.7 = 32.7$, the population increase in millions during the 1990s. In the same way, $a_6 = S_6 - S_5 = 308.7 - 281.4 = 27.3$, the population increase during the 2000s.

(iii) Using the answer to (ii), we have $a_6/10 = 27.3/10 = 2.73$, the average yearly population growth during the 2000s.

(b) We have

S_n = US population, in millions, n decades after 1950.
$a_n = S_n - S_{n-1}$ = growth in US population in millions, during the nth decade after 19540.
$a_n/10$ = Average yearly growth, in millions, during the nth decade after 1950.

37. Expanding both sums, we see

$$\sum_{i=1}^{15} i^3 - \sum_{j=3}^{15} j^3 = (1^3 + 2^3 + 3^3 + 4^3 + \cdots + 15^3) - (3^3 + 4^3 + \cdots + 15^3)$$
$$= 1^3 + 2^3 = 9.$$

41. We have $a_1 = 16$ and $d = 32$. At the end of n seconds, the object has fallen

$$S_n = \frac{1}{2}n(2 \cdot 16 + (n-1)32) = \frac{1}{2}n(32 + 32n - 32) = 16n^2.$$

The height of the object at the end of n seconds is

$$h = 1000 - S_n = 1000 - 16n^2.$$

When the object hits the ground, $h = 0$, so

$$0 = 1000 - 16n^2$$

$$n^2 = \frac{1000}{16} \quad \text{so} \quad n = \pm\sqrt{\frac{1000}{16}} = \pm 7.906 \text{ sec.}$$

Since n is positive, it takes 7.906 seconds, that is, nearly 8 seconds, for the object to hit the ground.

45. **(a)** Since there are 9 terms, we can group them into 4 pairs each totaling 66. The middle or fifth term, $a_5 = 33$, remains unpaired. This means that

$$\text{Sum of series} = 4(66) + 33 = 297.$$

This is the same answer we got by adding directly.

(b) We can use our formula derived for the sum of an even number of terms to add the first 8 terms. We have $a_1 = 5$, $n = 8$, and $d = 7$. This gives

$$\text{Sum of first 8 terms} = \frac{1}{2}(8)(2(5) + (8-1)(7)) = 236.$$

Adding the ninth term gives

$$\text{Sum of series} = 236 + 61 = 297.$$

This is the same answer we got by adding directly.

(c) Using the method from part (a), we add the first and last terms and obtain $a_1 + a_n$. Notice that $a_n = a_1 + (n-1)d$ and so this expression can be rewritten as $a_1 + a_1 + (n-1)d = 2a_1 + (n-1)d$. We then add the second and next to last terms and obtain $a_2 + a_{n-1}$. We know that $a_2 = a_1 + d$ and that $a_{n-1} = a_1 + (n-2)d$, and so this expression can be rewritten as $a_1 + d + a_1(n-2)d = 2a_1 + (n-1)d$. Continuing in this manner, we see that the sum of each pair is $2a_1 + (n-1)d$. The total number of such terms is given by $\frac{1}{2}(n-1)$. For instance, if there are 9 terms, the total number of pairs is $\frac{1}{2}(9-1) = 4$. Thus, the subtotal of these pairs is given by

$$\text{Subtotal of pairs} = \frac{1}{2}(n-1)(2a_1 + (n-1)d).$$

As you can check for yourself, the unpaired (middle) term is given by

$$\text{Unpaired (middle) term} = a_1 + \frac{1}{2}(n-1)d.$$

For instance, in the case of the series given in the question, the unpaired term is given by

$$5 + \frac{1}{2}(9-1)(7) = 5 + 4(7) = 33.$$

Therefore, the sum of the arithmetic series is given by

$$\begin{aligned}\text{Sum} &= \text{Subtotal of pairs} + \text{Unpaired term} \\ &= \underbrace{\frac{1}{2}(n-1)(2a_1 + (n-1)d)}_{\text{Subtotal of pairs}} + \underbrace{a_1 + \frac{1}{2}(n-1)d}_{\text{Unpaired term}}.\end{aligned}$$

One way to simplify this expression is to first factor out 1/2:

$$\text{Sum} = \frac{1}{2}\left[(n-1)(2a_1 + (n-1)d) + 2a_1 + (n-1)d\right].$$

The bracketed part of this expression involves $(n-1)$ terms equaling $2a_1+(n-1)d$, plus one more such term, for a total number of n such terms. We have

$$\text{Sum} = \frac{1}{2}n(2a_1+(n-1)d),$$

which is the same as the formula derived in the text.

Using the method from part (b), we see that since n is odd, $(n-1)$ is even, so we can use the formula derived for an even number of terms to add the first $(n-1)$ terms. Substituting $(n-1)$ for n in our formula, we have

$$\text{Sum of first } (n-1) \text{ terms} = \frac{1}{2}(n-1)(2a_1+(n-2)d).$$

The total sum is given by

$$\text{Sum} = \text{Sum of first } (n-1) \text{ terms} + n^{\text{th}} \text{ term}.$$

Since the nth term is given by $a_1+(n-1)d$, we have

$$\text{Sum} = \underbrace{\frac{1}{2}(n-1)(2a_1+(n-2)d)}_{\text{Sum of first } (n-1) \text{ terms}} + \underbrace{a_1+(n-1)d}_{n^{\text{th}} \text{ term}}.$$

To simplify this expression, we first factor out 1/2, as before:

$$\text{Sum} = \frac{1}{2}\left[(n-1)(2a_1+(n-2)d)+2a_1+2(n-1)d\right].$$

We can rewrite $2a_1+2(n-1)d$ as $2a_1+2nd-2d$, and then as $2a_1+nd-2d+nd$, and finally as $2a_1+(n-2)d+nd$. This gives

$$\text{Sum} = \frac{1}{2}\left[(n-1)(2a_1+(n-2)d)+2a_1+(n-2)d+nd\right].$$

We have $n-1$ terms each equaling $2a_1+(n-2)d$ plus one more such term plus a term equaling nd. This gives a total of n terms equaling $2a_1+(n-2)d$ plus the nd term:

$$\begin{aligned}\text{Sum} &= \frac{1}{2}\left[n(2a_1+(n-2)d)+nd\right]\\ &= \frac{1}{2}n\left[2a_1+(n-2)d+d\right] \qquad \text{factoring out } n\\ &= \frac{1}{2}n(2a_1+(n-1)d),\end{aligned}$$

which is the same answer as before.

Solutions for Section 13.3

Exercises

1. Since

$$\sum_{j=5}^{18} = 3\cdot 2^5+3\cdot 2^6+3\cdot 2^7+\cdots+3\cdot 2^{18},$$

there are $18-4=14$ terms in the series. The first term is $a=3\cdot 2^5$ and the ratio is $r=2$, so

$$\text{Sum} = \frac{3\cdot 2^5(1-2^{14})}{1-2} = 1{,}572{,}768.$$

5. This is a geometric series with a first term is $a = 1/125$ and a ratio of $r = 5$. To determine the number of terms in the series, use the formula, $a_n = ar^{n-1}$. Calculating successive terms of the series $(1/125)5^{n-1}$, we get

$$1/125, 1/25, 1/5, 1, 5, 25, 125, 625.$$

Thus we sum the first eight terms of the series:

$$S_8 = \frac{(1/125)(1-5^8)}{1-5} = \frac{97656}{125} = 781.248.$$

9. Yes, $a = 1$, ratio $= -1/2$.

13. These numbers are all power of 3, with signs alternating from positive to negative. We need to change the signs on an alternating basis. By raising -1 to various powers, we can create the pattern shown. One possible answer is: $\sum_{n=1}^{6}(-1)^{n+1}(3^n)$.

17. $\sum_{n=0}^{5} 3/(2^n) = 3 + 3/2 + 3/4 + 3/8 + 3/16 + 3/32 = 3(1 - 1/2^6)/(1 - 1/2)) = 189/32$

Problems

21. **(a)** Let a_n be worldwide oil consumption n years after 2011. Then, $a_1 = 88$, $a_2 = 88(1.007)$, and $a_n = 88(1.007)^{n-1}$. Thus, between 2012 and 2037,

$$\text{Total oil consumption } = \sum_{n=1}^{25} 88(1.007)^{n-1} \text{ billion barrels.}$$

(b) Using the formula for the sum of a finite geometric series, we have

$$\text{Total oil consumption } = \frac{88\left(1-(1.007)^{25}\right)}{1-1.007} = 2395.111 \text{ billion barrels.}$$

25. The answer to Problem 24 is given by

$$B_{20} = \frac{1000(1-(1.03)^{20})}{1-1.03} = 26{,}870.37.$$

(a) Replacing \$1000 by \$2000 doubles the answer, giving \$53,740.75.
(b) Doubling the interest rate to 6% by replacing 1.03 by 1.06 less than doubles the answer, giving \$36,785.60.
(c) Doubling the number of deposits to 40 by replacing $(1.03)^{20}$ by $(1.03)^{40}$ more than doubles the answer, giving \$75,401.26.

Solutions for Section 13.4

Exercises

1. Yes, $a = 1$, ratio $= -x$.

5. Yes, $a = e^x$, ratio $= e^x$.

9. Sum $= \dfrac{y^2}{1-y}$, $|y| < 1$.

13.

$$\sum_{i=4}^{\infty}\left(\frac{1}{3}\right)^i = \left(\frac{1}{3}\right)^4 + \left(\frac{1}{3}\right)^5 + \cdots = \left(\frac{1}{3}\right)^4\left(1 + \frac{1}{3} + \left(\frac{1}{3}\right)^2 + \cdots\right) = \frac{(\frac{1}{3})^4}{1-\frac{1}{3}} = \frac{1}{54}.$$

17. Since

$$\sum_{i=1}^{\infty} x^{2i} = x^2 + x^4 + x^6 \cdots,$$

the first term is $a = x^2$ and the ratio is $r = x^2$. Since $|x| < 1$,

$$\text{Sum} = \sum_{i=1}^{\infty} x^{2i} = \frac{a}{1-r} = \frac{x^2}{1-x^2}.$$

Problems

21. $0.122222\ldots = 0.1 + \frac{2}{100} + \frac{2}{1000} + \frac{2}{10000} + \frac{2}{100000} + \cdots$. Thus,

$$S = 0.1 + \frac{\frac{2}{100}}{1 - \frac{1}{10}} = \frac{1}{10} + \frac{2}{90} = \frac{11}{90}.$$

25. **(a)**

$$\begin{aligned}
P_1 &= 0 \\
P_2 &= 250(0.04) \\
P_3 &= 250(0.04) + 250(0.04)^2 \\
P_4 &= 250(0.04) + 250(0.04)^2 + 250(0.04)^3 \\
&\vdots \\
P_n &= 250(0.04) + 250(0.04)^2 + \cdots + 250(0.04)^{n-1}
\end{aligned}$$

(b) Factoring our formula for P_n, we see that it involves a geometric series of $n-2$ terms:

$$P_n = 250(0.04)\underbrace{\left[1 + 0.04 + (0.04)^2 + \cdots + (0.04)^{n-2}\right]}_{n-2 \text{ terms}}.$$

The sum of this series is given by

$$1 + 0.04 + (0.04)^2 + \cdots + (0.04)^{n-2} = \frac{1-(0.04)^{n-1}}{1-0.04}.$$

Thus,

$$\begin{aligned}
P_n &= 250(0.04)\left(\frac{1-(0.04)^{n-1}}{1-0.04}\right) \\
&= 10\left(\frac{1-(0.04)^{n-1}}{1-0.04}\right).
\end{aligned}$$

(c) In the long run, that is, as $n \to \infty$, we know that $(0.04)^{n-1} \to 0$, and so

$$P_n = 10\left(\frac{1-(0.04)^{n-1}}{1-0.04}\right) \to 10\left(\frac{1-0}{1-0.04}\right) = 10.417.$$

Thus, P_n gets closer to 10.417 and Q_n gets closer to 260.42. We'd expect these limits to differ because one is right before taking a tablet and one is right after. We'd expect the difference between them to be exactly 250 mg, the amount of ampicillin in one tablet.

29.

$$\begin{aligned}\text{Present value of first coupon} &= \frac{100}{1.02}\\ \text{Present value of second coupon} &= \frac{100}{(1.02)^2}, \text{etc.}\end{aligned}$$

$$\begin{aligned}\text{Total present value} &= \underbrace{\frac{100}{1.024} + \frac{100}{(1.02)^2} + \cdots + \frac{100}{(1.02)^{10}}}_{\text{coupons}} + \underbrace{\frac{10{,}000}{(1.02)^{10}}}_{\text{principal}}\\ &= \frac{100}{1.02}\left(1 + \frac{1}{1.02} + \cdots + \frac{1}{(1.02)^9}\right) + \frac{10{,}000}{(1.02)^{10}}\\ &= \frac{100}{1.02}\left(\frac{1-\left(\frac{1}{1.02}\right)^{10}}{1-\frac{1}{1.02}}\right) + \frac{10{,}000}{(1.04)^{10}}\\ &= 898.259 + 8203.480\\ &= \$9101.74.\end{aligned}$$

Solutions for Chapter 13 Review

Exercises

1. We have $n = 18$, $a_1 = 8$, $d = 3$. To find the sum of the series we use the formula $S = (1/2)n(2a_1 + (n-1)d)$. So

$$S_{18} = \frac{18}{2}(2(8) + 17(3)) = 603.$$

To find the 18th term, we use the formula $a_n = a_1 + (n-1)d$. Thus,

$$a_{18} = 8 + 17(3) = 59.$$

5. No. Ratio between successive terms is not constant: $\dfrac{x^2/3}{x/2} = \dfrac{2x}{3}$, while $\dfrac{x^3/4}{x^2/3} = \dfrac{3x}{4}$.

9. This is an arithmetic series. We have $n = 9$, $a_1 = 7$, $d = 7$. Using our formula, $S_n = (1/2)n(2a_1 + (n-1)d)$, we have

$$S_9 = \frac{9}{2}(2\cdot 7 + (9-1)7) = 315.$$

Problems

13. We know $S_8 = 108$, $a_1 = 24$ and $n = 8$. We need to find d. Substituting in the formula $S = (1/2)n(2a_1 + (n-1)d)$, we get

$$\begin{aligned}108 &= \frac{1}{2}(8)(2(24) + 7d)\\ 108 &= 4(48 + 7d)\\ 108 &= 192 + 28d\\ d &= -3.\end{aligned}$$

He can make 8 rows of cans, starting with 24 cans on the bottom. Each row of cans will have 3 fewer cans than the row underneath it.

17.

$$\begin{aligned}\text{Total present value, in dollars} &= 1000 + 1000e^{-0.04} + 1000e^{-0.04(2)} + 1000e^{-0.04(3)} + \cdots \\ &= 1000 + 1000(e^{-0.04}) + 1000(e^{-0.04})^2 + 1000(e^{-0.04})^3 + \cdots\end{aligned}$$

This is an infinite geometric series with $a = 1000$ and $x = e^{(-0.04)}$, and sum

$$\text{Total present value, in dollars} = \frac{1000}{1 - e^{-0.04}} = 25{,}503.33.$$

21. **(a)** Let h_n be the height of the nth bounce after the ball hits the floor for the nth time. Then from Figure 13.1,

$$\begin{aligned}h_0 &= \text{height before first bounce } = 10 \text{ feet},\\ h_1 &= \text{height after first bounce } = 10\left(\frac{3}{4}\right) \text{feet},\\ h_2 &= \text{height after second bounce } = 10\left(\frac{3}{4}\right)^2 \text{ feet}.\end{aligned}$$

Generalizing this gives

$$h_n = 10\left(\frac{3}{4}\right)^n.$$

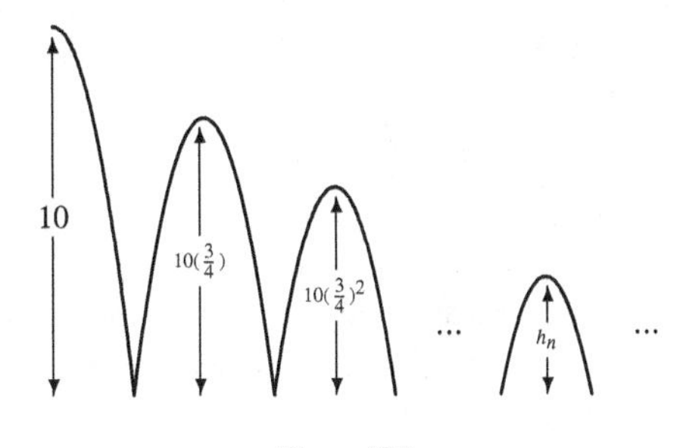

Figure 13.1

(b) When the ball hits the floor for the first time, the total distance it has traveled is just $D_1 = 10$ feet. (Notice that this is the same as $h_0 = 10$.) Then the ball bounces back to a height of $h_1 = 10\left(\frac{3}{4}\right)$, comes down and hits the floor for the second time. The total distance it has traveled is

$$D_2 = h_0 + 2h_1 = 10 + 2 \cdot 10\left(\frac{3}{4}\right) = 25 \text{ feet}.$$

Then the ball bounces back to a height of $h_2 = 10\left(\frac{3}{4}\right)^2$, comes down and hits the floor for the third time. It has traveled

$$D_3 = h_0 + 2h_1 + 2h_2 = 10 + 2 \cdot 10\left(\frac{3}{4}\right) + 2 \cdot 10\left(\frac{3}{4}\right)^2 = 25 + 2 \cdot 10\left(\frac{3}{4}\right)^2 = 36.25 \text{ feet}.$$

Similarly,

$$\begin{aligned}D_4 &= h_0 + 2h_1 + 2h_2 + 2h_3\\ &= 10 + 2 \cdot 10\left(\frac{3}{4}\right) + 2 \cdot 10\left(\frac{3}{4}\right)^2 + 2 \cdot 10\left(\frac{3}{4}\right)^3\\ &= 36.25 + 2 \cdot 10\left(\frac{3}{4}\right)^3\\ &\approx 44.688 \text{ feet}.\end{aligned}$$

(c) When the ball hits the floor for the nth time, its last bounce was of height h_{n-1}. Thus, by the method used in part (b), we get

$$D_n = h_0 + 2h_1 + 2h_2 + 2h_3 + \cdots + 2h_{n-1}$$

$$= \underbrace{10 + 2\cdot 10\left(\frac{3}{4}\right) + 2\cdot 10\left(\frac{3}{4}\right)^2 + 2\cdot 10\left(\frac{3}{4}\right)^3 + \cdots + 2\cdot 10\left(\frac{3}{4}\right)^{n-1}}_{\text{finite geometric series}}$$

$$= 10 + 2\cdot 10\cdot\left(\frac{3}{4}\right)\left(1 + \left(\frac{3}{4}\right) + \left(\frac{3}{4}\right)^2 + \cdots + \left(\frac{3}{4}\right)^{n-2}\right)$$

$$= 10 + 15\left(\frac{1-\left(\frac{3}{4}\right)^{n-1}}{1-\left(\frac{3}{4}\right)}\right)$$

$$= 10 + 60\left(1-\left(\frac{3}{4}\right)^{n-1}\right).$$

STRENGTHEN YOUR UNDERSTANDING

1. True. $a_1 = (1)^2 + 1 = 2$.

5. True. The differences between successive terms are all 1.

9. True. The first partial sum is just the first term of the sequence.

13. True. The sum is n terms of 3. That is, $3 + 3 + \cdots + 3 = 3n$.

17. False. The terms of the series can be negative so partial sums can decrease.

21. True. If $a = 1$ and $r = -\frac{1}{2}$, it can be written $\sum_{i=0}^{5}(-\frac{1}{2})^i$.

25. False. If payments are made at the end of each year, after 20 years, the balance at 1% is about \$44,000, while at 2% it would be about \$48,600. If payments are made at the start of each year, the corresponding figures are \$44,500 and \$49,600.

29. False. The series does not converge since the odd terms (Q_1, Q_3, etc.) are all -1.

33. False. An arithmetic series with $d \neq 0$ diverges.

CHAPTER FOURTEEN

Solutions for Section 14.1

Exercises

1. We use a parameter t so that when $t = 0$ we have $x = 1$ and $y = 3$. One possible parameterization is $x = 1 + 2t$, $y = 3 + t$, $0 \leq t \leq 1$.

5. True. Eliminating t gives $x = 3(y - 3)$. Thus we have the straight line $y = 3 + \frac{x}{3}$. We must also check the end points $t = 0$ and $t = 1$. When $t = 0$ we have $x = 0$ and $y = 3$, and when $t = 1$ we have $x = 3$ and $y = 3 + 1 = 4$.

9. The graph of the parametric equations is in Figure 14.1.

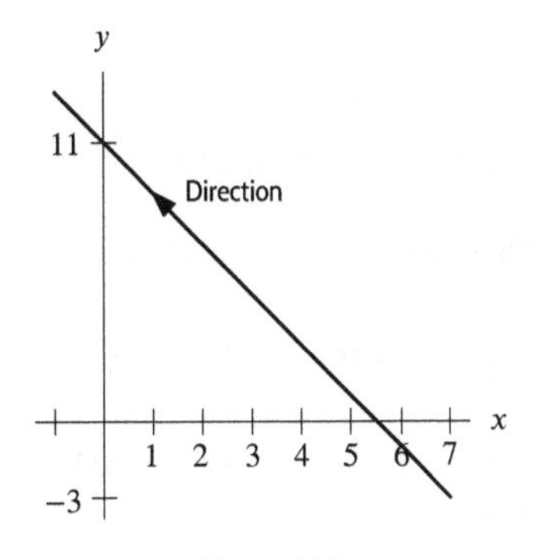

Figure 14.1

It is given that $x = 5 - 2t$, thus $t = (5 - x)/2$. Substitute this into the second equation:

$$y = 1 + 4t$$
$$y = 1 + 4\frac{(5 - x)}{2}$$
$$y = 11 - 2x.$$

13. The graph of the parametric equations is in Figure 14.2.

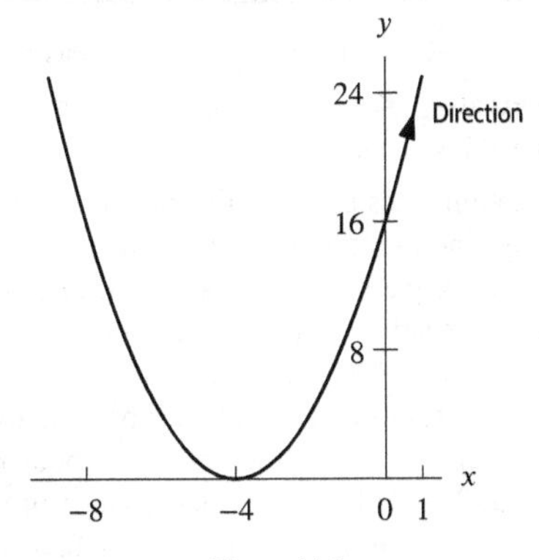

Figure 14.2

Since $x = t - 3$, we have $t = x + 3$. Substitute this into the second equation:

$$\begin{aligned} y &= t^2 + 2t + 1 \\ y &= (x+3)^2 + 2(x+3) + 1 \\ &= x^2 + 8x + 16 = (x+4)^2. \end{aligned}$$

17. The graph of the parametric equations is in Figure 14.3.

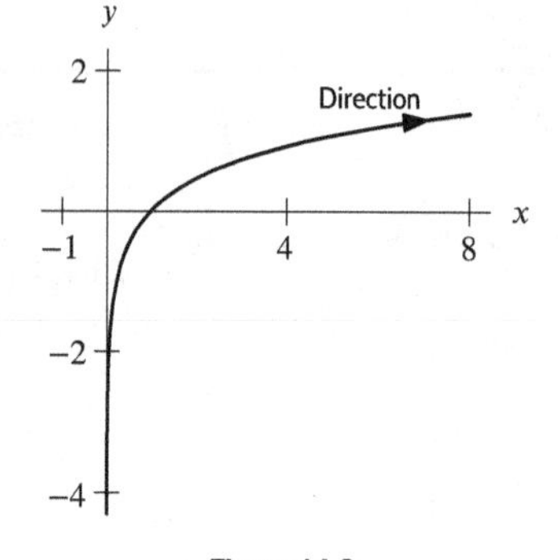

Figure 14.3

Since $x = t^3$, we take the natural log of both sides and get $\ln x = 3 \ln t$ or $\ln t = 1/3 \ln x$. We are given that $y = 2 \ln t$; thus,

$$y = 2\left(\frac{1}{3}\ln x\right) = \frac{2}{3}\ln x.$$

21. This is like Example 5 on page 570 of the text, except that the x-coordinate goes all the way to 2 and back. So the particle traces out the rectangle shown in Figure 14.4.

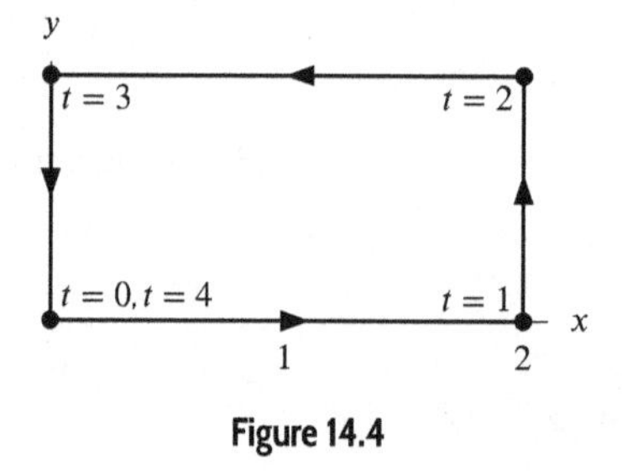

Figure 14.4

Problems

25. The particle moves clockwise: For $0 \le t \le \frac{\pi}{2}$, starting at $(1, 0)$ as t increases, we have $x = \cos t$ decreasing and $y = -\sin t$ decreasing. Similarly, for the time intervals $\frac{\pi}{2} \le t \le \pi, \pi \le t \le \frac{3\pi}{2}$, and $\frac{3\pi}{2} \le t \le 2\pi$, we see that the particle moves clockwise. The same is true for all $-\infty < t < +\infty$.

29. In all three cases, $y = x^2$, so that the motion takes place on the parabola $y = x^2$.

In case (a), the x-coordinate always increases at a constant rate of one unit distance per unit time, so the equations describe a particle moving to the right on the parabola at constant horizontal speed.

In case (b), the x-coordinate is never negative, so the particle is confined to the right half of the parabola. As t moves from $-\infty$ to $+\infty$, $x = t^2$ goes from ∞ to 0 to ∞. Thus the particle first comes down the right half of the parabola, reaching the origin $(0, 0)$ at time $t = 0$, where it reverses direction and goes back up the right half of the parabola.

In case (c), as in case (a), the particle traces out the entire parabola $y = x^2$ from left to right. The difference is that the horizontal speed is not constant. This is because a unit change in t causes larger and larger changes in $x = t^3$ as t approaches $-\infty$ or ∞. The horizontal motion of the particle is faster when it is farther from the origin.

33. For $0 \le t \le 2\pi$, the graph is in Figure 14.5.

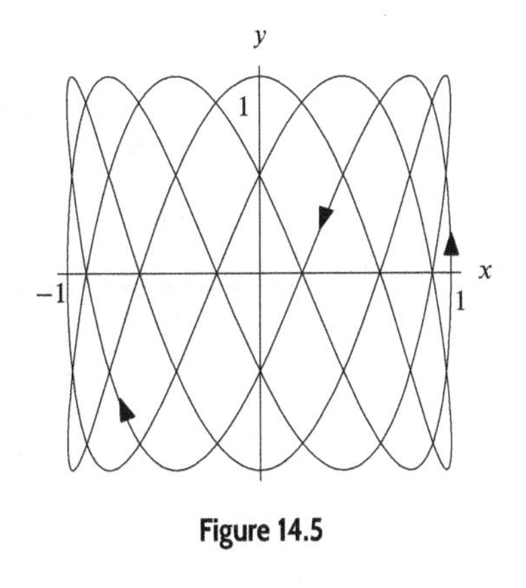

Figure 14.5

37. **(a)** Since the x-coordinate and the y-coordinate are always the same (they both equal t), the bug follows the path $y = x$.

(b) The bug starts at $(1, 0)$ because $\cos 0 = 1$ and $\sin 0 = 0$. Since the x-coordinate is $\cos x$, and the y-coordinate is $\sin x$, the bug follows the path of a unit circle, traveling counterclockwise. It reaches the starting point of $(1, 0)$ when $t = 2\pi$, because $\sin t$ and $\cos t$ are periodic with period 2π.

(c) Now the x-coordinate varies from 1 to -1, while the y-coordinate varies from 2 to -2; otherwise, this is much like part (b) above. If we plot several points, the path looks like an ellipse, which is a circle stretched out in one direction.

41. The particle moves back and forth between -1 and 1. See Figure 14.6.

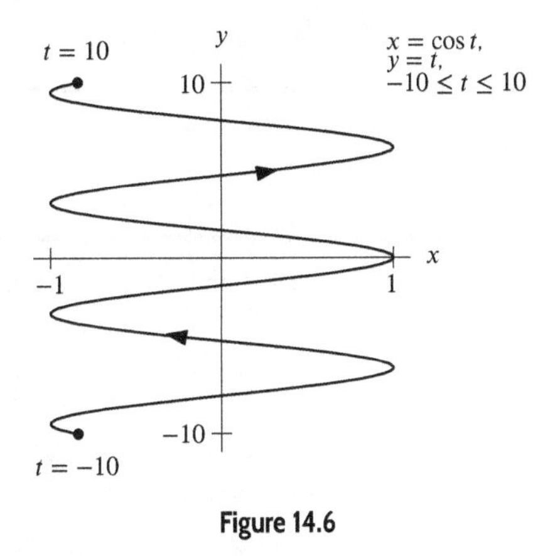

Figure 14.6

Solutions for Section 14.2

Exercises

1. Explicit. For each x we can write down the corresponding values of y.

5. Implicit. This is a quadratic in y so solving for y by completing the square, we obtain $y = 1 + \sqrt{2 + x}$ and $y = 1 - \sqrt{2 + x}$.

9. Multiplying by $(y - 4)^2$ gives

$$\begin{aligned} 4 - (x-4)^2 - (y-4)^2 &= 0 \\ (x-4)^2 + (y-4)^2 &= 4. \end{aligned}$$

Thus, the center is $(4, 4)$, and the radius is 2.

13. We can use $x = 5\sin t$, $y = -5\cos t$ for $0 \le t \le 2\pi$.

17. We consider $x = -2 + \sqrt{5}\sin t$, $y = 1 + \sqrt{5}\cos t$ for $0 \le t \le 2\pi$. However, as t increases from 0, the value of x increases, so this starts at the correct point but goes clockwise. We can use

$$x = -2 - \sqrt{5}\sin t, \quad y = 1 + \sqrt{5}\cos t, \quad \text{for } 0 \le t \le 2\pi.$$

Problems

21. **(a)** Center is $(2, -4)$ and radius is $\sqrt{20}$.

(b) Rewriting the original equation and completing the square, we have

$$\begin{aligned} 2x^2 + 2y^2 + 4x - 8y &= 12 \\ x^2 + y^2 + 2x - 4y &= 6 \\ (x^2 + 2x + 1) + (y^2 - 4y + 4) - 5 &= 6 \\ (x+1)^2 + (y-2)^2 &= 11. \end{aligned}$$

So the center is $(-1, 2)$, and the radius is $\sqrt{11}$.

25. Since $y - 3 = \sin t$ and $x = 4\sin^2 t$, this parameterization traces out the parabola $x = 4(y-3)^2$ for $2 \le y \le 4$.

29. Explicit: $y = \sqrt{4 - x^2}$
Implicit: $y^2 = 4 - x^2$ or $x^2 + y^2 = 4$, $y > 0$
Parametric: $x = 4\cos t$, $y = 4\sin t$, with $0 \le t \le \pi$.

Solutions for Section 14.3

Exercises

1. **(a)** The center is at the origin. The diameter in the x-direction is 4 and the diameter in the y-direction is $2\sqrt{5}$.

(b) The equation of the ellipse is

$$\frac{x^2}{2^2} + \frac{y^2}{(\sqrt{5})^2} = 1 \quad \text{or} \quad \frac{x^2}{4} + \frac{y^2}{5} = 1.$$

5. We consider $x = -2\cos t$, $y = 5\sin t$, for $0 \le t \le 2\pi$. However, in this parameterization, y increases as t increases from 0, so it traces clockwise. Thus, we take $x = -2\cos t$, $y = -5\sin t$, for $0 \le t \le 2\pi$.

9. The fact that the parameter is called s, not t, makes no difference. The minus sign means that the ellipse is traced out in the opposite direction. The graph of the ellipse in the xy-plane is the same as the ellipse in the example, and it is traced out once as s increases from 0 to 2π.

Problems

13. Completing the square on $x^2 - 2x$ and $y^2 + 4y$:

$$\begin{aligned} \frac{1}{4}(x^2 - 2x) + y^2 + 4y + \frac{13}{4} &= 0 \\ \frac{1}{4}((x-1)^2 - 1) + (y+2)^2 - 4 + \frac{13}{4} &= 0 \\ \frac{1}{4}(x-1)^2 - \frac{1}{4} + (y+2)^2 - 4 + \frac{13}{4} &= 0 \\ \frac{(x-1)^2}{4} + (y+2)^2 &= 1. \end{aligned}$$

The center is $(1, -2)$, and $a = 2$, $b = 1$.

17. Factoring out the 9 from $9x^2 + 9x = 9(x^2 + x)$ and the 4 from $4y^2 - 4y = 4(y^2 - y)$ and completing the square on $x^2 + x$ and $y^2 - y$:

$$\begin{aligned}
9(x^2 + x) + 4(y^2 - y) &= \frac{131}{4} \\
9\left(\left(x + \frac{1}{2}\right)^2 - \frac{1}{4}\right) + 4\left(\left(y - \frac{1}{2}\right)^2 - \frac{1}{4}\right) &= \frac{131}{4} \\
9\left(x + \frac{1}{2}\right)^2 - \frac{9}{4} + 4\left(y - \frac{1}{2}\right)^2 - 1 &= \frac{131}{4} \\
9\left(x + \frac{1}{2}\right)^2 + 4\left(y - \frac{1}{2}\right)^2 &= \frac{144}{4} = 36.
\end{aligned}$$

Dividing by 36 to get 1 on the right:

$$\begin{aligned}
\frac{9\left(x + \frac{1}{2}\right)^2}{36} + \frac{4\left(y - \frac{1}{2}\right)^2}{36} &= \frac{36}{36} \\
\frac{\left(x + \frac{1}{2}\right)^2}{4} + \frac{\left(y - \frac{1}{2}\right)^2}{9} &= 1.
\end{aligned}$$

The center is $(-\frac{1}{2}, \frac{1}{2})$, and $a = 2$, $b = 3$.

21. (a) Clearing the denominator, we have

$$\begin{aligned}
r(1 - \epsilon\cos\theta) &= r_0 \\
r - r\epsilon\cos\theta &= r_0 \\
r - \epsilon x &= r_0 && \text{because } x = r\cos\theta \\
r &= \epsilon x + r_0 \\
r^2 &= (\epsilon x + r_0)^2 && \text{squaring both sides} \\
x^2 + y^2 &= \epsilon^2 x^2 + 2\epsilon r_0 x + r_0^2 && \text{because } r^2 = x^2 + y^2 \\
x^2 - \epsilon^2 x^2 - 2\epsilon r_0 x + y^2 &= r_0^2 && \text{regrouping} \\
(1 - \epsilon^2)x^2 - 2\epsilon r_0 x + y^2 &= r_0^2 && \text{factoring} \\
Ax^2 - Bx + y^2 &= r_0^2,
\end{aligned}$$

where $A = 1 - \epsilon^2$ and $B = 2\epsilon r_0$. From Problem 20, we see that this is the formula of an ellipse.

(b) In the fraction $r_0/(1 - \epsilon\cos\theta)$, the numerator is a constant. Thus, the fraction's value is largest when the denominator is smallest, and smallest when the denominator is largest. The largest value of the denominator occurs at $\cos\theta = -1$, that is, at $\theta = \pi$, and so the minimum value of r is

$$r = \frac{r_0}{1 - \epsilon\cos\pi} = \frac{r_0}{1 + \epsilon}.$$

The smallest value of the denominator occurs at $\cos\theta = 1$, that is, at $\theta = 0$, and so the minimum value of r is

$$r = \frac{r_0}{1 - \epsilon\cos 0} = \frac{r_0}{1 - \epsilon}.$$

(c) See Figure 14.7. We know from part (b) that at $\theta = 0$, $r = r_0/(1 - \epsilon) = 6/(1 - 1/2) = 12$, and that at $\theta = \pi$, $r = r_0/(1 + \epsilon) = 6/(1 + 1/2) = 4$. In addition, at $\theta = \pi/2$ and $\theta = 3\pi/2$, we see that $\cos\theta = 0$ and so $r = r_0 = 6$. This gives us the four points labeled in the figure. By symmetry, the center of the ellipse must lie between the points $(4, \pi)$ and $(12, 0)$, or, in Cartesian coordinates, the points $(-4, 0)$ and $(12, 0)$. Thus, the center is at $(8, 0)$.

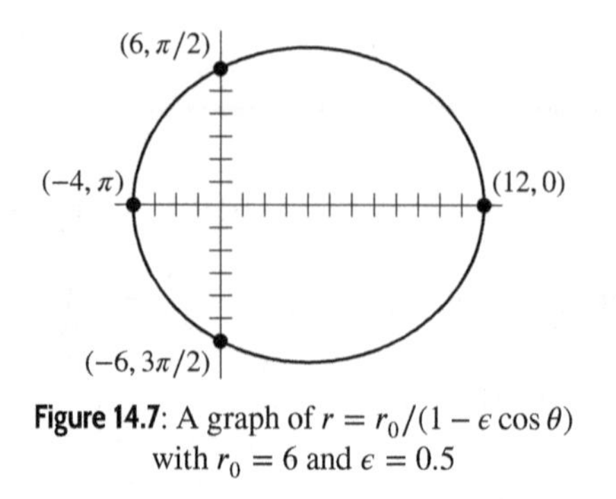

Figure 14.7: A graph of $r = r_0/(1 - \epsilon\cos\theta)$ with $r_0 = 6$ and $\epsilon = 0.5$

(d) The length of the horizontal axis is given by the minimum value of r (at $\theta = \pi$) plus the maximum value of r (at $\theta = 0$):

$$\begin{aligned}\text{Horizontal axis length} &= r_{\text{min}} + r_{\text{max}} \\ &= \frac{r_0}{1+\epsilon} + \frac{r_0}{1-\epsilon} \\ &= \frac{r_0(1-\epsilon) + r_0(1+\epsilon)}{(1-\epsilon)(1+\epsilon)} \\ &= \frac{2r_0}{1-\epsilon^2}.\end{aligned}$$

(e) At $\epsilon = 0$, we have $r = r_0/(1-0) = r_0$, which is the equation (in polar coordinates) of a circle of radius r_0. In our formula in part (d), we see that as ϵ gets closer to 1, the denominator gets closer to 0, and so the length of the horizontal axis increases rapidly. However, the y-intercepts at $(r_0, \pi/2)$ and $(r_0, 3\pi/2)$ do not change. For this to be true, the ellipse must be getting longer and longer along the horizontal axis as the eccentricity gets close to 1, but not be changing very much along the vertical axis.

Solutions for Section 14.4

Exercises

1. **(a)** The vertices are at $(0, 7)$ and $(0, -7)$. The center is at the origin.

(b) The asymptotes have slopes $7/2$ and $-7/2$. The equations of the asymptotes are

$$y = \frac{7}{2}x \quad \text{and} \quad y = -\frac{7}{2}x.$$

(c) The equation of the hyperbola is

$$\frac{y^2}{7^2} - \frac{x^2}{2^2} = 1 \quad \text{or} \quad \frac{y^2}{49} - \frac{x^2}{4} = 1.$$

5. The hyperbola is centered at the origin, and $a = 2$, $b = 7$. We can use $x = 2\tan t$, $y = 7\sec t = 7/\cos t$.
If $0 < t < \pi/2$, then $x > 0$, $y > 0$, so we have Quadrant I.
If $\pi/2 < t < \pi$, then $x < 0$, $y < 0$, so we have Quadrant III.
If $\pi < t < 3\pi/2$, then $x > 0$, $y < 0$, so we have Quadrant IV.
If $3\pi/2 < t < 2\pi$, then $x < 0$, $y > 0$, so we have Quadrant II.
So the upper half is given by $0 \le t < \pi/2$ together with $3\pi/2 < t < 2\pi$.

Problems

9. Factoring out -1 from $-y^2 + 4y = -(y^2 - 4y)$ and completing the square on $x^2 - 2x$ and $y^2 - 4y$ gives

$$\frac{1}{4}(x^2 - 2x) - (y^2 - 4y) = \frac{19}{4}$$

$$\begin{aligned}\frac{1}{4}((x-1)^2-1)-((y-2)^2-4)&=\frac{19}{4}\\ \frac{(x-1)^2}{4}-\frac{1}{4}-(y-2)^2+4&=\frac{19}{4}\\ \frac{(x-1)^2}{4}-(y-2)^2&=1.\end{aligned}$$

The center is $(1,2)$, the hyperbola opens right-left, and $a=2$, $b=1$.

13. Factoring out 4 from $4x^2-8x=4(x^2-2x)$ and 36 from $36y^2-36y=36(y^2-y)$ and completing the square on x^2-2x and y^2-y gives

$$\begin{aligned}4(x^2-2x)&=36(y^2-y)-31\\ 4((x-1)^2-1)&=36\left(\left(y-\frac{1}{2}\right)^2-\frac{1}{4}\right)-31\\ 4(x-1)^2-4&=36\left(y-\frac{1}{2}\right)^2-9-31\\ 4(x-1)^2&=36\left(y-\frac{1}{2}\right)^2-36.\end{aligned}$$

Moving $36(y-\frac{1}{2})^2$ to the left and dividing by -36 to get 1 on the right:

$$\begin{aligned}\frac{4(x-1)^2}{-36}-\frac{36\left(y-\frac{1}{2}\right)^2}{-36}&=-\frac{36}{-36}\\ -\frac{(x-1)^2}{9}+\left(y-\frac{1}{2}\right)^2&=1\\ \left(y-\frac{1}{2}\right)^2-\frac{(x-1)^2}{9}&=1.\end{aligned}$$

The center is $(1,\frac{1}{2})$, the hyperbola opens up-down, and $a=3$, $b=1$.

17. **(a)** The center is $(-5,2)$, vertices are $(-5+\sqrt{6},2)$ and $(-5-\sqrt{6},2)$. The asymptotes are $y=\pm\frac{2}{\sqrt{6}}(x+5)+2$. Figure 14.8 shows the hyperbola.

(b) Rewriting the equation and competing the square, we have

$$\begin{aligned}x^2-y^2+2x&=4y+17\\ x^2-y^2+2x-4y&=17\\ (x^2+2x+1)-(y^2+4y+4)+3&=17\\ (x+1)^2-(y+2)^2&=14\\ \frac{(x+1)^2}{14}-\frac{(y+2)^2}{14}&=1.\end{aligned}$$

Thus the center is $(-1,-2)$; vertices are $(-1-\sqrt{14},-2)$ and $(-1+\sqrt{14},-2)$; asymptotes are $y=\pm(x+1)-2$, that is, $y=x-1$ and $y=-x-3$. Figure 14.9 shows the hyperbola.

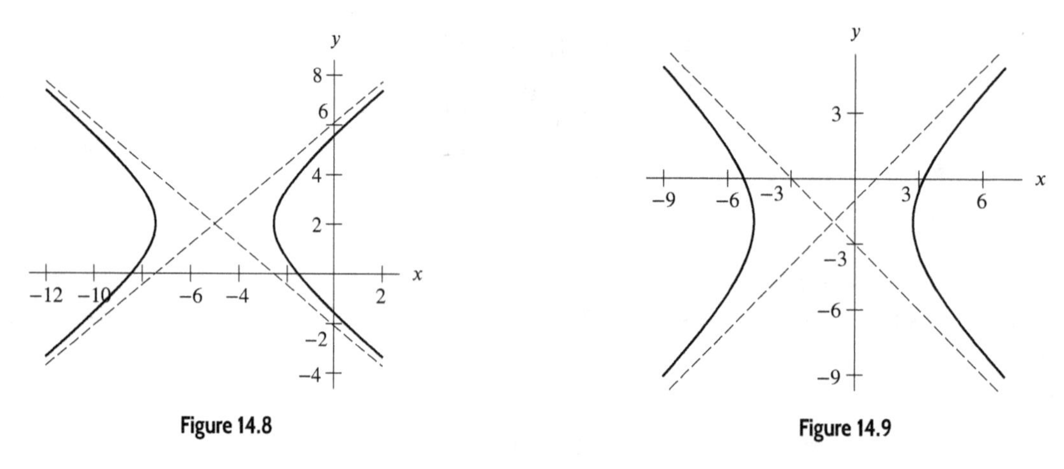

Figure 14.8

Figure 14.9

Solutions for Section 14.5

Exercises

1. This is the standard form of an ellipse. Since $a > b$, the major axis is the x-axis.

5. Subtract the x^2 term from both sides to get the standard form of a hyperbola. Since the y^2 term is positive, the vertices are on the y-axis.

9. The two focal points lie on the major axis, and are equidistant from the center. Therefore, the major axis is vertical and the other focal point is at $(0, -2)$.

13. In the form

$$\frac{x^2}{a^2} - \frac{y^2}{b^2} = 1$$

we see that $a^2 = b^2 = 1$. The focal points lie on the same axis as the positive squared term. Therefore the focal points are at $(\pm c, 0)$, where $c = \sqrt{a^2 + b^2} = \sqrt{1+1} = \sqrt{2}$. The two focal points are $(\sqrt{2}, 0)$ and $(-\sqrt{2}, 0)$.

Problems

17. The reflective properties of a parabola are somewhat helpful, because the transmission can be focused to go in only one direction. However, the location of the transmission will be perfectly clear to anyone noticing the transmission. A series of elliptical mirrors would be much more complicated and expensive to build, and by following the signal, someone could figure out where the transmission was coming from without your being aware, because surveying a series of mirrors is complicated. A hyperbolic reflection allows you to survey a small mirror near your camp while at the same time making it appear that your camp is somewhere else (at the other focus of the hyperbola). So you should follow the mess sergeant's advice.

21. From the figure, we see the vertex has coordinates $(1, 0)$ and the equation must be of the form

$$x = ay^2 + 1$$

The distance, c from the vertex to the focus is 2. Using $c = 1/(4a)$, we solve for a in $2 = 1/(4a)$ and find $a = 1/8$. The equation is $x = (1/8)y^2 + 1$. The directrix is the line $x = -1$, since it is 2 units to the left of the vertex.

25. Rewrite the formula as

$$(x^2 + 4x) + 2(y^2 - 6y) = 3.$$

Complete the square of both the x and y expressions,

$$(x^2 + 4x + 4) + 2(y^2 - 6y + 9) = 3 + 4 + 2(9),$$

and simplify to

$$(x+2)^2 + 2(y-3)^2 = 25$$

or

$$\frac{(x+2)^2}{25} + \frac{(y-3)^2}{(25/2)} = 1$$

This is an ellipse with center at $(-2, 3)$, and $a = 5$ and $b = \sqrt{25/2}$. Since $a > b$, the distance c, from the center to a focus point is found from

$$c = \sqrt{a^2 - b^2} = \sqrt{25 - 25/2} = \sqrt{25/2} = 5/\sqrt{2}.$$

With $c = 5/\sqrt{2}$ the focal points are at $(-2 + 5/\sqrt{2}, 3)$ and $(-2 - 5/\sqrt{2}, 3)$.

29. First rewrite the equation in the form,

$$\frac{x^2}{8} - \frac{y^2}{16} = 1.$$

Because the x^2 term is positive and the y^2 term is negative this hyperbola has vertices on a horizontal line through its center. The center is at the origin. The length from the center to the vertex is the distance $a = 2\sqrt{2}$ which is read from the equation since $a^2 = 8$. Thus the vertices are at the points $(2\sqrt{2}, 0)$ and $(-2\sqrt{2}, 0)$. Since $b^2 = 16$ and $a^2 = 8$, we find $c^2 = a^2 + b^2 = 24$, and $c = \sqrt{24}$. The focal points are $(\sqrt{24}, 0)$ and $(-\sqrt{24}, 0)$.

33. After passing through the second focal point it is reflected back to the original focal point.

37. For a cross-section of the dish centered at the origin, the equation of the parabola is $y = ax^2$. We know that the rim of the dish is the point $(12, 4)$, thus $4 = a(12)^2$ and solving for a we find $a = 1/36$. Using the equation $c = 1/(4a)$, we find

$$c = \frac{1}{4(1/36)} = 9.$$

The end of the arm is placed 9 inches above the center of the dish to be the focus of the incoming signal.

41. **(a)** The major axis has length $88 + 5250 = 5338$ million km.

(b) Since $2a = 5338$, we find $a = 2669$. Let $(\pm c, 0)$ be the two focal points. Since $a = 2669 = c + 88$, we find $c = 2581$. To find b we use the formula $c^2 = a^2 - b^2$ and find $b^2 = 2669^2 - 2581^2 = 462{,}000$, so $b = 680$.

Shifting the axis so the sun is at the origin, we have

$$\frac{(x-2581)^2}{2669^2} + \frac{y^2}{680^2} = 1.$$

(c) Use the parametric equations of an ellipse centered at (h, k),

$$x = h + a\cos t, \quad y = k + b\sin t, \quad 0 \le t \le 2\pi$$

to obtain

$$x = 2581 + 2669\cos t, \quad y = 680\sin t, \quad 0 \le t \le 2\pi.$$

45. Consider the set of points (x, y) so that the difference of the distances to two focal points $(\pm c, 0)$ is constant. Note that the x-intercepts $(\pm a, 0)$ satisfy this condition.

Using the distance formula, we see that the distance from (x, y) to $(c, 0)$ is $\sqrt{(x-c)^2 + y^2}$ and the distance from (x, y) to $(-c, 0)$ is $\sqrt{(x+c)^2 + y^2}$. The distance from $(a, 0)$ to $(-c, 0)$ is $c + a$ and the distance from $(a, 0)$ to $(c, 0)$ is $c - a$ since $c > a$. We have:

$$\begin{aligned}
\text{Difference of distances from } (x, y) \text{ to focal points} &= \text{Difference of distances from } (a, 0) \text{ to focal points} \\
\sqrt{(x-c)^2 + y^2} - \sqrt{(x+c)^2 + y^2} &= (c + a) - (c - a) \\
\sqrt{(x-c)^2 + y^2} - \sqrt{(x+c)^2 + y^2} &= 2a \\
\sqrt{(x-c)^2 + y^2} &= 2a + \sqrt{(x+c)^2 + y^2}.
\end{aligned}$$

We square both sides and simplify:

$$\begin{aligned}
(x-c)^2 + y^2 &= 4a^2 + 4a\sqrt{(x+c)^2 + y^2} + ((x+c)^2 + y^2) \\
x^2 - 2cx + c^2 + y^2 &= 4a^2 + 4a\sqrt{(x+c)^2 + y^2} + x^2 + 2cx + c^2 + y^2.
\end{aligned}$$

We cancel x^2, c^2, and y^2 from both sides, and subtract $2cx$ from both sides to obtain:

$$-4cx = 4a^2 + 4a\sqrt{(x+c)^2 + y^2}.$$

We divide through by 4:

$$-cx = a^2 + a\sqrt{(x+c)^2 + y^2},$$

and then isolate the square root:

$$-a\sqrt{(x+c)^2 + y^2} = a^2 + cx.$$

We square both sides again to obtain:

$$\begin{aligned} a^2((x+c)^2 + y^2) &= a^4 + 2cxa^2 + c^2x^2 \\ a^2(x^2 + 2cx + c^2 + y^2) &= a^4 + 2cxa^2 + c^2x^2 \\ a^2x^2 + 2cxa^2 + a^2c^2 + a^2y^2 &= a^4 + 2cxa^2 + c^2x^2 \\ a^2x^2 + a^2c^2 + a^2y^2 &= a^4 + c^2x^2. \end{aligned}$$

Since $c > a$, there is a positive number b such that $c^2 = a^2 + b^2$:

$$\begin{aligned} a^2x^2 + a^2(a^2 + b^2) + a^2y^2 &= a^4 + (a^2 + b^2)x^2 \\ a^2x^2 + a^4 + a^2b^2 + a^2y^2 &= a^4 + a^2x^2 + b^2x^2 \\ a^2b^2 + a^2y^2 &= b^2x^2 \\ b^2x^2 - a^2y^2 &= a^2b^2. \end{aligned}$$

Dividing through by a^2b^2, we arrive at the equation for a hyperbola given in Section 14.3:

$$\begin{aligned} \frac{b^2x^2}{a^2b^2} - \frac{a^2y^2}{a^2b^2} &= \frac{a^2b^2}{a^2b^2} \\ \frac{x^2}{a^2} - \frac{y^2}{b^2} &= 1. \end{aligned}$$

This is the equation for a hyperbola.

Solutions for Section 14.6

Exercises

1. We use $x = \sinh t$, $y = \cosh t$, for $-\infty < t < \infty$.

5. We use $x = 1 + 2\sinh t$, $y = -1 - 3\cosh t$, for $-\infty < t < \infty$.

9. Divide by 36 to rewrite the equation as

$$\frac{(y+3)^2}{4} - \frac{(x+1)^2}{9} = 1, \quad y > -3,$$

and use $x = -1 + 3\sinh t$, $y = -3 + 2\cosh t$, for $-\infty < t < \infty$.

13. The graph of $\sinh x$ in the text suggests that

$$\begin{aligned} &\text{As } x \to \infty, &\sinh x \to \tfrac{1}{2}e^x \\ &\text{As } x \to -\infty, &\sinh x \to -\tfrac{1}{2}e^{-x}. \end{aligned}$$

Using the facts that

$$\begin{aligned} &\text{As } x \to \infty, &e^{-x} \to 0, \\ &\text{As } x \to -\infty, &e^x \to 0, \end{aligned}$$

we can predict the same results algebraically:

$$\text{As } x \to \infty, \qquad \sinh x = \tfrac{e^x - e^{-x}}{2} \to \tfrac{1}{2}e^x$$
$$\text{As } x \to -\infty, \qquad \sinh x = \tfrac{e^x - e^{-x}}{2} \to -\tfrac{1}{2}e^{-x}.$$

Problems

17. Factoring out 2 from $2x^2 - 12x = 2(x^2 - 6x)$ and 4 from $4y^2 + 4y = 4(y^2 + y)$ and completing the square on $x^2 - 6x$ and $y^2 + y$ gives

$$\begin{aligned}
25 + 2(x^2 - 6x) &= 4(y^2 + y), \quad y > -\frac{1}{2} \\
25 + 2((x-3)^2 - 9) &= 4\left(\left(y + \frac{1}{2}\right)^2 - \frac{1}{4}\right), \quad y > -\frac{1}{2} \\
25 + 2(x-3)^2 - 18 &= 4\left(y + \frac{1}{2}\right)^2 - 1, \quad y > -\frac{1}{2} \\
2(x-3)^2 &= 4\left(y + \frac{1}{2}\right)^2 - 8, \quad y > -\frac{1}{2}.
\end{aligned}$$

Then, moving $4(y + \frac{1}{2})^2$ to the left and dividing by -8 to get 1 on the right,

$$\begin{aligned}
\frac{2(x-3)^2}{-8} - \frac{4\left(y + \frac{1}{2}\right)^2}{-8} &= 1, \quad y > -\frac{1}{2} \\
\frac{\left(y + \frac{1}{2}\right)^2}{2} - \frac{(x-3)^2}{4} &= 1, \quad y > -\frac{1}{2},
\end{aligned}$$

so we use $x = 3 + 2\sinh t$, $y = -\frac{1}{2} + \sqrt{2}\cosh t$ for $-\infty < t < \infty$.

21. Yes. First, we observe that

$$\cosh 2x = \frac{e^{2x} + e^{-2x}}{2}.$$

Now, using the fact that $e^x \cdot e^{-x} = 1$, we calculate

$$\begin{aligned}
\cosh^2 x &= \left(\frac{e^x + e^{-x}}{2}\right)^2 \\
&= \frac{(e^x)^2 + 2e^x \cdot e^{-x} + (e^{-x})^2}{4} \\
&= \frac{e^{2x} + 2 + e^{-2x}}{4}.
\end{aligned}$$

Similarly, we have

$$\begin{aligned}
\sinh^2 x &= \left(\frac{e^x - e^{-x}}{2}\right)^2 \\
&= \frac{(e^x)^2 - 2e^x \cdot e^{-x} + (e^{-x})^2}{4} \\
&= \frac{e^{2x} - 2 + e^{-2x}}{4}.
\end{aligned}$$

Thus, to obtain $\cosh 2x$, we need to add (rather than subtract) $\cosh^2 x$ and $\sinh^2 x$, giving

$$\cosh^2 x + \sinh^2 x = \frac{e^{2x} + 2 + e^{-2x} + e^{2x} - 2 + e^{-2x}}{4}$$

$$= \frac{2e^{2x} + 2e^{-2x}}{4}$$
$$= \frac{e^{2x} + e^{-2x}}{2}$$
$$= \cosh 2x.$$

Thus, we see that the identity relating $\cosh 2x$ to $\cosh x$ and $\sinh x$ is

$$\cosh 2x = \cosh^2 x + \sinh^2 x.$$

25. We know that $\sinh(iz) = i \sin z$, where z is real. Substituting $z = ix$, where x is real so z is imaginary, we have

$$\begin{aligned} \sinh(iz) &= i \sin z \\ \sinh(i \cdot ix) &= i \sin(ix) \qquad \text{substituting } z = ix \\ \sinh(-x) &= i \sin(ix). \end{aligned}$$

But $\sinh(-x) = -\sinh(x)$; thus we have

$$-\sinh x = i \sin(ix).$$

Multiplying both sides by i gives

$$-i \sinh x = -1 \sin(ix).$$

Thus,

$$i \sinh x = \sin(ix).$$

Solutions for Chapter 14 Review

Exercises

1. The coefficients of x^2 and y^2 are equal, so this is a circle with center $(0, 3)$ and radius $\sqrt{5}$.

5. Dividing by 36 to get 1 on the right gives

$$\frac{9(x-5)^2}{36} + \frac{4y^2}{36} = \frac{36}{36}$$
$$\frac{(x-5)^2}{4} + \frac{y^2}{9} = 1.$$

This is an ellipse centered at $(5, 0)$, with $a = 2$, $b = 3$.

9. One possible answer is $x = 3\cos t, y = -3\sin t, 0 \le t \le 2\pi$.

13. The ellipse $x^2/25 + y^2/49 = 1$ can be parameterized by $x = 5\cos t,\ y = 7\sin t, 0 \le t \le 2\pi$.

17. In the form

$$\frac{x^2}{a^2} + \frac{y^2}{b^2} = 1$$

we see that $a > b$. Therefore the focal points are at $(\pm c, 0)$, where $c = \sqrt{a^2 - b^2} = \sqrt{25 - 4} = \sqrt{21}$. The two focal points are $(\sqrt{21}, 0)$ and $(-\sqrt{21}, 0)$.

Problems

21. A parametric equation for the circle is

$$x = \cos t, y = \sin t.$$

As t increases from 0, we have x increasing and y decreasing, which is a counterclockwise movement, so this parameterization is correct.

25. Factoring out 6 from $6x^2 - 12x = 6(x^2 - 2x)$ and 9 from $9y^2 + 6y = 9(y^2 + \frac{2}{3}y)$, and completing the square on $x^2 - 2x$ and $y^2 + \frac{2}{3}y$ gives

$$6(x^2 - 2x) - 9\left(y^2 + \frac{2}{3}y\right) + 1 = 0$$
$$6((x-1)^2 - 1) + 9\left(\left(y + \frac{1}{3}\right)^2 - \frac{1}{9}\right) + 1 = 0$$
$$6(x-1)^2 - 6 + 9\left(y + \frac{1}{3}\right)^2 - 1 + 1 = 0$$
$$6(x-1)^2 + 9\left(y + \frac{1}{3}\right)^2 = 6.$$

(Alternatively, you may recognize $9y^2 + 6y + 1$ as the perfect square $(3y+1)^2$.) Dividing by 6 to get 1 on the right

$$(x-1)^2 + \frac{9\left(y + \frac{1}{3}\right)^2}{6} = 1$$
$$(x-1)^2 + \frac{3\left(y + \frac{1}{3}\right)^2}{2} = 1.$$

This is an ellipse centered at $(1, -\frac{1}{3})$ with $a = 1$, $b^2 = 2/3$, so $b = \sqrt{2/3}$.

29. **(a)** Since P moves in a circle we have

$$x = 10\cos t$$
$$y = 10\sin t.$$

This completes a revolution in time 2π.

(b) First, consider the planet as stationary at (x_0, y_0). Then the equations for M are

$$x = x_0 + 3\cos 8t$$
$$y = y_0 + 3\sin 8t.$$

The factor of 8 is inserted because for every $2\pi/8$ units of time, $8t$ covers 2π, which is one orbit. But since P moves, we must replace (x_0, y_0) by the position of P. So we have

$$x = 10\cos t + 3\cos 8t$$
$$y = 10\sin t + 3\sin 8t.$$

(c) See Figure 14.10.

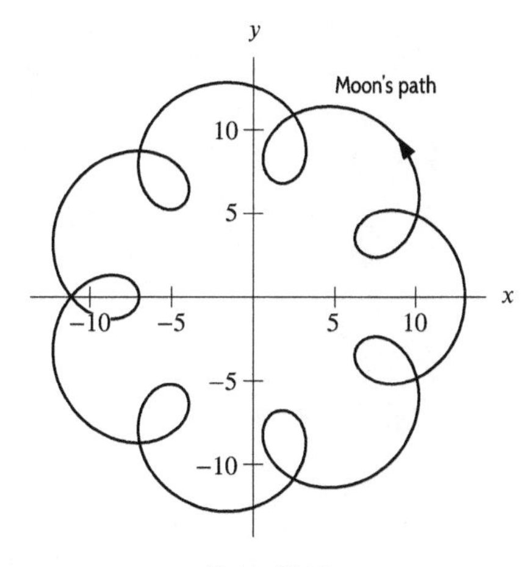

Figure 14.10

STRENGTHEN YOUR UNDERSTANDING

1. True. The path is on the line $3x - 2y = 0$.

5. False. Starting at $t = 0$ the object is at $(0, 1)$. For $0 \le t \le \pi$, $x = \sin(t/2)$ increases and $y = \cos(t/2)$ decreases, thus the motion is clockwise.

9. False. The standard form of the circle is $(x + 4)^2 + (y + 5)^2 = 141$. The center is $(-4, -5)$.

13. False. There are many parameterizations; $x = \cos t, y = \sin t$ and $x = \sin t, y = \cos t$ are two of them.

17. True. Since

$$\frac{x^2}{4} + y^2 = \frac{(2\cos t)^2}{4} + (\sin t)^2 = \cos^2 t + \sin^2 t = 1,$$

these equations parameterize the ellipse $(x^2/4) + y^2 = 1$.

21. False. It is defined as $\sinh x = \dfrac{e^x - e^{-x}}{2}$.

25. True. We have $\cosh(-x) = (e^{-x} + e^{-(-x)})/2 = (e^x + e^{-x})/2 = \cosh x$, so the hyperbolic cosine is an even function.

29. False. We have

$$\sinh \pi = \frac{e^\pi}{2} - \frac{e^{-\pi}}{2} = 11.549 \ne 0.$$

We do have $\sin \pi = 0$.

33. False. The two asymptotes are entirely in the region between the two branches of the hyperbola. The two focal points are outside this region. To move from an asymptote to a focal point you must cross the hyperbola.

CPSIA information can be obtained
at www.ICGtesting.com
Printed in the USA
FSHW022001140320
68133FS

9 781118 941638